UNITEXT for Physics

Series Editors

Michele Cini, University of Rome Tor Vergata, Roma, Italy

Attilio Ferrari, University of Turin, Turin, Italy

Stefano Forte, University of Milan, Milan, Italy

Guido Montagna, University of Pavia, Pavia, Italy

Oreste Nicrosini, University of Pavia, Pavia, Italy

Luca Peliti, University of Napoli, Naples, Italy

Alberto Rotondi, Pavia, Italy

Paolo Biscari, Politecnico di Milano, Milan, Italy

Nicola Manini, University of Milan, Milan, Italy

Morten Hjorth-Jensen, University of Oslo, Oslo, Norway

UNITEXT for Physics series publishes textbooks in physics and astronomy, characterized by a didactic style and comprehensiveness. The books are addressed to upper-undergraduate and graduate students, but also to scientists and researchers as important resources for their education, knowledge, and teaching.

More information about this series at http://www.springer.com/series/13351

Giovanni Carraro

Astrophysics
of the Interstellar Medium

Springer

Giovanni Carraro
Dipartimento di Fisica e Astronomia
Università di Padova
Padova, Italy

ISSN 2198-7882   ISSN 2198-7890   (electronic)
UNITEXT for Physics
ISBN 978-3-030-75295-8   ISBN 978-3-030-75293-4   (eBook)
https://doi.org/10.1007/978-3-030-75293-4

This Springer imprint is published by the registered company Springer Nature Switzerland AG
The registered company address is: Gewerbestrasse 11, 6330 Cham, Switzerland

Preface

This book collects the material of the course of *Astrophysics of the Inter-Stellar Medium* for the master degree in Astrophysics and Cosmology at Padova University. It has been translated and adapted from the original hardcopy paper-notes in Italian of the course of *Mezzo Inter-stellare* given by Prof. Guido Barbaro until the academic year 2012–2013. The course (of 48 hours) has been retaken with the same syllabus after a few years hiatus by this writer in the academic year 2018–2019. Several chapters have been largely improved with the inclusion of new, modern material. As such, this publication effort is heart-fully dedicated to the memory of Prof. Guido Barbaro.

As brilliantly and effectively iconised by Donald Osterbrok, a pioneer of Inter-Stellar Medium (ISM) studies and author of numerous papers and books on the subject, the ISM is *everything not in stars.*

The modern picture of the ISM is of a multi-component, multi-phase, turbulent, and magneto-hydro-dynamic system. Each component is characterized by a different particle number density and mean temperature. From the cold molecular phase, where stars and planetary systems form, to the very hot coronal gas which surrounds galaxies and galaxy clusters, the ISM is everywhere. Studying the properties of the ISM is vital for the exploration of virtually any field of modern research in Astronomy and Cosmology.

The ISM is a violent system, where energy is generated in spectacular events like Supernovae explosions, spiral density wave shocks and galaxy mergers or pumped in the medium by strong UV flux and stellar winds from massive stars and/or Active Galactic Nuclei.

These notes bring the student through all this with a coherent and accurate mathematical and physical approach, with continuous references to the real ISM in galaxies.

The course (and the book) is naturally divided into three parts.

Part I (Chaps. 1 to 6) describes the ISM neglecting electrical and magnetic fields. Being the ISM a hydro-dynamic system, this part starts setting the scene by introducing the equations of the fluid dynamics (the conservation equations) for a system at rest. Then, acoustic waves are introduced, which represent the main avenue to the study of instabilities and out-of-equilibrium situations. The real ISM is then described by exploring the role of thermal conduction and viscosity. This first part is concluded by discussing shock waves and turbulence, two ubiquitous and essential ingredients of the thermal balance of the ISM.

In Part II (Chaps. 7 to 9), the electro-magnetic field is switched on and its role in modulating shock waves and contrasting gravity is studied.

The third and last part (Chaps. 10 to 16) is phenomenological. First, dust and its properties are described, then a study of the stellar main sources of energy (HII regions, stellar winds, and supernovae) are discussed. The various components of the ISM (HIM, WIM, WNM, and CNM) are discussed in Chap. 14, while entire Chap. 15 is dedicated to molecular clouds. Eventually, star formation is presented in the last Chapter.

This preface cannot end without mentioning the continuous contribution of the students who over the years attended this course. Two students in particular have provided an essential help for the preparation of this book: Nicola Gaspari, and Martina Lai, to whom much gratitude is expressed. Finally, nobody more than my old friend Butler Burton instilled interest for the ISM in me.

Padova, Italy   Giovanni Carraro

Contents

Chapter 1
Fundamental Equations for Ideal Fluids

1.1 Introduction: Fluids as Continuous Systems

The properties of matter in its different states of aggregation depend on the entity of intermolecular forces, on density, and on the degree of thermal agitation. Let us analyze the dependence of such forces on intermolecular distance, as illustrated in Fig. 1.1. Forces are repulsive when $d < d_0$, attractive when $d > d_0$ and in an equilibrium configuration when $d = d_0$.

The knowledge of the density of a substance allows us to determine the mean distance between molecules. Let us consider two examples. Air under standard conditions, i.e. at atmospheric pressure and $T = 15$ C, has a density of $\rho = 1.3 \cdot 10^{-2}$ g/cm^3 and therefore the number of molecules per unit volume is: $n = \rho/\mu H$, where $\mu = 30$ is its molecular weight. Hence $n = 2.6 \cdot 10^{20}$ cm^{-3}. Each molecule occupies a volume equal to $1/n$ so the mean distance is of the order of $d = n^{-1/3} = 1.6 \cdot 10^{-7}$ cm. Ethanol has a density of $\rho = 0.79$ g/cm^3 and a molecular weight $\mu = 34$, hence an intermolecular distance equal to $4 \cdot 10^{-8}$ cm.

The mean distance between molecules for a gaseous substance under standard conditions is roughly $10\, d_0$. At such distances the attractive forces are so small that they can be neglected and, because of thermal agitation, they also have high mobility: the perfect gas model is then a good approximation. For these reasons gases does not have shape nor volume, but adapt to the shape and volume of their container.

The mean distance between molecules in a solid or a liquid are of the order of d_0, which is 10^{-8} cm. At equilibrium each molecule is blocked by the molecular forces and so its mobility is limited. This limitation is particularly stringent in the case of crystalline solids, in which molecules oscillate around a stable configuration; in this case the thermal energy is the energy associated to oscillationa plus the translational energy due to any free electron. The increase in thermal agitation can break the bonds leading to the melting of the solid.

Solids have their own shape and undergo minimal variations of volume when subject to external forces (even if the forces are considerable): this happens because

© The Author(s), under exclusive license to Springer Nature Switzerland AG 2021
G. Carraro, *Astrophysics of the Interstellar Medium*,
UNITEXT for Physics, https://doi.org/10.1007/978-3-030-75293-4_1

Fig. 1.1 Dependence of the
inter-molecular force as a
function of inter-molecular
distance

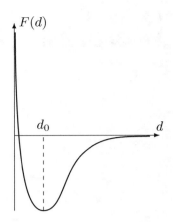

a decrease in the mean distance between molecules leads to the development of
repulsive forces.

The properties of liquids are more difficult to be interpreted. We have already
seen that their intermolecular distances are of the same order of those in the solids,
but that the thermal agitation prevents their organisation into a structure. Rather, it
seems that molecules tend to organise into clusters that can move with respect to
each other. In other words liquids are characterised by both ordered and disordered
phenomena. Therefore the difference between solids and liquids lies in the level of
mobility of the crystalline structure constituents. Because of this mobility, liquids
does not hold a shape but they resist volume variations.

An indication of the organised behavior of liquids is the atomic heat of monoatomic
liquids, such as mercury or liquefied argon, that is around 6 cal/mole K. This value
is twice the one expected for free atoms (i.e. a perfect gas) for which each degree
of freedom contributes with $\mathcal{R}/2$ cal/mole K = 1 cal/mole K, where $\mathcal{R}$ is the gas
constant. In fact in a lattice the components vibrate around an equilibrium position:
the three degrees of freedom related to the vibrational motion contribute together
with $3\mathcal{R}/2$ cal/mole K. We can understand why the specific heat is twice the value
due to the vibrational degrees of freedom by considering the presence of three other
degrees of freedoms related to the translational motion, which accounts for the other
half of the specific heat.

Gases and liquids are characterised by *fluidity*, that is the relative mobility of their
parts: because of this they are together called fluids. The study of relative motions
of the various portions of these systems is the subject of fluid dynamics.

The main difference between liquids and gases is their compressibility. The *coefficient of compressibility* is defined as:

$$\beta = -\frac{1}{V}\frac{d\,V}{d\,p}\,. \tag{1.1}$$

Let us consider water and a perfect gas as examples. For water at ambient temperature and atmospheric pressure we have $\beta = 4.84 \cdot 10^{-10}$ Pa^{-1}. For the perfect gas, using the equation of state, we have:

$$\beta = \frac{1}{p} \tag{1.2}$$

hence in standard conditions $\beta \approx 10^{-5}$ Pa^{-1}. The low degree of compressibility of liquids comes from the fact that small variations in the molecular distances trigger strong repulsive forces; for gases these repulsive forces cannot develop due to the larger intermolecular distances.

In the analysis of the dynamical behaviour is usually not necessary to distinguish the fluid nature or to recognise whether it is a liquid or a gas. In fact motions in a fluid are usually followed by such small pressure variations that, following (1.1), gases and liquids behave in the same way, since density variations are negligible (i.e. they're approximatively incompressible):

$$\frac{\delta V}{V} = -\beta \Delta p \approx 0 . \tag{1.3}$$

In the case of gases, molecules are separated by voids that have a linear dimension much larger than the dimension of the molecules; even in the case of solids and liquids mass is almost completely concentrated in the nuclei, which occupy a very small portion of the volume occupied by the whole system. Hence, all the matter at a microscopic level has a discrete structure. However fluid dynamics deals with the macroscopic behaviour of the motion, thus at a scale that is much larger than molecular dimensions and inter-molecular distances. In these conditions it is not necessary to consider the discrete structure and a continuous model can be exploited and used.

Let us analyse the main characteristics of this model by considering an infinitesimal volume element. This must have dimensions much smaller than the global dimensions of the system, in order to provide a precise local description, but at the same time much larger than the intermolecular distances, to provide a statistically meaningful estimates of the local value of any thermodynamical quantity.

At a given instant a fluid element harbours a huge number of molecules with different velocities. The velocity of each molecule $\mathbf{v}_i$ can be represented as the sum of two vectors: the vector $\mathbf{v}$, which is the velocity averaged over all the molecules enclosed in the volume and the vector $\mathbf{u}_i$, which is the velocity of the ith molecules with respect to the point with velocity $\mathbf{v}$. From the macroscopic point of view the infinitesimal volume can be considered as a particle moving with velocity $\mathbf{v}$ and having a density defined by:

$$\rho = \frac{\sum_i m_i}{V} \tag{1.4}$$

where m_i is the mass of the ith molecule and V is the fluid element volume. The fluid element absolute temperature is also known, since, according to the kinetic theory of gases, it is proportional to the mean value of u_i^2. Finally, let us consider a plane impermeable membrane, of area ΔA, immersed in the fluid at the point P and with the velocity $\mathbf{v}$ evaluated at P. Each side of the membrane is then bombarded by molecules, each with its velocity $\mathbf{u}_i$. The fast succession of impulses received by the membrane produces a force ΔF and we define the quantity:

$$\lim_{\Delta A \to 0} \frac{\Delta F}{\Delta A} \tag{1.5}$$

as the pressure p at point P.

If the fluid element volume is too small (such that only a few molecules are enclosed) then the quantities evaluated this way do not represent the fluid behaviour; on the other side, if the volume of the fluid element is too large, then ρ and $\mathbf{v}$ do not provide a local representation of the fluid but just mean values.

In the study of the dynamics of fluids this continuous scheme is routinely adopted and the state of the system under consideration is characterised by the velocity $\mathbf{v}(x, y, z, t)$, the pressure $p(x, y, z, t)$ and the density $\rho(x, y, z, t)$, all expressed as functions of spatial and time coordinates.

The description of the system is closed with the equation of state, from which, given p and ρ, one can obtain the gas temperature $T(x, y, z, t)$. This in most cases coincides with the temperature inferred from pure thermal agitation. All these quantities are continuous functions of both position and time.

The continuous model fails when the mean free path of molecules has the same order of magnitude of the smallest linear dimension that is significant in the problem. In this circumstance the action of single molecules or groups of molecules becomes significant and a different model needs to be constructed.

In the framework of this continuous model we can adopt two different perspectives, or points of view:

(a) the **eulerian** approach, in which we analyse the evolution of the fluid in a point fixed in space: this way we don't refer to a particular physical portion of fluid;
(b) the **lagrangian** approach, in which we analyse the evolution of the single particles of fluid, in the sense that we study how kinematics quantities related to a single element of fluid (as defined above) evolve.

Let's then consider a fluid: let C_0 be its configuration at the instant $t = 0$, which is taken as the reference configuration. Let P_0 be the point of the space occupied by a fluid element in C_0 at $t = 0$. With respect to an appropriate reference system, let (x_0, y_0, z_0) be the coordinates of P_0. If C is the configuration at the generic moment t and P the position occupied in C by the element which at $t = 0$ was in P_0, at each instant t there is a one-to-one (biunivocal) correspondence between the points P_0 of C_0 and the points P of C. If (x, y, z) are the coordinates of P instantly t this correspondence is expressed by the relations:

$$x = x(x_0, y_0, z_0, t) \qquad y = y(x_0, y_0, z_0, t) \qquad z = z(x_0, y_0, z_0, t) \ . \qquad (1.6)$$

From (1.6) the fluid element velocities can be derived:

$$\dot{x} = \frac{\partial x}{\partial t} = v_x(x_0, y_0, z_0, t) \ , \qquad (1.7)$$

$$\dot{y} = \frac{\partial y}{\partial t} = v_y(x_0, y_0, z_0, t) \ , \qquad (1.8)$$

$$\dot{z} = \frac{\partial z}{\partial t} = v_z(x_0, y_0, z_0, t) \ . \qquad (1.9)$$

If we want to pass to the Eulerian representation, that is, we want to study how the kinematic quantities related to the motion of the fluid vary in a given point of space, we need to eliminate the parameters (x_0, y_0, z_0) using the (1.6).

The two points of view are linked by the formula of the *Lagrangian* or *molecular derivation*. Let f be a typical thermodynamical function of the fluid, for example the density ρ or the pressure p. The variations of f can occur in two ways: variations due to time $\partial f / \partial t$ and variations consequent to changes in the spatial position: the global variation, which is indicated with df/dt will be a consequence of the superposition of the two individual modes of variation. This global variation is the one of interest for the description of the motion of a particular element of the fluid. Generally f will be a function of the coordinates (x, y, z) and time: $f = f(x, y, z, t)$. The differential of f is:

$$df = \frac{\partial f}{\partial x} dx + \frac{\partial f}{\partial y} dy + \frac{\partial f}{\partial z} dz + \frac{\partial f}{\partial t} dt. \qquad (1.10)$$

If (dx, dy, dz) is the infinitesimal displacement of a volume element in the time interval dt it will be:

$$dx = v_x \, dt \qquad dy = v_y \, dt \qquad dz = v_z \, dt \qquad (1.11)$$

and hence:

$$df = \left(\frac{\partial f}{\partial x} v_x + \frac{\partial f}{\partial y} v_y + \frac{\partial f}{\partial z} v_z + \frac{\partial f}{\partial t} \right) dt \qquad (1.12)$$

and therefore in vector notation:

$$\frac{df}{dt} = \frac{\partial f}{\partial t} + \mathbf{v} \cdot \mathrm{grad}\, f \ . \qquad (1.13)$$

The term df/dt is the quantity that must be used in the Lagrangian description, while $\partial f / \partial t$ is appropriate for the Eulerian description. This latter is a measure of the non-stationary character of the motion, which is instead stationary if this term is identically null. The term $\mathbf{v} \cdot \mathrm{grad}\, f$ is instead indicative of non-uniformity in the various points of the flow field. It is also referred to as *advection* of a scalar field. In the following we will refer predominantly to the Eulerian point of view.

1.2   Continuity Equation

The continuity equation is the mathematical formulation of the conservation of matter: *the variation of the mass contained in a certain fixed volume V_0 equals the quantity of matter passing through the surface encapsulating the volume.*

Let us consider a fixed volume V_0: the fluid mass contained in this volume is $\int_{V_0} \rho \, d V$. Let $d\sigma$ be an infinitesimal element of the contour surface S_0 of V_0: the fluid mass that comes out per unit of time through $d\sigma$ is $\rho \mathbf{v} \cdot d\sigma$ where $d\sigma$ is a vector with magnitude $d\sigma$, oriented as the outgoing normal of V_0.

Conservation of matter therefore implies:

$$\int_{S_0} \rho \mathbf{v} \cdot d\sigma = -\frac{\partial}{\partial t} \int_{V_0} \rho \, d V \,. \tag{1.14}$$

The surface integral can be transformed into a volume integral using the Gauss formula:

$$\int_{S_0} \rho \mathbf{v} \cdot d\sigma = \int_{V_0} \mathrm{div}\,(\rho \mathbf{v}) \, d V \tag{1.15}$$

and from (1.14) we have:

$$\frac{\partial}{\partial t} \int_{V_0} \rho \, d V + \int_{V_0} \mathrm{div}\,(\rho \mathbf{v}) \, d V = 0 \,. \tag{1.16}$$

Since V_0 is fixed, the derivation with respect to time can be brought inside the integral. Given the arbitrariness of V_0, the cancellation of the integral implies that of the integrand function:

$$\frac{\partial \rho}{\partial t} + \mathrm{div}\,(\rho \mathbf{v}) = 0 \,. \tag{1.17}$$

This is the continuity equation in the Eulerian form. It can be given a different form by recalling (1.13):

$$\frac{d \rho}{d t} + \rho \, \mathrm{div}\, \mathbf{v} = 0 \,. \tag{1.18}$$

When the density of a fluid is constant, i.e. $d\rho / d t = 0$, the fluid is called *incompressible* and in this case the continuity equation is reduced to:

$$\mathrm{div}\, \mathbf{v} = 0 \,. \tag{1.19}$$

We now derive the Lagrangian expression of the continuity equation. A material volume is a portion of the system consisting of a certain number of fluid particles. Those particles that occupy the space volume V at the time t, will at $t = 0$ occupy the volume V_0. If f is a function of coordinates and time, i.e. $f = f(x, y, z, t)$, the integral

$$\int_V f(x, y, z) \, d V \tag{1.20}$$

can be calculated both with respect to the variables (x, y, z) and to the variables (x_0, y_0, z_0), using the transformations defined by the (1.6). It turns out in fact that:

$$\int_V f \, d V = \int_{V_0} fJ \, d V_0 \tag{1.21}$$

being J the determinant of the Jacobian of the transformation. Hence, conservation of matter can be expressed as:

$$\int_V \rho \, d V = \int_{V_0} \rho_0 \, d V_0 \tag{1.22}$$

and, expressing the first member through the (1.21), we have:

$$\int_{V_0} (\rho J - \rho_0) \, d V_0 = 0 \tag{1.23}$$

from which, given the arbitrariness of V_0, the following expression holds:

$$\rho J = \rho_0 \, . \tag{1.24}$$

Note that it is an algebraic equation instead of a partial differential equation as in Eulerian coordinates.

By exploiting this relationship we can prove the *transport theorem*: if f is a function of coordinates (x, y, z) and time t, it results:

$$\frac{d}{dt} \int_V \rho f \, d V = \int_V \rho \frac{df}{dt} \, d V \, . \tag{1.25}$$

Indeed it is:

$$\frac{d}{dt} \int_V \rho f \, d V = \frac{d}{dt} \int_{V_0} \rho fJ \, d V_0 \tag{1.26}$$

and being $V_0 = $ const,

$$\frac{d}{dt} \int_{V_0} \rho f J \, d V_0 = \int_{V_0} \frac{d}{dt} (\rho fJ) \, d V_0 \, . \tag{1.27}$$

It is, obviously :

$$\frac{d}{dt}(\rho f J) = \frac{df}{dt}\rho J + f\frac{d}{dt}(\rho J) \tag{1.28}$$

and, since for (1.24) we have $\rho J = \rho_0 = $ const, the second term on the right hand side cancels out. Therefore, we have:

$$\frac{d}{dt}\int_V \rho f \, dV = \int_{V_0} \frac{d}{dt}(\rho f J) \, dV_0 = \int_{V_0} \rho J \frac{df}{dt} \, dV_0 \, . \tag{1.29}$$

The last integral can be expressed, according to (1.21), as an integral in the variables (x, y, z) and therefore the (1.25) is found.

1.3   Volume Forces and Surface Forces. Stresses

In the study of fluids, two types of forces are distinguished:

(1) Long-range forces that act on matter without direct contact and whose intensities slowly decrease as the distance between interacting elements increases; examples of this type of force are gravitational forces, electromagnetic forces and fictitious forces, such as centrifugal force, which occur when the reference system is in accelerated motion with respect to an inertial system. These forces act on each element of the fluid and are proportional to the mass of the element on which they act:

$$d\mathbf{f} = \mathbf{f} \, dm = \mathbf{f}\rho \, dV \tag{1.30}$$

where $\mathbf{f}$ is the force acting on the mass unit, while $\rho\mathbf{f}$ is the force acting per unit of volume.

(2) Short-range forces, of molecular origin, which decrease rapidly as the distance between the interacting elements increases and are important only when the distance is of the order of intermolecular distances; they are practically negligible unless there is direct contact between interacting elements. These forces are therefore related to the contact surface between the two elements taken into consideration, while the volume does not intervene directly. The force exerted through the surface element $d\sigma$ is expressed by:

$$\mathbf{\Sigma}(\hat{\mathbf{n}}, x, y, z, t) \, d\sigma \tag{1.31}$$

where $\hat{\mathbf{n}}$ is the versor of the normal to $d\sigma$ properly oriented and the force is intended to be exerted by the fluid on the side of the surface in the positive direction of $\hat{\mathbf{n}}$.

The force $\mathbf{\Sigma}$, which represents the force acting on the surface unit, is called *stress*. Of this type are the forces that develop between different elements of a fluid.

Let's consider a fluid in equilibrium. The resultant of the surface forces that all parts of the body exert on an arbitrary volume V is zero. If the fluid is deformed, it is out of its state of equilibrium and forces, called internal stresses, arise which tend to bring the body back into its state of equilibrium. We choose any volume V in the body and let S be the surface enclosing it. The resultant of the surface forces acting on this volume is given by:

$$\int_V \mathbf{F} \, dV = \int_S \mathbf{\Sigma} \, d\sigma \tag{1.32}$$

and results from the contribution of the external elements acting on V and from that of the internal elements; however the forces that are exerted between elements inside V cancel each other out by virtue of the principle of action and reaction and therefore the net force acting on the volume is the result of the forces with which the volume is stimulated by the external entities only. These forces act through the boundary of V and therefore their resultant can be expressed as a surface integral.

So far we have denoted with (x, y, z) the three coordinates of the reference system; for the sake of an easier and more compact notation in the rest of this section we will make use of tensors, and will switch to the notation x_i $(i = 1, 2, 3)$.

The integral of a scalar extended to a volume can be transformed into a surface integral when the scalar is the divergence of a vector; in our case the integrand function in (1.32) is a vector; we can express it as the divergence of a second order tensor π_{ij}:

$$F_i = \frac{\partial \pi_{ij}}{\partial x_j} \tag{1.33}$$

called stress tensor.

For the ith component of the resulting force we will then have:

$$\int_V F_i \, dV = \int_V \frac{\partial \pi_{ij}}{\partial x_j} \, dV = \int_S \pi_{ij} \hat{n}_j \, d\sigma \, , \tag{1.34}$$

where $d\sigma$ is the vector that has the surface element $d\sigma$ as its magnitude and is directed as the external normal and $\hat{n}_j$ is the versor of that particular direction. The general versor $\hat{n}$ declines in $\hat{n}_j$, with $j = 1, 2$, or 3, depending on the axis. The quantity

$$\pi_{ij} \, d\sigma_j = \pi_{ij} \hat{n}_j \, d\sigma \, , \tag{1.35}$$

is the ith component of the force acting on the surface element $d\sigma$.

Choosing the surface elements perpendicular to the axes x_1, x_2, x_3 one can see that the component π_{ij} of the stress tensor is the ith component of the force acting on the unit of surface perpendicular to x_j. For example taking $d\sigma$ perpendicular to the axis x_1 and placing $d\sigma = 1$ we have:

$$\pi_{ij} n_j = \pi_{i1} \tag{1.36}$$

being $\hat{\boldsymbol{n}} = (1, 0, 0)$; π_{i1} is the ith component of the force acting on the normal surface unit at x_1.

The surface integral:

$$\int_S \pi_{ij}\hat{\boldsymbol{n}}_j \,\mathrm{d}\boldsymbol{\sigma} \tag{1.37}$$

is the ith component of the force the volume V is subject to; conversely, the ith component of the force that this volume exerts on the surface that surrounds it is given by:

$$-\int_S \pi_{ij}\hat{\boldsymbol{n}}_j \,\mathrm{d}\boldsymbol{\sigma} \;. \tag{1.38}$$

We now distinguish two particular cases.

(A) Case of uniform compression.

We define as *ideal* those fluids in which viscosity and thermal conductivity are negligible, that is, those fluids in which there is no production of heat by internal friction nor heat exchange between portions of fluid or between the fluid and external systems. The ideal fluid approximation somewhat simplifies the equations that describe the fluid dynamic behaviour, but can correctly describe the behaviour of real fluids only in particular situations.

In ideal fluids, each unit surface, regardless of its orientation, is subjected to a pressure of equal intensity directed along the normal internal surface. Denoting this pressure with p, the force acting on the surface element $\mathrm{d}\,\sigma_i$ will be equal to $-p\,\mathrm{d}\,\sigma_i = -\mathrm{pn}_i\,\mathrm{d}\,\sigma$.

On the other hand, this force can be expressed through the stress tensor and there will be:

$$\pi_{ij}\,\mathrm{d}\,\sigma_j = -\mathrm{p}\,\mathrm{d}\,\sigma_i = -\mathrm{p}\delta_{ij}\,\mathrm{d}\,\sigma_j \tag{1.39}$$

and hence:

$$\pi_{ij} = -p\delta_{ij} \;. \tag{1.40}$$

This is the expression of the stress tensor in the case of an ideal fluid.

(B) General case of an arbitrary deformation.

In this case the non-diagonal components (they are called shear terms) of the stress tensor will not be null:

$$\pi_{ij} = -p\delta_{ij} + \varepsilon_{ij} \;. \tag{1.41}$$

This means that, in addition to the normal force, tangential forces also act on each surface element inside the body, which have a tendency to slide the elements parallel to their common surface.

1.4 Equation of Motion (Euler Equation) for Ideal Fluids

A further condition for the study of the dynamic behaviour of ideal fluids can be obtained from the fundamental law of dynamics: *the rate of change of the momentum of a material volume V of fluid is equal to the resultant of the forces acting on the volume*. We will apply this law to ideal fluids, neglecting for now the magnetic forces on which we will fix our attention starting from Chap. 7.

The conservation of the ith component of the momentum of the fluid contained in V is then expressed by:

$$\frac{d}{dt} \int_V \rho v_i \, dV = \int_V \rho f_i \, dV + \int_S \pi_{ij} n_j \, d\sigma \tag{1.42}$$

where $\mathbf{f}$ is a generic long-range force per unit of mass.

Taking into account the (1.40) the previous relationship becomes:

$$\frac{d}{dt} \int_V \rho v_i \, dV = \int_V \rho f_i \, dV - \int_S p n_i \, d\sigma \tag{1.43}$$

and in vector terms:

$$\frac{d}{dt} \int_V \rho \mathbf{v} \, dV = \int_V \rho \mathbf{f} \, dV - \int_S p \hat{\mathbf{n}} \, d\sigma \tag{1.44}$$

The last second member term can be transformed into a volume integral by applying the Gauss theorem to a vector $\mathbf{u} = F\hat{\mathbf{w}}$ where the scalar F is a function of coordinates and $\hat{\mathbf{w}}$ is a constant versor. It is:

$$\hat{\mathbf{w}} \cdot \int_S F \, d\boldsymbol{\sigma} = \hat{\mathbf{w}} \cdot \int_V \operatorname{grad} F \, dV \tag{1.45}$$

and the (1.44) becomes:

$$\frac{d}{dt} \int_V \rho \mathbf{v} \, dV = \int_V \rho \mathbf{f} \, dV - \int_V \operatorname{grad} p \, dV . \tag{1.46}$$

Using the transport theorem, the previous equation changes to:

$$\int_V \left(\rho \frac{d\mathbf{v}}{dt} - \rho \mathbf{f} + \operatorname{grad} p \right) dV = 0 \tag{1.47}$$

and, given the arbitrariness of V, the cancellation of the integrated function follows. Keeping also in mind the molecular derivation rule (1.13) we have:

$$\frac{\partial \mathbf{v}}{\partial t} + (\mathbf{v} \cdot \operatorname{grad}) \mathbf{v} = \mathbf{f} - \frac{1}{\rho} \operatorname{grad} p \tag{1.48}$$

called Euler's equation in eulerian form. The second term of the left hand side member of this equation is also referred to as the *advection* of a vectorial field.

1.5 Conservation of Total Energy and Thermal Energy

For an ideal fluid the conservation of energy is expressed as follows: *the rate of change of the total energy of a material volume is equal to the work of the forces done per unit of time (power).* The mathematical formulation of this law relative to the material volume V is:

$$\frac{\mathrm{d}}{\mathrm{d}\,t} \int_V \rho \left(\frac{1}{2} v^2 + \epsilon \right) \mathrm{d}\,V = \int_V \rho \mathbf{f} \cdot \mathbf{v}\,\mathrm{d}\,V - \int_S p \mathbf{v} \cdot \mathrm{d}\,\sigma \qquad (1.49)$$

where ϵ is the internal energy of the unit mass and S is the contour of V. Recalling the transport theorem and applying the Gauss formula we deduce the:

$$\frac{\mathrm{d}}{\mathrm{d}\,t} \left(\frac{1}{2} v^2 + \epsilon \right) = \mathbf{f} \cdot \mathbf{v} - \frac{1}{\rho} \operatorname{div}\,(p\mathbf{v})\ . \qquad (1.50)$$

This expression of energy conservation is in Lagrangian form. It can be shown that this formula, if the work of gravitational forces is neglected, is equivalent to:

$$\frac{\partial}{\partial t} \left(\frac{1}{2} \rho v^2 + \rho\epsilon \right) = - \operatorname{div} \left[\rho \mathbf{v} \left(\frac{1}{2} v^2 + w \right) \right] \qquad (1.51)$$

where $w = \epsilon + p/\rho$ is the enthalpy of the mass unit. The expression (1.51) is an Eulerian formulation.

The physical meaning of the (1.51) can be seen from the integral on a fixed volume V_0

$$\int_{V_0} \frac{\partial}{\partial t} \left(\frac{1}{2} \rho v^2 + \rho\epsilon \right) \mathrm{d}\,V = - \int_{V_0} \operatorname{div} \left[\rho \mathbf{v} \left(\frac{1}{2} v^2 + w \right) \right] \mathrm{d}\,V\ . \qquad (1.52)$$

Since V_0 is fixed, the derivation sign can be exchanged with the integration sign. Furthermore, applying the Gauss formula to the second member, we have:

$$\frac{\partial}{\partial t} \int_{V_0} \left(\frac{1}{2} \rho v^2 + \rho\epsilon \right) \mathrm{d}\,V = - \int_{S_0} \rho \left(\frac{1}{2} v^2 + w \right) \mathbf{v} \cdot \mathrm{d}\,\sigma\ . \qquad (1.53)$$

where S_0 is the contour of V_0.

The fact that w appears in the second member of the equation depends on the requirement that pressure forces work must appear in the energy budget balance; in fact it is:

$$\int_{S_0} \rho \left(\frac{1}{2}v^2 + w \right) \mathbf{v} \cdot d\boldsymbol{\sigma} = \int_{S_0} \rho \left(\frac{1}{2}v^2 + \epsilon \right) \mathbf{v} \cdot d\boldsymbol{\sigma} + \int_{S_0} p\mathbf{v} \cdot d\boldsymbol{\sigma} \ . \quad (1.54)$$

Then we have:

$$\frac{\partial}{\partial t} \int_{V_0} \left(\frac{1}{2}\rho v^2 + \rho\epsilon \right) dV = -\int_{S_0} \rho \left(\frac{1}{2}v^2 + \epsilon \right) \mathbf{v} \cdot d\boldsymbol{\sigma} - \int_{S_0} p\mathbf{v} \cdot d\boldsymbol{\sigma} \quad (1.55)$$

Let's now compare the (1.55) with the (1.49) in which the term of the work of the volume forces is suppressed: the derivation sign is different, in fact in (1.55) the volume is fixed while in (1.49) it is a material volume. Furthermore, in the first the term for the transport of energy shows up; this term is not necessary in the other formula.

The (1.53) is a particular form of a general energy conservation equation:

$$\frac{\partial U}{\partial t} = -\operatorname{div} \mathbf{g} \quad (1.56)$$

where U is the energy of the volume unit and $\mathbf{g} = \rho\mathbf{v}(v^2/2 + w)$ is the energy flow density. From the relation (1.50) we can subtract a relation expressing the conservation of mechanical energy, obtained from the Euler equation: this way a conservation law for thermal energy is derived. From Euler's equation multiplying scalarly by $\mathbf{v}$ we have:

$$\mathbf{v}\frac{d\mathbf{v}}{dt} = \mathbf{f} \cdot \mathbf{v} - \frac{1}{\rho}\mathbf{v} \cdot \operatorname{grad} p \ . \quad (1.57)$$

Subtracting the (1.57) from the (1.50) one obtains:

$$\frac{d\epsilon}{dt} + \frac{p}{\rho} \operatorname{div} \mathbf{v} = 0 \ . \quad (1.58)$$

From the continuity equation we have:

$$\operatorname{div} \mathbf{v} = -\frac{1}{\rho}\frac{d\rho}{dt} \ , \quad (1.59)$$

and, introducing the specific volume $V = 1/\rho$, we deduce:

$$\frac{d\epsilon}{dt} + p\frac{dV}{dt} = 0 \ . \quad (1.60)$$

On the other hand from thermodynamics we have

$$T\frac{ds}{dt} = \frac{dQ}{dt} = \frac{d\epsilon}{dt} + p\frac{dV}{dt} \quad (1.61)$$

and hence we have:

$$\frac{d s}{d t} = \frac{\partial s}{\partial t} + \mathbf{v} \cdot \text{grad } s = 0 \tag{1.62}$$

where s is the entropy of the mass unit. We find the constancy of entropy in the absence of dissipation, thermal conduction and production or subtraction of energy due to the interactions between matter and radiation.

1.6   Vorticity in Ideal Fluids. Potential Motion

We introduce this topic by demonstrating first the Kelvin theorem of the conservation of the velocity circulation: *for an ideal fluid in isentropic motion the velocity circulation along a closed material line is constant in time*. A motion is called isentropic when the entropy density is uniform. By closed material line we mean a closed line which remains such during motion and which always consists of the same fluid particles.

It must be shown that the temporal derivative of the integral

$$\oint_l \mathbf{v} \cdot d\mathbf{l} , \tag{1.63}$$

where the integration is extended to the closed material line l, is null.

Let $\delta\mathbf{l}$ be an element of the material line l and let $\mathbf{v}$ the velocity of the fluid at an extreme P of $\delta\mathbf{l}$. The velocity at the other extreme Q, at the same instant, will be:

$$\mathbf{v}' = \mathbf{v} + (\delta\mathbf{l} \cdot \text{grad })\mathbf{v} . \tag{1.64}$$

In the interval $d t$ the length of the line element will vary by:

$$d \delta\mathbf{l} = (\mathbf{v}' - \mathbf{v}) d t = (\delta\mathbf{l} \cdot \text{grad })\mathbf{v} d t \tag{1.65}$$

and the variation in the unit of time will be:

$$\frac{d}{d t}\delta\mathbf{l} = (\delta\mathbf{l} \cdot \text{grad })\mathbf{v} = d \mathbf{v} . \tag{1.66}$$

Consider the line integral extended to an open portion of the material line between the points P and Q. We will then have for the time variation:

$$\frac{d}{d t}\int_P^Q \mathbf{v} \cdot \delta\mathbf{l} = \int_P^Q \frac{d \mathbf{v}}{d t} \cdot \delta\mathbf{l} + \int_P^Q \mathbf{v} \cdot \frac{d \delta\mathbf{l}}{d t} \tag{1.67}$$

and taking into account (1.66):

$$\frac{d}{dt} \int_P^Q \mathbf{v} \cdot \delta \mathbf{l} = \int_P^Q \frac{d\mathbf{v}}{dt} \cdot \delta \mathbf{l} + \int_P^Q \mathbf{v} \cdot d\mathbf{v} \tag{1.68}$$

introducing Euler's equation and calculating the integral along the closed material line we have:

$$\frac{d}{dt} \oint_l \mathbf{v} \cdot d\mathbf{l} = \oint_l \left(-\frac{1}{\rho} \operatorname{grad} p - \operatorname{grad} \Phi \right) \cdot d\mathbf{l} + \oint_l \mathbf{v} \cdot d\mathbf{v} \tag{1.69}$$

where Φ is the potential of a conservative mass force, for example the gravitational potential. It is:

$$dw = d\epsilon + p\,dv + V\,dp = T\,ds + \frac{1}{\rho}\,dp \tag{1.70}$$

from which:

$$\operatorname{grad} w = T \operatorname{grad} s + \frac{1}{\rho} \operatorname{grad} p \tag{1.71}$$

and being $\operatorname{grad} s = 0$ (isentropic fluid) we have:

$$\frac{d}{dt} \oint_l \mathbf{v} \cdot d\mathbf{l} = -\oint_l \operatorname{grad}(w + \Phi) \cdot d\mathbf{l} + \oint_l \mathbf{v} \cdot d\mathbf{v}. \tag{1.72}$$

In both integrals in the second member of the equation the integrand function is an exact differential and therefore the integrals cancel each other out.

Let us suppose now to consider a motion characterised by a velocity field $\mathbf{v} = \mathbf{v}(\mathbf{r}, t)$ and to introduce a new field defined by:

$$\boldsymbol{\omega} = \operatorname{rot} \mathbf{v} \tag{1.73}$$

The vector $\boldsymbol{\omega}$ thus defined is called *vorticity*; the field lines are called vortex lines and the equations that describe them are:

$$\frac{dx}{\omega_x} = \frac{dy}{\omega_y} = \frac{dz}{\omega_z}. \tag{1.74}$$

Being:

$$\operatorname{div} \boldsymbol{\omega} = \operatorname{div} \operatorname{rot} \mathbf{v} \equiv 0 \tag{1.75}$$

the vector field $\boldsymbol{\omega}$ is solenoidal. Consider a flow tube of this field: the flow of the vector $\boldsymbol{\omega}$ through any section of the tube is constant and constitutes the *intensity* of the flow tube. Again due to the solenoid character, the vortex lines are closed lines or leave and arrive at the contour of the fluid.

Let's try to derive a physical meaning for vorticity. Consider the motion in the surroundings of a point. It can be shown that the most general local displacement in this motion is the overlap of (a) a rigid displacement, which generally consists of a translation and a rotation, and (b) a deformation along each of the three reference frame axes. From Stokes' theorem we have:

$$\oint \mathbf{v} \cdot \mathrm{d}\mathbf{l} = \int_S \mathrm{rot}\,\mathbf{v} \cdot \hat{\boldsymbol{n}}\,\mathrm{d}\sigma\ . \tag{1.76}$$

Taking for l a circumference of radius a and as surface S that of the corresponding circle we have:

$$2\pi a v_\perp = \pi a^2\,\mathrm{rot}\,\mathbf{v} \cdot \hat{\boldsymbol{n}} \tag{1.77}$$

from which the angular velocity Ω is obtained:

$$\Omega = \frac{v_\perp}{a} = \frac{1}{2}\boldsymbol{\omega} \cdot \hat{\boldsymbol{n}} \tag{1.78}$$

therefore a non-zero vorticity means that the fluid has a rotating motion component.

There are velocity fields characterised by high vorticity values only in the vicinity of a vortex line while elsewhere the vorticity is zero: in this case we speak of a linear vortex. The current lines, that is, the flow lines of the vector $\mathbf{v}$ are circumferences contained in planes perpendicular to the vortex line and centred on the intersection of these planes with the line. Examples of this type are cyclones and tornadoes, vortices that form in the water, smoke trails.

A motion for which $\boldsymbol{\omega} \equiv 0$ is called *irrotational*; in this case being identically $\mathrm{rot}\,\mathbf{v} \equiv 0$ it will be:

$$\mathbf{v} = \mathrm{grad}\,\psi\ . \tag{1.79}$$

The scalar quantity ψ is the *velocity potential*. An irrotational motion has particular characteristics and can be comfortably integrated. In the case of irrotational motion of incompressible fluid, in addition to (1.51), $\mathrm{div}\,\mathbf{v} = 0$ holds, which together give:

$$\nabla^2 \psi = 0 \tag{1.80}$$

which is Laplace's equation, whose solutions are harmonic functions.

A motion is said to be *rotational* if $\boldsymbol{\omega}$ is not identically null.

We now prove that: *if a motion of an ideal fluid is irrotational at a certain instant, it is always irrotational*. Suppose that at time t is:

$$\boldsymbol{\omega} = \mathrm{rot}\,\mathbf{v} = 0\ . \tag{1.81}$$

On the basis of Stokes' theorem, we find that the circulation of $\mathbf{v}$ is instantaneously null:

$$\oint \mathbf{v} \cdot \mathrm{d}\mathbf{l} = \int_S \mathrm{rot}\,\mathbf{v} \cdot \mathrm{d}\boldsymbol{\sigma} = 0\ . \tag{1.82}$$

But, by Kelvin's theorem, the circulation is always zero and therefore ω is also always null.

From a physical point of view, this implies that, under the conditions assumed by us, vorticity cannot be produced. In reality this is not true: for example, moving an oar can produce a vortex in the water. Obviously the hypotheses adopted by us are not realistic; we will go back to the problem in Chap. 3.

We now derive the equation which vorticity obeys in the case of ideal fluids in isentropic motion. If we make explicit the Lagrangian derivative and make use of the vector identity:

$$(\mathbf{v} \cdot \text{grad })\mathbf{v} = \frac{1}{2} \text{grad } v^2 - \mathbf{v} \times \text{rot } \mathbf{v} . \tag{1.83}$$

by virtue of (1.73). Euler's equation becomes:

$$\frac{\partial \mathbf{v}}{\partial t} - \mathbf{v} \times \text{rot } \mathbf{v} = - \text{grad } w - \frac{1}{2} \text{grad } v^2 - \text{grad } \Phi . \tag{1.84}$$

Taking the curl of this expression we obtain:

$$\frac{\partial}{\partial t} \text{rot } \mathbf{v} - \text{rot } (\mathbf{v} \times \text{rot } \mathbf{v}) = \text{rot grad} \left(-w - \frac{1}{2}v^2 - \Phi \right) = 0 \tag{1.85}$$

from which:

$$\frac{\partial \omega}{\partial t} = \text{rot } (\mathbf{v} \times \omega) . \tag{1.86}$$

From this equation we find the conservation of the condition of irrotationality.

The Eq. (1.86) expresses the freezing condition of the vortex lines in the fluid: in an ideal fluid a vortex tube moves with the fluid and, since its intensity remains constant, where the section decreases the vorticity increases. If the section of the vortex tube is squeezed to zero, a vortex line is obtained and we can conclude that a material line initially coinciding with a vortex line will continue to be superimposed on it.

Let's highlight another aspect contained in the Eq. (1.86). Using the vector identity:

$$\text{rot } (\mathbf{v} \times \omega) = \mathbf{v} \text{ div } \omega - \omega \text{ div } \mathbf{v} - (\mathbf{v} \cdot \text{grad })\omega + (\omega \cdot \text{grad })\mathbf{v} \tag{1.87}$$

where the first addend in the second member is canceled by the solenoid character of ω. Equation (1.86) thus becomes:

$$\frac{\partial \omega}{\partial t} = -\omega \text{ div } \mathbf{v} + (\omega \cdot \text{grad })\mathbf{v} - (\mathbf{v} \cdot \text{grad })\omega . \tag{1.88}$$

This can be written in the form:

$$\frac{d\,\omega}{d\,t} + \omega \, div \, \mathbf{v} = (\omega \cdot grad\,)\mathbf{v} \qquad (1.89)$$

the div $\mathbf{v}$ can be eliminated by recalling the continuity equation, and we obtain:

$$\frac{d\,\omega}{d\,t} - \omega\frac{1}{\rho}\frac{d\,\rho}{d\,t} = (\omega \cdot grad\,)\mathbf{v} \,. \qquad (1.90)$$

Dividing by ρ and performing some simple algebraic manipulations we obtain the equation:

$$\frac{d}{d\,t}\left(\frac{\omega}{\rho}\right) = \left(\frac{\omega}{\rho} \cdot grad\,\right)\mathbf{v} \,. \qquad (1.91)$$

Let's now consider a material line that is also a vortex line: it will always be a vortex line. Furthermore, if we consider an infinitesimal element on it δl which has the points P and Q as extremes, and if $\mathbf{v}$ and $\mathbf{v}'$ are the velocities at P and at Q, it will be:

$$\mathbf{v}' = \mathbf{v} + (\delta\mathbf{l} \cdot grad\,)\mathbf{v} \qquad (1.92)$$

and therefore:

$$d\,\delta\mathbf{l} = (\mathbf{v}' - \mathbf{v})\,d\,t = (\delta\mathbf{l} \cdot grad\,)\mathbf{v}\,d\,t \qquad (1.93)$$

that is

$$\frac{d\,\delta\mathbf{l}}{d\,t} = (\delta\mathbf{l} \cdot grad\,)\mathbf{v} \qquad (1.94)$$

which is the same equation as (1.91). If initially ω/ρ and $\delta\mathbf{l}$ have the same direction, they will always be parallel and it will also be:

$$\frac{\omega}{\rho} \propto \delta\mathbf{l} \qquad (1.95)$$

If the fluid is incompressible, since $\omega \propto \delta l$, if the vortex lines stretches (δl increases) ω will grow and, on the contrary, ω will decrease in case of a vortex line contraction (δl decreases).

1.7   Numerical Solutions of Hydrodynamical Equations

When dealing with ideal fluids, the system of equations of the hydrodynamics consists of two scalar (continuity and energy conservation) and one vectorial equations (Euler). Therefore we have a system of 5 scalar equations with 5 independent variables: the three velocity components, pressure, and density. All of them are functions of two independent variables: space and time. In very rare cases this system admits an analytic solution. Routinely, the equations of hydrodynamics are solved numerically.

A brief summary of solutions methods are given here. For more details, the student is referred to Muller (1994). Essentially, there are two general different schemes, which take advantage of the two approaches one can adopt to write down gas-dynamical equations: the Eulerian and the Lagrangian schemes. The two schemes are related by the so-called formula of Lagrangian or molecular derivation, namely Eq. (1.13).

Eulerian schemes, by definition, employ meshes. Meshes can be fixed or adaptive, in the sense that, depending on the complexity of the problem, grids can be opened up into several level of sub-grids. The values of the dependent variables are then estimated at the corners, or at the center of each grid, and then evolved in time. Numerical diffusion, which originates from the advection term, is limited using high order Taylor expansion of the functions of interest. Many methods have been devised (implicit, explicit, Godunov methods, Piecewise Parabolic Methods, Adaptive Mesh Refinement, and so forth) over the years, since Eulerian methods have a very long tradition.

Lagrangian methods have been developed later than Eulerian methods. The obvious advantage here is that there is no need to set up a grid system, and that numerical diffusion is less important, since no advection is present in the Lagrangian form of the Euler equation. Lagrangian methods employ particles to sample the fluid properties. Therefore at particles positions fluid dynamic functions are known exactly. In locations of the fluids where particles are absent, an interpolation method is used. One of the most widely used Lagrangian method is SPH (Smoothed Particle Hydrodynamics), and in this method particles are smoothed according to a kernel function which depends on a smoothing length. This way the hydrodynamical functions can be evaluated everywhere. These methods fail to work effectively when multi-dimensional problems are to be treated.

1.8 Synthetic Summary

Fluids are characterised by the mobility of their parts. The study of the relative motions of their portions constitutes the topic of fluid dynamics, which treats liquids and gases indifferently; they are essentially distinguished by their different compressibility only.

The model used for the study of a fluid is a continuous one: the requirement for its applicability is that the average free path of the molecules is much smaller than the smallest typical size of the system under consideration.

There are two different descriptions of the fluid motion: the **Eulerian point of view** which consists in considering velocity, acceleration, density, etc. at a fixed point in space (the observer focuses on a point and watches the fluid passing through) and the **Lagrangian point of view** which follows the motion of the material elements of the fluid (the observer moves together with the fluid element). The two points of view are connected by the expression of the Lagrangian or molecular derivative.

The physical state of a fluid is described by the velocity field $\mathbf{v}(x, y, z, t)$ and the pressure fields $p(x, y, z, t)$, of the density $\rho(x, y, z, t)$ and the temperature

$T(x, y, z, t)$. Bearing in mind that the latter can be derived from the equation of state, the dynamic and thermodynamic behaviour of a fluid is defined by a vector function and two scalars. These can be determined by means of differential equations inferred from the conservation principles:

(1) from the conservation of matter we obtain the **continuity equation**,
(2) from the conservation of the momentum, and in the hypothesis that the fluid is ideal (absence of dissipation phenomena by friction and heat conduction) we deduce the **Euler equation**,
(3) from the conservation of the total energy a differential equation is obtained and eventually, through the Euler equation, the entropy density conservation is obtained.

In Euler's equation there are forces that can be of two types: long-range forces or volume forces and short-range forces or surface forces. The latter can be expressed by means of a tensor: the stress tensor. In the case of an ideal fluid, only the components of the main diagonal that are equal to $-p$ are different from 0, which implies that the force exerted on the surface unit is perpendicular to it and is independent of the orientation of the surface itself. For a real fluid, an additional tensor must be considered: the viscous stress tensor.

Given a velocity field **v**, vorticity can be defined. The fact that vorticity shows up in a given motion does not necessarily mean that there is a rotational component in that motion.

For an ideal fluid in isentropic motion (grad $s = 0$) conservation of the circulation of velocity along a closed material line (Kelvin's theorem) holds. From this property it follows that if a motion is irrotational at a certain instant, it will always be: in other words, vorticity cannot be created in an ideal fluid. The fact that eddies can occur in the water means then that the model of the ideal fluid simply has strong limitations.

The conservation of the circulation implies the freezing of the vortex lines to the fluid: there is no relative motion with respect to the vortex lines in a direction perpendicular to them, that is, the vortex lines are material lines. In an incompressible fluid, stretching of the vortex lines leads to an increase in vorticity.

Chapter 2
Acoustic Waves

2.1 Acoustic Waves in an Ideal Fluid at Rest

Let us consider the motion induced in an ideal fluid by a small perturbation. An object moving in the air exerts an action on (in other words, perturbs) the surrounding medium; if its motion is slow the air slides along it but, if the speed is high, air gets compressed and the pressure variation generates oscillations propagating around it in the form of a wave. One would argue that as a consequence of this molecules simply move from denser to less dense areas without necessarily wobbling. What is needed for the oscillations to occur is that the effects of the collisions between the molecules maintain and this requires that the dimensions L of the region in which the variations in density and pressure materialise are much larger than the mean free path λ of the molecules. Otherwise, molecules move freely and cancel the oscillation out very quickly. If we take the characteristic length L as oscillation wavelength, it will be necessary that this is larger than the average free path of the molecules in the medium.

We can expect the temperature to rise in a compression region, while it decreases in the rarefaction regions, heat conduction being negligible. Newton was the first to calculate the speed of sound propagation; he assumed instead that, due to conduction, the temperature could remain unchanged. The argument that the speed of sound can be calculated under isothermal conditions was actually wrong; the correct deduction indeed assumes that p and ρ vary adiabatically.

Let us suppose that the fluid is compressible but the density variations are small so that terms of higher order than the first can be neglected (method of small perturbations). As will be seen, such perturbation generates small oscillations in the fluid: *this oscillatory motion of small amplitude in a compressible fluid is called an acoustic wave.*

For the moment, we neglect the effect of the gravitational field and suppose that the unperturbed fluid is at rest and is uniform; let ρ_0 and p_0 be the constant values of density and pressure before the perturbation switches on. We are going to use the continuity and Euler equations and the condition $s = \text{cost}$.

© The Author(s), under exclusive license to Springer Nature Switzerland AG 2021
G. Carraro, *Astrophysics of the Interstellar Medium*,
UNITEXT for Physics, https://doi.org/10.1007/978-3-030-75293-4_2

Let's indicate with ρ and p the values of a small density and pressure perturbation, and with $\mathbf{v}$ the perturbed velocity. With the procedure described in Appendix A we arrive at the following equation in p:

$$\nabla^2 p - \frac{1}{c^2}\frac{\partial^2 p}{\partial t^2} = 0 \tag{2.1}$$

where it is:

$$c^2 = \left(\frac{\partial p}{\partial \rho}\right)_s \tag{2.2}$$

It can be verified that equations equal to (2.1) can be obtained for ρ and for the components of $\mathbf{v}$ and for the displacement ξ of the fluid elements.

The solutions of the equation (2.1) represent waves. Let us consider the case of a plane wave propagating in the direction of the x axis: the solution will be a function of the x coordinate and of the t time only. The equation becomes in this case:

$$\frac{\partial^2 p}{\partial x^2} - \frac{1}{c^2}\frac{\partial^2 p}{\partial t^2} = 0 \tag{2.3}$$

It can be shown that the more general solution has the form:

$$p(x, t) = f_1(x - ct) + f_2(x + ct) \tag{2.4}$$

where f_1 and f_2 are arbitrary functions.

Let us suppose $f_2 \equiv 0$; the solution $p = f_1(x - ct)$ represents a plane wave that propagates in the positive direction of the x axis with velocity c. Solutions of the same type would be obtained for the equations for ρ and for the velocity v_x: only this component is different from zero and therefore the velocity of the fluid has the same direction of the perturbation and the wave is said to be longitudinal (the proof is provided in Appendix B). This result is general and applies to all acoustic waves.

In a plane wave the velocity $v_x = v$ is related to the perturbation of the pressure p and the density ρ by a simple relation. Suppose p is described by $p = f(x - ct) := f(w)$. From Euler's equation, perturbed and linearised, we have:

$$\frac{\partial v}{\partial t} = -\frac{1}{\rho_0}\frac{\partial p}{\partial x} = -\frac{1}{\rho_0}f' \tag{2.5}$$

where f' is the derivative of the function $f(w)$, i.e. $\mathrm{d}f/\mathrm{d}w$. By integrating this equation we obtain:

$$v = -\frac{1}{\rho_0}\int f'\,\mathrm{d}t = \frac{p}{\rho_0 c} \tag{2.6}$$

Recalling (2.2) we have:

$$v = c\frac{\rho}{\rho_0} \tag{2.7}$$

From this last relation it can be seen that, being:

$$\frac{\rho}{\rho_0} \ll 1, \tag{2.8}$$

(compression and rarefaction are small) the velocity of the fluid elements is much lower than the speed of sound propagation.

The condition $v \ll c$ is therefore a necessary condition for incompressibility. On the other hand, the analysis of the properties of motion when $v \geq c$ shows that in such conditions the fluid is compressible and the effects of the compression are relevant (see Chap. 4). We will return to this topic in Chap. 5.

A detailed analysis of the equations in their general form would lead to the conclusion that if, within the fluid, a pressure perturbation is generated, when the cause that produced it turns off, the perturbation is extinguished by the propagation of sound waves and the time required to restore the fluid equilibrium in a region of dimension l is $t = l/c$. If the system undergoes an evolution in which a pressure gradient is generated in a region of dimension l and the characteristic time of evolution is $\tau > l/c$, the system can reach a condition of dynamic equilibrium; if, instead, the time during which the system conditions change is $\tau < l/c$ the system is not able to settle in an equilibrium configuration.

From thermodynamics we know a relationship between the adiabatic compressibility coefficient and the isothermal compressibility coefficient:

$$\left(\frac{\partial p}{\partial \rho}\right)_s = \gamma \left(\frac{\partial p}{\partial \rho}\right)_T \tag{2.9}$$

from which

$$c = \sqrt{\gamma \left(\frac{\partial p}{\partial \rho}\right)_T} \tag{2.10}$$

In the case of an ideal gas, the equation of state is $p = (k/\mu H)\rho T$, hence

$$\left(\frac{\partial p}{\partial \rho}\right)_T = \frac{kT}{\mu H} \tag{2.11}$$

and therefore:

$$c = \sqrt{\gamma \frac{kT}{\mu H}} \tag{2.12}$$

where μ is the average molecular weight. Introducing the values of the constants, with $\gamma = \frac{5}{3}$, we have:

$$c = 1.17 \cdot 10^4 \left(\frac{T}{\mu}\right)^{\frac{1}{2}} \text{cm/s} \tag{2.13}$$

Table 2.1 Typical sound speed velocity in different ISM components

Gas condition	T (K)	μ	c (km/s)
Cloud of H_2	10	2	0.4
HI Region	10^2	1	1.2
HII Region	10^4	1	10
Hot gas	$\geq 10^6$	1	≥ 100
Air	273	30	0.36

Recalling that, from the kinetic theory of perfect gases, the following holds:

$$kT = \frac{1}{2}\mu H < v >^2 \tag{2.14}$$

it is obtained from (2.12):

$$c^2 = \frac{1}{3}\frac{\gamma}{\mu H}\mu H < v >^2 = \frac{\gamma}{3}\langle < v >^2\rangle \tag{2.15}$$

which implies

$$c = \left(\frac{\gamma}{3}\right)^{\frac{1}{2}} < v > \tag{2.16}$$

where $< v >$ is the mean square velocity (Table 2.1). Therefore the sound propagation speed is about $3/4$ of the mean square velocity of the molecules. In other words, the speed of sound is of the same order of magnitude of the speed of molecules. This result is not unexpected because a disturbance (perturbation) such as a pressure variation is propagated by the motion of the molecules. Note that v is not the speed of molecules but that of a fluid volume element oscillating around its equilibrium position.

As will be shown later (Chap. 3) in the interstellar medium there are interaction processes between matter and radiation that involve the acquisition and loss of thermal energy by the gas: if these processes occur more rapidly than the variations due to the motion, the temperature is determined by these processes and the gas behaves as isothermal: this also happens for the propagation of sound waves and therefore the propagation speed is evaluated with the (2.15) setting $\gamma = 1$.

When the quantities that define a wave are harmonic functions of time, the wave is called monochromatic. Usually such functions are written as a real part of a complex quantity, for example

$$p = \Re[p_0(x, y, z)\exp(-i\omega t)] \tag{2.17}$$

where ω is the frequency of the wave. Substituting (2.17) into (2.1) we find the equation that p_0 must satisfy

$$\nabla^2 p_0 + \frac{\omega^2}{c^2} p_0 = 0 \qquad (2.18)$$

Let us consider a monochromatic plane wave moving in the positive direction of the x axis. All quantities are functions only of $(x - ct)$ and p can be written in the form:

$$p = \Re \left\{ A \exp \left[-i\omega \left(t - \frac{x}{c} \right) \right] \right\} \qquad (2.19)$$

where A is a complex constant. Putting $A = a \exp(-i\alpha)$ we get:

$$p = a \cos \left[\omega \left(t - \frac{x}{c} \right) + \alpha \right] = a \cos \left[\omega t - \frac{\omega x}{c} + \alpha \right] \qquad (2.20)$$

where a is the wave amplitude and the cosine argument is the phase. Indicating with $\hat{\mathbf{n}}$ the versor of the direction of propagation, the vector:

$$\mathbf{k} = \frac{\omega}{c} \hat{\mathbf{n}} = \frac{2\pi}{\lambda} \hat{\mathbf{n}} \qquad (2.21)$$

it is called the wave vector and wave number is its modulus, while λ is the wavelength.

The importance of monochromatic waves derives from the fact that, through Fourier analysis, any wave can be thought of as a superposition of monochromatic waves of different frequencies.

2.2 Velocity of Propagation in a Dispersive Medium

The velocity c that appears in the relation (2.21), valid for harmonic waves, is called phase velocity. In the case of a non-harmonic wave, the concept of wave propagation speed must be clarified. Let us consider the case of a wave having a finite amplitude at a given instant only in a limited region of space (wave packet). Fourier analysis teaches us that this wave is the superposition of harmonic waves whose frequency and wavelength vary continuously in a narrow range, e.g.

$$P(x, t) = \int_{k_0 - dk}^{k_0 + dk} A(k) \exp[i(kx - \omega t)] \, dk \qquad (2.22)$$

$A(k)$ is the amplitude that can be expressed as a Fourier transform of P.

Suppose that ω is a known function of k, for k belonging to the range $(k_0 - dk, k_0 + dk)$: ω will not differ much from $\omega_0 = \omega(K_0)$. Expanding in series and neglecting the terms of higher order than the first, we have:

$$\omega(k) = \omega_0 + \frac{d\omega}{dk} \bigg|_{k_0} (k - k_0) \qquad (2.23)$$

Substituting in (2.22) we have:

$$P(x, t) = \int_{k_0 - dk}^{k_0 + dk} A(k) \exp\left[i\left(kx - \omega_0 t - \left.\frac{d\omega}{dk}\right|_{k_0}(k - k_0)t\right)\right] dk \quad (2.24)$$

If in the exponential argument we add and remove $k_0 x$ we get:

$$P(x, t) = \int_{k_0 - dk}^{k_0 + dk} A(k) \exp[i(k_0 x - \omega_0 t)] \exp[i(x - \omega'|_{k_0}t)(k - k_0)] dk \quad (2.25)$$

where we have placed $\omega' = d\omega/dk$, which can be written in the form:

$$P(x, t) = P_0(x, t) \exp[i(k_0 x - \omega_0 t)] \quad (2.26)$$

with

$$P_0(x, t) = \int_{k_0 - dk}^{k_0 + dk} A(k) \exp[i(x - \omega'|_{k_0}t)(k - k_0)] dk \quad (2.27)$$

where P_0 is the width of the packet. This amplitude is constant on the surfaces:

$$x - \left.\frac{d\omega}{dk}\right|_{k_0} t = \text{cost.} \quad (2.28)$$

i.e. the packet propagates with the speed:

$$v_g = \left.\frac{d\omega}{dk}\right|_{k_0} \quad (2.29)$$

From (2.21) we have:

$$dc = \frac{k\, d\omega - \omega\, dk}{k^2} \quad (2.30)$$

from this and from (2.29) we get:

$$v_g = \frac{d\omega}{dk} = \frac{\omega}{k} + k\frac{dc}{dk} \quad (2.31)$$

If there is no dispersion, i.e. if the phase velocity of a harmonic wave is independent of the wave number and therefore is the same for all components, i.e. $c(k) = \text{cost.}$, the group velocity v_g is equal to the phase velocity of the component harmonic waves and the packet does not disperse.

If the medium is dispersive, the harmonic components of the packet propagate with different speeds, albeit slightly. This tends to produce the scattering of the packet which will however be limited: the condition that the frequencies of the component harmonic waves are kept in a narrow range allows us to define the group velocity, which is given by (2.31).

The relationship $\omega = c \cdot k$ between the frequency and the wave number is valid only for a monochromatic sound wave that propagates in a medium at rest; this relation also loses its validity when the medium is at rest but the effect of its gravitational field is important (see Sect. 2.4).

It is not difficult to derive an analogous relationship for a wave that propagates in a moving medium and is observed in a fixed coordinate system. Let us consider a homogeneous fluid moving with velocity v; take a fixed K coordinate system (x, y, z) and a K' coordinate system (x', y', z') tied to the fluid, which therefore moves with velocity v with respect to K. In the K' system the fluid is at rest and a monochromatic wave has the usual form:

$$p = a \cos(\mathbf{k} \cdot \mathbf{r'} - kct) \tag{2.32}$$

The vector radius $\mathbf{r'}$ in system K' is connected to the vector radius $\mathbf{r}$ in system K by:

$$\mathbf{r} = \mathbf{r'} + \mathbf{v}t \tag{2.33}$$

In the K system the wave has the form:

$$p = a \cos[\mathbf{k} \cdot \mathbf{r} - (\mathbf{k} \cdot \mathbf{c} + \mathbf{k} \cdot \mathbf{v})] \tag{2.34}$$

Since the coefficient of t in the exponent is the frequency ω of the wave, this is related to the wave vector in the reference system K by:

$$\omega = \mathbf{c} \cdot \mathbf{k} + \mathbf{v} \cdot \mathbf{k} \tag{2.35}$$

and the propagation speed is:

$$\frac{c}{k}\mathbf{k} + \mathbf{k} \tag{2.36}$$

and is the sum vector of the velocity $\mathbf{c}$ in the direction of $\mathbf{k}$ and the velocity $\mathbf{v}$ with which the sound is transported by the fluid.

In Appendix C the waves propagating in a liquid are presented as an example of wave propagation in a dispersive medium.

2.3 Acoustic Waves in a Patchy Medium at Rest

If the medium is initially not homogeneous, the analysis of Sect. 2.1 can be repeated taking into account that $\mathrm{grad}p_0 = 0$. However, $\mathrm{grad}p_0 \neq 0$ cannot be set because otherwise there would be no mechanical equilibrium. By perturbing the continuity equation and taking into account the adiabatic condition we obtain:

$$\frac{\partial p}{\partial t} + \rho_0 c^2 \operatorname{div} \mathbf{v} = 0 \tag{2.37}$$

Combining this with Euler's perturbed equation, and eliminating a variable, we arrive at the sound wave equation:

$$\operatorname{div}\left(\frac{1}{\rho_0} \operatorname{grad} p\right) - \frac{1}{\rho_0 c^2} \frac{\partial^2 p}{\partial t^2} = 0 \tag{2.38}$$

putting $\rho_0 = \text{cost}$. From this we get the Eq. (2.1).

At each separation surface between two media with different density the incident wave undergoes the phenomena of reflection and refraction. Therefore, in passing through a medium with a density gradient, the transmitted wave is attenuated.

2.4 Acoustic Waves in a Gravity-Dominated Medium

Let's consider the case of acoustic waves when the self-gravitation field cannot be neglected. Euler's equation is:

$$\frac{\partial \mathbf{v}}{\partial t} + (\mathbf{v} \cdot \operatorname{grad})\mathbf{v} = -\operatorname{grad}\Phi - \frac{1}{\rho} \operatorname{grad} p \tag{2.39}$$

where the gravitational potential satisfies the Poisson equation:

$$\nabla^2 \Phi = 4\pi G \rho \tag{2.40}$$

Suppose the medium is at rest and is uniform. Let ρ_0, p_0, Φ_0 be the constant values of density, pressure and gravitational potential. By applying the method of small perturbations (see Appendix D), with a procedure similar to that followed to obtain the equation of sound waves in the absence of gravity, the following equation is obtained:

$$\nabla^2 \rho - \frac{1}{c^2} \frac{\partial^2 \rho}{\partial t^2} = -\frac{4\pi G \rho_0}{c^2} \rho \tag{2.41}$$

For simplicity we suppose that the problem depends only on the spatial coordinate x as well as on time. We are looking for a solution such as:

$$\rho = A \exp[i(kx - \omega t)] \tag{2.42}$$

By imposing that ρ is the solution of the equation derived from (2.41)

$$\frac{\partial^2 \rho}{\partial x^2} - \frac{1}{c^2} \frac{\partial^2 \rho}{\partial t^2} = -\frac{4\pi G \rho_0}{c^2} \rho \tag{2.43}$$

we obtain an equation in ω:

$$\omega^2 = k^2 c^2 - 4\pi G \rho_0 \tag{2.44}$$

which is referred to as *dispersion relation*.

Let's define:

$$k_J^2 = \frac{4\pi G \rho_0}{c^2} \tag{2.45}$$

where k_J has the dimensions of the inverse of a length and:

$$\lambda_J = \frac{2\pi}{k_J} = \sqrt{\frac{\pi c^2}{G \rho_0}} \, , \tag{2.46}$$

it is called the wavelength of Jeans. The dispersion relation becomes:

$$\omega^2 = c^2 (k^2 - k_J^2) \tag{2.47}$$

If $k > k_J$ ($\lambda < \lambda_J$) it results $\omega^2 > 0$ and the solution represents a gravitationally modified sound wave whose propagation speed is:

$$s = \frac{d\omega}{dk} = \frac{ck}{\sqrt{k^2 - k_J^2}} = \frac{k}{k_J} \frac{c}{\sqrt{(k/k_J)^2 - 1}} \tag{2.48}$$

whose trend is in Fig. 2.1. If $k \to \infty$ and then $s \to c$ we have the case of ordinary sound waves. If $k < k_J$ ($\lambda > \lambda_J$) the solution is exponential and the perturbation grows over time with a time-scale:

$$\tau = \frac{1}{|\omega|} = \frac{1}{\sqrt{4\pi G \rho_0}} \frac{1}{\sqrt{1 - (k/k_J)^2}} \tag{2.49}$$

whose trend is in Fig. 2.2. Therefore wavelengths are unstable such that:

$$\lambda > \lambda_J \tag{2.50}$$

We can summarize the results obtained as follows:

- the effect of self-gravitation on solutions representing sound waves is to alter the relationship between ω and k and to make the medium dispersive by modifying the propagation speed according to (2.48);
- self-gravitation reduces the elasticity of the medium and consequently perturbations of fairly large scale density are amplified and cause gravitational instability;
- if the system has finite dimensions, let's think, for example, to a sphere with radius R and density ρ_0, any perturbation can be expanded in Fourier series and can be expressed as the superposition of harmonic waves with $0 \leq \lambda \leq 2R$, so if $2R < \lambda_J$

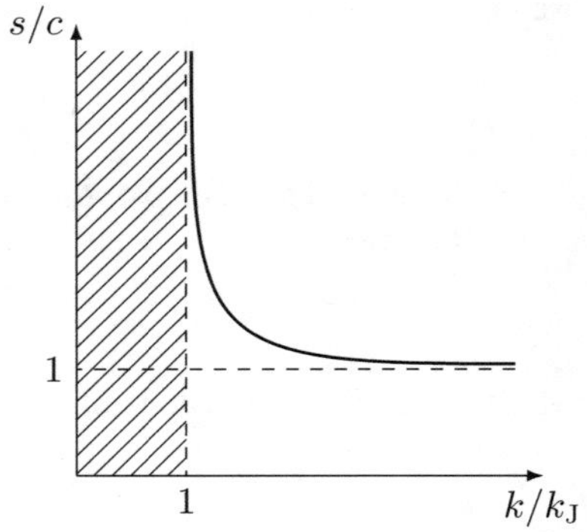

Fig. 2.1 Trend of sound
speed versus Jean wave
number

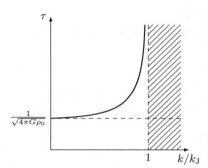

Fig. 2.2 Time-scale
dependence of perturbation
growth

all perturbations are stable, while if $R > \lambda_J/2$ the perturbations with $\lambda_J \leq \lambda \leq 2R$ are unstable and lead to the prevalence of self gravitation forces.

The critical mass (Jeans mass) for gravitational instability is

$$M_J = \frac{4\pi}{3} \left(\frac{\lambda_J}{2} \right)^3 \rho_0 = 1.13 \left[\frac{\pi k T}{G \mu H} \right]^{\frac{3}{2}} \rho_0^{\frac{1}{2}} \tag{2.51}$$

This approach has the great advantage of simplicity but the solution found does not correspond to any real physical solution. In fact, the Euler equation for a homogeneous medium with zero speed implies $\nabla^2 \Phi_0 = 0$ while from the Poisson equation it follows that if $\rho_0 \neq 0$ it will be $\nabla^2 \Phi_0 \neq 0$.

In particular physical situations (i.e. for systems with a different geometry and with different initial conditions) the Jeans criterion can be correct but in general it is not valid *a priori*. Its validity, at least at the level of orders of magnitude, can be argued with other justifications, for example with the virial theorem, which we will deal with in the next section.

2.5 Virial Theorem

Let's consider Euler's equation and multiply it scalarly by $\mathbf{r}$, then let us integrate on the material volume V, thus obtaining:

$$\int_V \rho \mathbf{r} \cdot \frac{d\mathbf{v}}{dt} \, dV = -\int_V \mathbf{r} \cdot \text{gradp} \, dV - \int_V \rho \mathbf{r} \cdot \text{grad}\Phi \, dV \qquad (2.52)$$

With the procedure described in Appendix E the (2.52) transforms into:

$$\frac{1}{2} \frac{d^2 I}{dt^2} = 2K + 2U - \int_S p\mathbf{r} \cdot d\boldsymbol{\sigma} + W \qquad (2.53)$$

where it is:

$$I = \int_V r^2 \rho \, dV \qquad (2.54)$$

is the moment of inertia of the material enclosed within V, K the kinetic energy of the macroscopic motion of the system, U the internal energy, W the gravitational potential energy and S is the boundary of V.

The relation (2.53) constitutes the *virial theorem* and, for an equilibrium system, it is reduced to:

$$2K + 2U + W - \int_S p\mathbf{r} \cdot d\boldsymbol{\sigma} = 0 \qquad (2.55)$$

If the first-member quantity is negative, the system is unstable and collapses. We apply the virial theorem to two particular cases.

(A) Isothermal and homogeneous sphere in equilibrium without external pressure. We can assume that the sphere is at rest from the macroscopic point of view, hence $K = 0$. Furthermore, the integral in (2.55) vanishes because on the surface S it is $p = 0$. Therefore we obtain:

$$2U + W = 0 \qquad (2.56)$$

Introducing the appropriate expressions for internal and gravitational energy we have:

$$2\left(\frac{3}{2}\frac{kTM}{\mu H}\right) - \frac{3}{5}\frac{GM^2}{R} = 0 \qquad (2.57)$$

where M and R are the mass and radius of the gas sphere respectively. By eliminating R as a function of ρ and solving for mass we have:

$$M = 0.98 \left(\frac{\pi kT}{G\mu H}\right)^{\frac{3}{2}} \rho^{-\frac{1}{2}} \qquad (2.58)$$

Fig. 2.3 Pressure profile of
an homogeneous isothermal
sphere subject to an external
pressure p_o

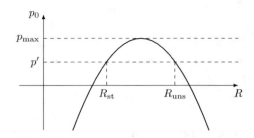

A mass greater than this cannot be in equilibrium and in fact in correspondence
it is $\mathrm{d}^2 I / \mathrm{d} t^2 < 0$. The mass defined by (2.58) is the analogue of the critical
mass of Jeans for gravitational instability; the dependence on ρ and T is the
same while the numerical coefficient differs slightly in the two expressions. The
Jeans criterion is therefore confirmed by the virial theorem.

(B) Homogeneous isothermal sphere immersed in a uniform medium with pressure
p_0. If we consider the states of equilibrium, the virial theorem applied to this
system gives:

$$2U + W - \int_S p\mathbf{r} \cdot \mathrm{d}\boldsymbol{\sigma} = 0 \tag{2.59}$$

Expanding the different terms we arrive at:

$$3\frac{kTM}{\mu H} - \frac{3}{5}\frac{GM^2}{R} = 4\pi R^3 p_0 \tag{2.60}$$

This relation gives, for a given mass and a given temperature, the radius of the
equilibrium configurations at an assigned external pressure p_0. For $p_0 < p_{max}$
(see p' in Fig. 2.3) the solutions are two and represent two equilibrium configurations; an analysis of their stability reveals that the one corresponding to the
greater radius (see R_{st}) is stable while the one corresponding to the smaller radius
(see R_{ins}) is unstable. For $p_0 > p_{max}$ there are no equilibrium configurations and
in such circumstances the external compression causes the sphere to collapse
even when its mass is less than the Jeans mass.

The value p_{max} is determined by the condition $\mathrm{d} p_0 / \mathrm{d} R = 0$ and we find:

$$p_{max} = 3.15 \left(\frac{kT}{\mu H}\right)^4 \frac{1}{G^3 M^2} \tag{2.61}$$

2.6 Gravitational Collapse of a Sphere

A gaseous cloud that becomes unstable under the action of self gravitating force or with the help of an external compression tends to rapidly evolve towards a collapse situation in which the only force acting is gravitation (free fall) . The gravitational energy that is released would tend to transform into thermal energy and therefore to increase the pressure that would then contrast the collapse. But the collisions between atoms produce ionisations and excitations and therefore the disappearance of this thermal energy. In fact, the de-excitation or recombination of the ions releases radiant energy which, in turn, is lost to the system since the average free path of the photons is greater than the size of the cloud. Subsequently, because of the increase of the density and consequently of the opacity, the radiant energy produced remains trapped and contribute to increasing the thermal energy and the pressure until the system evolves through states that differ slightly from states of hydrostatic equilibrium.

A first approximation to treat this collapse consists in considering the motion as a free fall in which all the forces are negligible except for self-gravitation ; this way a correct estimate of the time scale is obtained. Let us consider a sphere of mass M, initially at rest, with an initially uniform or decreasing density profile towards the outside; we consider a spherical shell (shell) initially placed at a distance of a from the center. Denoting by $M(a)$ the mass enclosed within the sphere of radius a the equation of motion for this shell in Lagrangian notation is:

$$\frac{d^2 r}{d t^2} = -\frac{GM(a)}{r^2} \tag{2.62}$$

The various shells do not overlap during motion. The initial conditions for a shell are $t = 0, r = a, \ dr/dt = 0$. The equation can be integrated with the substitution:

$$r = a \cos^2 \theta \tag{2.63}$$

and one then finds:

$$\theta + \frac{1}{2} \sin 2\theta = \left(\frac{2GM(a)}{a^2} \right)^{\frac{1}{2}} t \tag{2.64}$$

The shells reach the center when $\theta = \pi/2$. The time taken for a shell to fall back on the center is the free fall time (or free-fall time) τ_{ff} and is obtained from (2.64) by setting $\theta = \pi/2$

$$t_{ff} = \left(\frac{3\pi}{32G\langle\rho\rangle_a} \right)^{\frac{1}{2}} = 1.92\frac{1}{\sqrt{4\pi G\langle\rho\rangle_a}} \tag{2.65}$$

where it is:

$$\langle\rho\rangle_a = \frac{3M(a)}{4\pi a^2} \tag{2.66}$$

is the average density below the considered shell. Compare the free-fall time with the time given by (2.49) relative to the growth of the perturbations when $k = 0$.

2.7   Synthetic Summary

Sound waves are generated by perturbations in a fluid that propagate as longitudinal waves. In order for wave propagation to occur, the wavelength must be much larger than the average free path of the molecules.

Wave propagation occurs under adiabatic conditions (in the interstellar medium there can be isothermal propagation if there are cooling processes (see Chap. 4) with a cooling time shorter than the wave period).

The study of the propagation of a small amplitude wave in a uniform medium at rest, assuming the action of the gravitational field to be negligible, is done starting from the fundamental equations of fluids with the method of small perturbations. It consists in supposing that a perturbation small enough to be able to neglect the second order terms is introduced. This means that the resulting equations are linear in the perturbations.

If in the group of equations we eliminate all the variables except one, we obtain the equation that satisfies this variable, which is the wave equation.

From the condition that the perturbation of the density is small compared to the initial density it follows that the velocity of the elements of the fluid is much lower than the speed of sound. Viceversa supersonic motions imply a strong compression.

The action of an impulsive force, which alters the equilibrium in a fluid, determines a propagation of sound waves through which the equilibrium is restored: to restore the equilibrium in a region of dimensions L it will take a time $\tau = L/c$. If the perturbation lasts $T > \tau$ the system can constantly adapt to the new situation and evolve through equilibrium states. If, on the other hand, $T < \tau$ the system fails to bring itself to equilibrium.

Harmonic waves have a particular importance in wave phenomena due to the fact that any wave can, through Fourier analysis, be thought of as a superposition of harmonic waves. An example of this is offered by the wave packet which is a wave with a non-zero amplitude in a limited region of space. It can be thought of as a superposition of harmonic waves in which pulsation and wave number vary over a narrow range.

The wave packet propagates with a *group speed* which is expressed by (2.31). If the phase velocity $c(k)$ of each harmonic component independent of k, the medium is called non-dispersive and the group velocity coincides with the phase velocity. If the phase velocity $c(k)$ depends on k the medium is said to be dispersive and during propagation the packet tends to gradually deform.

The introduction of the self gravitation force makes a fluid dispersive with respect to the propagation of sound waves. The study of small perturbations is done as in the absence of gravity: in this case a further equation must be introduced to describe the evolution of the perturbation of the gravitational potential. By perturbing, linearising the equations and eliminating all the variables except one, we arrive at the Eq. (2.41); of this we look for a solution that represents a plane wave.

By introducing this expression we obtain the dispersion relation (2.44). From this we deduce that if $k > k_0$ the solution represents a sound wave altered by the action of gravity so that the phase velocity is a function of k.

If instead $k < k_0$ the solution is no longer a wave but describes the progressive growth of the perturbation: we are in the presence of a gravitational instability that triggers a collapse in free fall (Jeans criterion).

Although this collapse criterion is not correct in the case considered, it is valid for other geometric configurations and its correctness is proved by the virial theorem. This theorem can be deduced from Euler's equation and is in integral form. It establishes conditions for the dynamic equilibrium of a system when it is imposed that the second derivative with respect to the time of the moment of inertia of the system is zero.

If applied to a homogeneous and isothermal sphere not subject to external pressure, it reproduces the Jeans criterion.

The time scale of gravitational collapse is inversely proportional to the square root of the density.

Appendices

Appendix A: Derivation of the Sound Wave Equation

We denote by ρ and p the perturbation values of the density and pressure and by $\mathbf{v}$ the perturbed velocity. Suppose the perturbation is small so that we can consider ρ, p and $\mathbf{v}$ infinitesimal with respect to ρ_0, p_0 and neglect terms of higher order than the first.

The equations that describe the behaviour of the system are:

$$\frac{\partial}{\partial t}(\rho_0 + \rho) = -\operatorname{div}\left[(\rho_0 + \rho)\mathbf{v}\right] \tag{2.67}$$

$$\frac{\partial \mathbf{v}}{\partial t} + (\mathbf{v} \cdot \operatorname{grad})\mathbf{v} = -\frac{1}{\rho_0 + \rho}\operatorname{grad}(p_0 + p) \tag{2.68}$$

By exploiting the fact that ρ_0 and p_0 satisfy the unperturbed equations of motion and therefore that:

$$\partial \rho_0 / \partial t = 0 \tag{2.69}$$

$$\operatorname{grad} p_0 = 0 \tag{2.70}$$

and ignoring terms of higher order than the first in p and ρ we obtain the following:

$$\frac{\partial \rho}{\partial t} + \rho_0 \operatorname{div} \mathbf{v} = 0 \tag{2.71}$$

$$\frac{\partial \mathbf{v}}{\partial t} = -\frac{1}{\rho_0} \operatorname{grad} p \tag{2.72}$$

A relationship between p and ρ can be derived from the adiabatic conditions:

$$p = \left(\frac{\partial p}{\partial \rho}\right)_s \rho = c^2 \rho \tag{2.73}$$

Removing ρ from (2.73) and (2.71) gives:

$$\frac{\partial p}{\partial t} + c^2 \rho_0 \operatorname{div} \mathbf{v} = 0 \tag{2.74}$$

The two equations (2.72) and (2.74) give the description of the sound wave. We can reduce ourselves to a single equation in an unknown: to obtain it we apply the operator $\partial/\partial t$ to the (2.74) and the operator div to the (2.72) and multiply the latter for ρ_0. You get:

$$\frac{1}{c^2} \frac{\partial^2 p}{\partial t^2} = -\rho_0 \frac{\partial}{\partial t} \operatorname{div} \mathbf{v} \tag{2.75}$$

$$\rho_0 \operatorname{div} \frac{\partial \mathbf{v}}{\partial t} = -\operatorname{div} \operatorname{grad} p \tag{2.76}$$

Subtracting member from member we have:

$$\nabla^2 p - \frac{1}{c^2} \frac{\partial^2 p}{\partial t^2} = 0 \tag{2.77}$$

Appendix B: Sound Waves Are Longitudinal Waves

We verify this circumstance in the case of plane waves moving in the positive direction of the x axis. The solution in this case is given by $p = p(x - ct)$, $\rho = \rho(x - ct)$ and $\mathbf{v} = (v_x(x - ct), v_y(x - ct), v_z(x - ct))$.

From Euler's equation, perturbed and linearised, we obtain:

$$\frac{\partial v_x}{\partial t} = -\frac{1}{\rho_0}\frac{\partial p}{\partial x} \neq 0 \tag{2.78}$$

$$\frac{\partial v_y}{\partial t} = -\frac{1}{\rho_0}\frac{\partial p}{\partial y} = 0 \tag{2.79}$$

$$\frac{\partial v_z}{\partial t} = -\frac{1}{\rho_0}\frac{\partial p}{\partial z} = 0 \tag{2.80}$$

from which it follows that v_y and v_z are constant since $\partial v_i/\partial t \approx d\,v_i/d\,t$. A reference system can then be chosen in which v_y and v_z are null and the only non-zero component is v_x: in this case the oscillation is perpendicular to the direction of propagation of the phase.

Appendix C: Waves in a Liquid

The general expression of the phase velocity of the harmonic waves propagating on the surface of a liquid is given by:

$$v = \left[\left(\frac{g\lambda}{2\pi} + \frac{2\pi\tau}{\rho\lambda}\right)\tanh\frac{2\pi h}{\lambda}\right]^{\frac{1}{2}} \tag{2.81}$$

where h is the depth of the liquid, ρ the density and τ the surface tension, g is the acceleration due to gravity and λ the wavelength.

Let's consider some special situations.

(1) The depth is very large relative to the wavelength so that $\tanh(2\pi h/\lambda) \approx 1$ (surface waves). The (2.81) reduces to:

$$v = \left(\frac{g\lambda}{2\pi} + \frac{2\pi\tau}{\rho\lambda}\right)^{\frac{1}{2}} \tag{2.82}$$

Within the range of validity of this approximation we can consider two further approximations.

(a) The wavelength is large enough to neglect the second term with respect to the first wave:

$$v = \sqrt{\frac{g\lambda}{2\pi}} \tag{2.83}$$

A wave of this type is called a gravity wave and its propagation speed is independent of the nature of the medium as the parameters of the liquid do not appear in the expression of v. The larger λ, the faster the propagation. For this reason a strong and fast wind produces longer waves than a slow wind. For a harmonic wave the relation $\omega = vk$ holds. Therefore, combining with the (2.83) we have:

$$\omega = \sqrt{gk} \tag{2.84}$$

and the group speed will be:

$$v_g = \frac{d\omega}{dk} = \frac{1}{2}\sqrt{\frac{g}{k}} = \frac{1}{2}v \tag{2.85}$$

(b) The wavelength is small so the dominant term is the second. Then you will have:

$$v = \sqrt{\frac{2\pi\tau}{\rho\lambda}} \tag{2.86}$$

These waves are called capillary waves; they are observed when a very light wind blows or when the container containing the liquid is subjected to vibrations of high frequency and small amplitude. The speed of propagation increases as the wavelength decreases. Yes has:

$$\omega = kv = \sqrt{\frac{k^3\tau}{\rho}} \tag{2.87}$$

and the group speed is:

$$v_g = \frac{3}{2}\sqrt{\frac{k\tau}{\rho}} \tag{2.88}$$

In the two types of waves considered there is dispersion. If a wave motion results from the superposition of harmonic waves of different frequencies, it distorts because each component propagates with a different speed.

(2) When the depth is small compared to the wavelength it results $2\pi h/\lambda \ll 1$ and we have:

$$\tanh\frac{2\pi h}{\lambda} \approx \frac{2\pi h}{\lambda} \tag{2.89}$$

therefore the phase velocity of a harmonic wave is:

$$v = \sqrt{\frac{g\lambda}{2\pi} + \frac{2\pi\tau}{\rho\lambda}\frac{2\pi h}{\lambda}} = \sqrt{gh + \frac{4\pi^2\tau h}{\rho\lambda^2}} \tag{2.90}$$

Furthermore, since $h \ll \lambda$ we can neglect the second addend and therefore we have:

$$v = \sqrt{gh} \tag{2.91}$$

The phase velocity is independent of the wavelength and therefore there is no dispersion.

Appendix D: Acoustic Waves in a Gravity-Dominated Medium

Suppose that the medium is uniform and at rest and that the density, pressure and gravitational potential are constant ρ_0, p_0, Φ_0. These quantities must satisfy the equations of motion, in steady state, of a homogeneous medium:

$$\frac{\partial \rho_0}{\partial t} = 0 \qquad \text{grad}\,\Phi_0 = 0 \tag{2.92}$$

$$\text{grad}\,p_0 = 0 \qquad \nabla^2 \Phi_0 = 4\pi G \rho_0 \tag{2.93}$$

Assuming that there is a small perturbation $\mathbf{v}$, p, ρ, Φ the linearised equations to which the perturbed quantities satisfy are obtained taking into account the equations (2.92) and (2.93) and ignoring higher order terms. They are:

$$\frac{\partial \rho}{\partial t} + \rho_0\,\text{div}\,\mathbf{v} = 0 \tag{2.94}$$

$$\frac{\partial \mathbf{v}}{\partial t} = -\,\text{grad}\,\Phi - \frac{1}{\rho_0}\,\text{grad}\,p \tag{2.95}$$

$$\nabla^2 \Phi = 4\pi G \rho \tag{2.96}$$

to which the adiabatic condition $p = c^2 \rho$ is added where $c^2 = (\partial p / \partial \rho)_0$, is the square of the speed of sound in a non-autogravitating medium that obeys to the same equation of state of the medium that we consider.

Eliminating p with the help of this last relation in (2.95) we obtain:

$$\frac{\partial \mathbf{v}}{\partial t} = -\,\text{grad}\,\Phi - \frac{c^2}{\rho_0}\,\text{grad}\,\rho \tag{2.97}$$

deriving the (2.94) with respect to time and applying the div operator to the (2.97) and multiplying it by ρ_0 and subtracting it from the (2.94) we obtain:

$$\frac{\partial^2 \rho}{\partial t^2} - \rho_0 \nabla^2 \Phi - c^2 \nabla^2 \rho = 0 \tag{2.98}$$

By eliminating the potential Φ between this and (2.96), we obtain at the end:

$$\nabla^2 \rho - \frac{1}{c^2}\frac{\partial^2 \rho}{\partial t^2} = -\frac{4\pi G \rho_0}{c^2}\rho \tag{2.99}$$

Appendix E: The Virial Theorem

(a) Let us consider the integral a first member of (2.52). As it results:

$$\frac{d^2 r^2}{d t^2} = \frac{d}{d t}(2\mathbf{r} \cdot \mathbf{v}) = 2\mathbf{r} \cdot \frac{d\mathbf{v}}{d t} + 2\mathbf{v} \cdot \mathbf{v} \tag{2.100}$$

and therefore:

$$\mathbf{r} \cdot \frac{d\mathbf{v}}{d t} = \frac{1}{2}\frac{d^2 r^2}{d t^2} - v^2 \tag{2.101}$$

you will have:

$$\int_V \rho \mathbf{r} \cdot \frac{d\mathbf{v}}{d t}\,d V = \frac{1}{2}\frac{d^2}{d t^2}\int_V \rho r^2\,d V - \int_V \rho v^2\,d V \tag{2.102}$$

is

$$\int_V \rho \mathbf{r} \cdot \frac{d\mathbf{v}}{d t}\,d V = \frac{1}{2}\frac{d^2 I}{d t^2} - 2K \tag{2.103}$$

where K is the kinetic energy of the macroscopic motion.

(b) Keeping in mind the identity:

$$\mathrm{div}\,(p\mathbf{r}) = p\,\mathrm{div}\,\mathbf{r} + \mathbf{r} \cdot \mathrm{grad}p = 3p + \mathbf{r} \cdot \mathrm{grad}p \tag{2.104}$$

we have for the first integral to the second member of (2.52):

$$\int_V \mathbf{r} \cdot \mathrm{grad}p\,d V = \int_V \mathrm{div}\,(p\mathbf{r})\,d V - 3\int_V p\,d V \tag{2.105}$$

On the other hand, the integral

$$\int_V p\,d V \tag{2.106}$$

can be transformed using the equation of state:

$$\int_V p\,d V = \int_V nkT\,d V = \frac{2}{3}\int_M \frac{3}{2}\frac{kT}{\mu H}\,d M = \frac{2}{3}U \tag{2.107}$$

where U is the thermal energy of the system. Therefore the (2.52) becomes:

$$\int_V \mathbf{r} \cdot \mathrm{grad}p\,d V = \int_S p\mathbf{r} \cdot d\boldsymbol{\sigma} - 2U \tag{2.108}$$

(c) Considering the second integral in the second member we have:

$$\int_V \rho \mathbf{r} \cdot \operatorname{grad} \Phi \, d V = \int_M \mathbf{r} \cdot \operatorname{grad} \Phi \, d M \qquad (2.109)$$

We work on this integral to show that it is equal to a potential energy. We will do this demonstration limited to the case of spherical symmetry. We have:

$$\operatorname{grad} \Phi = \frac{d\,\Phi}{d\,r} = \frac{GM(r)}{r^2} \qquad (2.110)$$

where $M(r)$ is the mass enclosed within the sphere of radius r. The potential energy of a given matter distribution is defined as the work done by the system to form the distribution by carrying the diffused matter from infinity. Assuming you have already brought a quantity of matter $M(r)$ from infinity, the work required to add a spherical shell of thickness $d\,r$ is:

$$GM(r)\, d\, M(r) \int_\infty^r \frac{d\,r}{r^2} \qquad (2.111)$$

So the potential energy W of the spherical distribution of mass M and radius R is:

$$W = - \int_0^M \frac{GM(r)}{r}\, d\, M(r) \qquad (2.112)$$

We multiply and divide the integrating function by r:

$$\frac{GM(r)}{r} = r \frac{GM(r)}{r^2} = r \frac{d\,\Phi}{d\,r} = r \cdot \operatorname{grad} \Phi \,, \qquad (2.113)$$

therefore we have:

$$W = - \int_0^M \mathbf{r} \cdot \operatorname{grad} \Phi \, d\, M(r) \qquad (2.114)$$

Using the (2.103), (2.108), and (2.112), the (2.52) becomes:

$$\frac{1}{2} \frac{d^2 I}{d\,t^2} = 2K - \int_S p\mathbf{r} \cdot d\boldsymbol{\sigma} + 2U + W \qquad (2.115)$$

References

1. Landau, L.D., Lifshitz, E.M.: Fluid Mechanics, Chapter 8. Pergamon Press (1987)
2. Batchelor, G.K.: An Introduction to Fluid Dynamics, Chapters 10 and 13. Cambridge University Press (2012)

Chapter 3
Real Fluids

3.1 Transport Phenomena

In conditions of thermodynamic equilibrium, matter is characterised by the uniformity of the quantities that define the state of the system (pressure, temperature, chemical composition, etc.). If any of these quantities is not uniformly distributed, the system is not in equilibrium and exchanges occur of a mechanical or thermal nature between nearby elements until equilibrium is eventually achieved.

These phenomena are called transport phenomena and are related to the motion of thermal agitation (brownian motion) of the molecules and have the effect of transferring a certain quantity Q, when an associated quantity Θ, called intensity, is not uniform and the direction of the transfer is such as to homogenise its value. The nature of the process depends on the quantity Q taken into consideration, the molecular structure of the system considered, and the state of inhomogeneity but the essential characteristics are described by the same differential equation, of the first order with respect to time, which has the following form:

$$\frac{\partial \Theta}{\partial t} = D \nabla^2 \Theta \tag{3.1}$$

called diffusion equation, where D is a constant called diffusivity or diffusion coefficient. A more complicated equation may be necessary to describe particular phenomena (think for example of the ambipolar diffusion, which will be discussed in Sect. 8.6).

There are three basic types of transport phenomena, each characterised by the consideration of a particular quantity Q : there is (a) the standard, classical *diffusion*, in which the intensity Θ is the concentration of a given molecular species while the quantity Q is the flow of material of that molecular species, (b) the *thermal conduction*, where Q is the thermal energy and the corresponding intensity Θ is the temperature, and (c) *internal friction* where Q is the momentum and Θ is the

© The Author(s), under exclusive license to Springer Nature Switzerland AG 2021
G. Carraro, *Astrophysics of the Interstellar Medium*,
UNITEXT for Physics, https://doi.org/10.1007/978-3-030-75293-4_3

velocity. While we omit the description of diffusion, we will briefly analyse the other two phenomena.

3.2 Thermal Conduction

Thermal conduction is a phenomenon of thermal energy transport determined by the presence of a temperature gradient. The temperature gradient must be small otherwise another mechanism of thermal energy transfer occurs due to the macroscopic motion of matter (convection).

We denote with Φ_c the current density of thermal energy, i.e. the heat flux through the unit of surface in the unit of time. Φ_c is given by the Fourier's law:

$$\Phi_c = -K \text{ grad } T \tag{3.2}$$

where K is the thermal conductivity and the negative sign appears because the heat flows in the opposite direction to that of the gradient, which is that of increasing temperatures. Although the conduction mechanism may be different depending on the state of aggregation of the system, Fourier's law is still applicable.

Let us assume for simplicity that the temperature gradient is parallel to the x axis and that therefore the heat flow occurs in this direction. Let us consider a parallelepiped with its axis parallel to the x axis, with height dx and base area S. The heat entering from one of the bases is $\Phi_c S$, while what flows out from the other is $-\Phi'_c S$, and therefore the increase in thermal energy per unit time within the volume is:

$$\frac{\partial E_t}{\partial t} = \Phi_c S - \Phi'_c S = -\frac{\partial \Phi_c}{\partial x} S \, dx \tag{3.3}$$

This increase in thermal energy corresponds to an increase in temperature defined by:

$$\frac{\partial E_c}{\partial t} = \zeta \frac{\partial T}{\partial t} \rho S \, dx \tag{3.4}$$

where ζ is the specific heat. Combining the last two relations we obtain:

$$\rho \zeta \frac{\partial T}{\partial t} = -\frac{\partial \Phi_c}{\partial x} \tag{3.5}$$

and making use of Fourier's law we obtain:

$$\rho \zeta \frac{\partial T}{\partial t} = -\frac{\partial}{\partial x} \left(-K \frac{\partial T}{\partial x} \right) \tag{3.6}$$

If we assume that the thermal conductivity is constant we get the equation:

$$\frac{\partial T}{\partial t} = D_{\rm c} \frac{\partial^2 T}{\partial x^2} \tag{3.7}$$

with $D_{\rm c} = K/\rho \zeta$, which is identical to (3.1). In reality the thermal conductivity depends on the temperature and therefore it is not constant, but since the gradient is small, this approximation can generally be adopted.

Using diffusion theory we evaluate the thermal conductivity in the case of a plasma. The conductive flux can be expressed, as an order of magnitude, by:

$$\Phi_{\rm c} = -n_{\rm e} v_{\rm m} l \frac{{\rm d}\,E_{\rm t}}{{\rm d}\,x} \tag{3.8}$$

where $n_{\rm e}$ is the density of electrons, $v_{\rm m}$ their average velocity, l the average free path, and $E_{\rm t}$ the average kinetic energy of translation. Electron-electron collisions are neglected and the gas is assumed to be fully ionized. For non-degenerate non-relativistic electrons it is:

$$E_{\rm t} = \frac{1}{2} m v_{\rm m}^2 = \frac{3}{2} k_{\rm B} T \tag{3.9}$$

($k_{\rm B}$ is Boltzmann's constant) so:

$$\Phi_{\rm c} = -\frac{3}{2} n_{\rm e} v_{\rm m} l k_{\rm B} \frac{{\rm d}\,T}{{\rm d}\,x} \tag{3.10}$$

Comparing with the Fourier equation (3.2) we deduce the expression of thermal conductivity:

$$K = \frac{3}{2} n_{\rm e} v_{\rm m} l k_{\rm B} \tag{3.11}$$

The mean free path is given by $l = (n_{\rm i} \sigma)^{-1}$ where σ is the mean cross section for the collision of an electron with an ion, and $n_{\rm i}$ is the density of the ions. To evaluate σ it is assumed that the collision takes place when the electron approaches the ion to the point that the average kinetic energy of the electron is equal to its electrostatic potential energy in the ion field, i.e. when:

$$\frac{1}{2} m_{\rm e} v^2 = \frac{3}{2} k_{\rm B} T = \frac{1}{4\pi \varepsilon_0} \frac{Z e^2}{r_0} \tag{3.12}$$

where Ze is the average charge of the ions. Therefore:

$$r_0 = \frac{Z e^2}{6\pi \varepsilon_0 k_{\rm B} T} \tag{3.13}$$

is

$$\sigma = \pi r_0^2 = \frac{Z^2 e^4}{36\pi \varepsilon_0^2 k_{\rm B}^2 T^2} \tag{3.14}$$

It turns out then:

$$K = 93.53\pi k_B \frac{n_e}{n_i} \frac{\varepsilon_0^2}{\sqrt{m_e} Z^2 e^4} (k_B T)^{5/2} \tag{3.15}$$

For a fully ionised pure hydrogen gas $n_e = n_i$ e $Z = 1$, then:

$$K = 6 \cdot 10^{-7} T^{5/2} \ [\text{erg cm}^{-1} \text{ s}^{-1} \text{ K}^{-1}] \tag{3.16}$$

Despite the crudeness of the derivation procedure, this formula is correct within a factor of 2 or 3.

Given the strong dependence of K on temperature, it can be expected that conduction is important in an ionised gas only when the temperature is high; in this case, and if the magnetic field is negligible, the large mean free path of the electrons makes the thermal conductivity very high. The magnetic field reduces the mean free path of electrons in the direction transverse to the field and therefore reduces conductivity.

Usually when dealing with an ordinary plasma the mean free path l of the electrons is considerably smaller than the size of the plasma and in particular the scale

$$L_T = \frac{T}{|\text{ grad T}|} \tag{3.17}$$

on which there is an appreciable variation in temperature. In the case of astrophysical systems, on the other hand, it may happen that, due to the very low density, the mean free path becomes comparable to or greater than L_T. In these conditions the laws for the propagation of heat that we have described lose their validity and the conductive flux tends to become saturated. The flow of thermal energy then depends on the global temperature distribution because electrons can move over very large distances. Essentially the condition of saturation of the conductive flux depends on the fact that the electrons transfer the energy over distances over which they do not suffer collisions. In this case the flow is saturated at the value:

$$\Phi_{c,sat} = f \sqrt{\frac{2k_B T}{\pi m_e}} n_e k_B T \tag{3.18}$$

where f is a correction factor of the order of the unity.

3.3   Internal Friction

Let us consider a fluid in which, in addition to the thermal agitation of the molecules, a macroscopic motion occurs within the fluid. For simplicity we suppose that the fluid is incompressible and moves, in this motion, along the y axis with a velocity v_y, a function of the coordinate x (parallel motion); furthermore, suppose that the modulus of velocity is a decreasing function of x.

Molecules not only move parallel to the y axis because of thermal agitation. Let Σ be a flat surface inside the fluid, perpendicular to the x axis and therefore parallel to the y axis. There are molecules that, due to the effect of thermal agitation, cross this surface: those that move towards increasing x have a momentum, due to the overall motion, greater than that possessed by the molecules crossing Σ in the direction of decreasing x. The momentum current density is the momentum transferred in the unit of time through a unit area perpendicular to the direction in which the speed of the ensemble motion changes. The momentum current density Φ_p, is related to the gradient of the velocity of the whole motion by the experimental relation:

$$\Phi_p = -\eta \frac{\partial v_y}{\partial x} \tag{3.19}$$

analogous to Fourier's law; the quantity η is called the coefficient of dynamic viscosity and the negative sign is due to the fact that the transfer of the momentum takes place in the direction in which the speed of the overall motion decreases; this relation can be considered as an approximate expression of a more general law in which other powers of the gradient also appear. Since Φ_p is expressed in m^{-1} kg s^{-2}, the viscosity coefficient η is expressed in m^{-1} kg s^{-1}.

Since the fluid which, with respect to Σ, is on the side of the increasing x gains momentum while that which is on the side of the decreasing x looses momentum, it can be said that the first portion of fluid undergoes the action of a force parallel to the direction of flow and the second portion of fluid is subjected to an equal but opposite force. This force per unit area is called *tangential force* τ and according to the second principle of dynamics it results:

$$\tau = \Phi_p \tag{3.20}$$

Let us consider a parallelogram of length dx parallel to the axis x and base area S. The fluid contained in this volume element has a velocity of v_y. The molecules that enter by moving towards increasing x produce an increase in momentum per unit of time, parallel to the y axis, equal to $\Phi_p S$: it is said that there is a flow of incoming momentum of this entity. Similarly there will be an outgoing flow on the other side equal to $\Phi'_p S' = \Phi_p S$. Therefore the increase in the unit of time of the fluid momentum within the parallelepiped is equal to the difference between the incoming and outgoing momentum flow. Given p_y, the momentum per unit of volume, for the conservation of the momentum will be:

$$\frac{\partial p_y}{\partial t} S \, dx = -\frac{\partial \Phi_p}{\partial x} S \, dx \tag{3.21}$$

from which:

$$\frac{\partial p_y}{\partial t} = -\frac{\partial \Phi_p}{\partial x} \tag{3.22}$$

The term on the left represents a force per unit volume and is the viscous force $\mathbf{f}_V$. We can obtain for $\mathbf{f}_V$ a different expression by making use of (3.19) and assuming $\eta = \text{cost}$:

$$\mathbf{f}_{V,y} = \frac{\partial p_y}{\partial t} = \eta \frac{\partial^2 v_y}{\partial x^2} \tag{3.23}$$

Since $p_y = \rho v_y$, we obtain:

$$\frac{\partial v_y}{\partial t} = \frac{\eta}{\rho} \frac{\partial^2 v_y}{\partial x^2} \tag{3.24}$$

with $D_p = \eta/\rho = \nu$ which is identical to the Eq. (3.1).

This equation has been obtained in the hypothesis that the fluid is incompressible and has a small overall velocity, furthermore a parallel motion is assumed: in the general case the expression of the viscous force is more complicated and requires the use of a tensor .

An estimate of the viscosity coefficient η can be derived from the kinetic theory of gases. You get:

$$\eta = \frac{1}{3} nm \langle v \rangle l \tag{3.25}$$

where n is the particle density, $\langle v \rangle$ the mean velocity of the thermal agitation motion and l the mean free path.

3.4 Navier–Stokes Equation

To derive the equation of motion in the case of real fluids, let us refer to the considerations made in Chap. 1, Sect. 1.3. It has been seen that in the case of real fluids the stress tensor in addition to the term $-p\delta_{ij}$ contains a further term ε_{ij} representing the tangential stresses. So let's say:

$$\Pi_{ij} = -p\delta_{ij} + \varepsilon_{ij} \tag{3.26}$$

where π_{ij} is the stress tensor and in the absence of friction it is reduced to the pressure term only; ε_{ij} is the viscous stress tensor.

Let us first consider the case of an incompressible fluid. The form of ε_{ij} can be obtained from the general properties:

(a) internal friction occurs when there is a velocity gradient, i.e. ε_{ij} depends on the spatial derivatives of velocity;
(b) if the gradient is small we can assume that the momentum transfer depends only on the first derivatives of the velocity and that the dependence is linear;

(c) a further condition can be derived considering the fluid in rigid rotation since in this case there is no relative motion and ε_{ij} must vanish (this introduces a symmetry character in the expression of the tensor).

With these conditions the most general expression of ε_{ij} for an incompressible fluid is of the type:

$$\varepsilon_{ij} = \eta \left(\frac{\partial v_i}{\partial x_j} + \frac{\partial v_j}{\partial x_i} \right) \tag{3.27}$$

with $\eta = \text{cost}$.

The components of the frictional force per unit of volume are:

$$f_{V,i} = \frac{\partial \varepsilon_{ij}}{\partial x_j} = \eta \frac{\partial^2 v_i}{\partial x_j \partial x_j} + \eta \frac{\partial^2 v_j}{\partial x_i \partial x_j} = \eta \nabla^2 v_i + \eta \frac{\partial}{\partial x_i} \text{div } \mathbf{v} = \eta \nabla^2 v_i \tag{3.28}$$

The equation of motion for incompressible real fluids will be:

$$\frac{d\mathbf{v}}{dt} = \mathbf{F} - \frac{1}{\rho} \text{grad p} + \nu \nabla^2 \mathbf{v} \tag{3.29}$$

called Navier–Stokes equation, with $\nu = \eta/\rho$ (kinematic viscosity); F_i is a generic external force per unit of mass.

In the general case of a compressible fluid in the expression of the viscous stress tensor, a further term appears which describes internal stresses generated by a compression or an expansion. The viscous stress tensor is then:

$$\varepsilon_{ij} = \eta \left(\frac{\partial v_i}{\partial x_j} + \frac{\partial v_j}{\partial x_i} \right) + \zeta \delta_{ij} \text{ div } \mathbf{v} \tag{3.30}$$

Another constant coefficient ζ appears.

The corresponding viscous force per unit of volume is:

$$f_{V,i} = \frac{\partial \varepsilon_{ij}}{\partial x_j} = \eta \frac{\partial^2 v_i}{\partial x_j \partial x_j} + \eta \frac{\partial^2 v_j}{\partial x_i \partial x_j} + \zeta \delta_{ij} \frac{\partial}{\partial x_j} \text{div } \mathbf{v} = \eta \nabla^2 v_i + (\eta + \zeta) \frac{\partial}{\partial x_i} \text{div } \mathbf{v} \tag{3.31}$$

The equation of motion in vector form is:

$$\rho \frac{d\mathbf{v}}{dt} = \rho \mathbf{F} - \text{grad p} + \eta \nabla^2 \mathbf{v} + (\zeta + \eta) \text{ grad div } \mathbf{v} \tag{3.32}$$

The expression of the viscous stress tensor is greatly simplified in the case of parallel motion considered in the previous section. Keeping the same hypotheses on the direction of motion we obtain:

$$\varepsilon_{11} = \varepsilon_{22} = \varepsilon_{33} = 0 \qquad \varepsilon_{13} = \varepsilon_{23} = 0 \tag{3.33}$$

$$\varepsilon_{12} = \eta \frac{\partial v_2}{\partial x_1} \tag{3.34}$$

while the other components are obtained from the symmetry condition of the tensor; thus we find the expression (3.23).

3.5   Energy Dissipation by Internal Friction

Let's limit ourselves to the case of an incompressible fluid. The variation of the kinetic energy of the unit of volume is:

$$\frac{\partial}{\partial t} \left(\frac{1}{2} \rho v^2 \right) = \frac{1}{2} v^2 \frac{\partial \rho}{\partial t} + \rho v_i \frac{\partial v_i}{\partial t} \tag{3.35}$$

Using the continuity equation and the Navier–Stokes equation we obtain:

$$\frac{\partial}{\partial t} \left(\frac{1}{2} \rho v^2 \right) = -\frac{1}{2} v^2 \, \mathrm{div}\,(\rho \mathbf{v}) - \rho \mathbf{v} \cdot \mathrm{grad}\, \frac{v^2}{2} - \mathbf{v} \cdot \mathrm{grad}\, p + v_i \frac{\partial \varepsilon_{ij}}{\partial x_j} \tag{3.36}$$

The term $\mathbf{v} \cdot \mathrm{grad}\, p$, remembering the identities relating to the divergence, can be replaced with $\mathrm{div}\,(p\mathbf{v})$ and using the same relations the first two terms can be modified to second member:

$$\frac{\partial}{\partial t} \left(\frac{1}{2} \rho v^2 \right) = -\mathrm{div} \left[\rho \mathbf{v} \left(\frac{1}{2} v^2 + \frac{p}{\rho} \right) \right] + v_i \frac{\partial \varepsilon_{ij}}{\partial x_j} \tag{3.37}$$

We can write:

$$v_i \frac{\partial \varepsilon_{ij}}{\partial x_j} = \mathrm{div}\,(\mathbf{v} \cdot \boldsymbol{\varepsilon}) - \varepsilon_{ij} \frac{\partial v_i}{\partial x_j} \tag{3.38}$$

where $\mathbf{v} \cdot \boldsymbol{\varepsilon}$ is the vector whose components are $v_i \varepsilon_{ij}$.

With this result, (3.37) becomes:

$$\frac{\partial}{\partial t} \left(\frac{1}{2} \rho v^2 \right) = -\mathrm{div} \left[\rho \mathbf{v} \left(\frac{1}{2} v^2 + \frac{p}{\rho} \right) - \mathbf{v} \cdot \boldsymbol{\varepsilon} \right] - \varepsilon_{ij} \frac{\partial v_i}{\partial x_j} \tag{3.39}$$

The first term within the square bracket is the flow of energy due to the transport of the fluid as well as the effect of the work of the pressure forces, the second is the flow of energy produced by internal friction (a transport of momentum always involves also an energy transport).

Integrating on a fixed volume V_0 we have:

$$\frac{\partial}{\partial t}\int_{V_0}\frac{1}{2}\rho v^2\,\mathrm{d}V=-\int_{S_0}\left[\rho\mathbf{v}\left(\frac{1}{2}v^2+\frac{p}{\rho}\right)-\mathbf{v}\cdot\boldsymbol{\varepsilon}\right]\cdot\mathrm{d}\boldsymbol{\sigma}-\int_{V_0}\varepsilon_{ij}\frac{\partial v_i}{\partial x_j}\,\mathrm{d}V$$

$$(3.40)$$

The rate of change of the kinetic energy of the fluid contained in V_0 is equal to the flow of energy through the boundary of V_0 minus a further term whose meaning becomes clear if we stretch S_0 to infinity. In this case the surface integral vanishes and the integral:

$$\int_{V_0}\varepsilon_{ij}\frac{\partial v_i}{\partial x_j}\,\mathrm{d}V \qquad\qquad (3.41)$$

it therefore represents the kinetic energy dissipated per unit of time throughout the fluid as a result of internal friction.

3.6 Energy Balance in a Real Fluid

In the case of a real fluid, the equation (1.63) is generalized by including the effects of energy dissipation due to viscosity and those of thermal conduction.

In general, the conservation of energy can be expressed by the condition:

$$\frac{\partial U}{\partial t}+\operatorname{div}\mathbf{g}=0 \qquad\qquad (3.42)$$

where U is the total energy per unit volume and $\mathbf{g}$ is the energy flux density vector. In the case of real fluids the energy flux density requires the inclusion of two other terms than in the case of the ideal fluid:

(a) a term expressing the flow of energy produced by internal friction represented by the divergence of the vector $\mathbf{v}\cdot\boldsymbol{\varepsilon}$ having as components $v_i\varepsilon_{ij}$;
(b) a flow Φ_c due to the transport of heat by conduction from the hottest points to the coldest points of the fluid expressed by Fourier's law (3.2); it does not involve macroscopic motions and occurs even when the fluid is at rest.

The (3.42) then becomes:

$$\frac{\partial}{\partial t}\left(\frac{1}{2}\rho v^2+\rho\epsilon\right)=-\operatorname{div}\left[\rho\mathbf{v}\left(\frac{1}{2}v^2+w\right)-\mathbf{v}\cdot\boldsymbol{\varepsilon}-K\operatorname{grad}\mathrm{T}\right] \qquad (3.43)$$

where ϵ is the internal energy of the unit of mass.

We now derive another expression of the rate of change of total energy. The rate of change of kinetic energy per unit of volume is given by (3.36). Recalling the thermodynamic relationship:

$$\mathrm{d}w=T\,\mathrm{d}s+\frac{1}{\rho}\,\mathrm{d}p \qquad\qquad (3.44)$$

which translates into:

$$\text{grad p} = \rho \,\text{grad w} - \rho T \,\text{grad s} \tag{3.45}$$

and making use of known properties of the div and grad operators, the (3.36) becomes:

$$\frac{\partial}{\partial t}\left(\frac{1}{2}\rho v^2\right) = -\frac{1}{2}v^2 \,\text{div}\,(\rho \mathbf{v}) - \rho \mathbf{v} \cdot \text{grad}\left(\frac{1}{2}v^2 + w\right) + \rho T \mathbf{v} \cdot \text{grad s} + v_i \frac{\partial \varepsilon_{ij}}{\partial x_j} \tag{3.46}$$

The rate of change of internal energy per unit of volume is given by:

$$\frac{\partial}{\partial t}(\rho \epsilon) = \epsilon \frac{\partial \rho}{\partial t} + \rho \frac{\partial \epsilon}{\partial t} \tag{3.47}$$

which, with the thermodynamic relationship:

$$\frac{\partial \epsilon}{\partial t} = T \frac{\partial s}{\partial t} + \frac{p}{\rho^2}\frac{\partial \rho}{\partial t} \tag{3.48}$$

can be put in the form:

$$\frac{\partial}{\partial t}(\rho \epsilon) = -\left(\epsilon + \frac{p}{\rho}\right)\text{div}\,(\rho \mathbf{v}) + \rho T \frac{\partial s}{\partial t} \tag{3.49}$$

The rate of change of total energy per unit of volume is obtained by adding (3.46) and (3.49):

$$\frac{\partial}{\partial t}\left(\frac{1}{2}\rho v^2 + \rho \epsilon\right) = -\left(\frac{1}{2}v^2 + w\right)\text{div}\,(\rho \mathbf{v}) - \rho \mathbf{v} \cdot \text{grad}\left(\frac{1}{2}v^2 + w\right) + \rho T\left(\frac{\partial s}{\partial t} + \mathbf{v} \cdot \text{grad s}\right) + v_i \frac{\partial \varepsilon_{ij}}{\partial x_j} \tag{3.50}$$

or:

$$\frac{\partial}{\partial t}\left(\frac{1}{2}\rho v^2 + \rho \epsilon\right) = -\text{div}\left[(\rho \mathbf{v})\left(\frac{1}{2}v^2 + w\right)\right] + \rho T\left[\frac{\partial s}{\partial t} + \mathbf{v} \cdot \text{grad s}\right] + v_i \frac{\partial \varepsilon_{ij}}{\partial x_j} \tag{3.51}$$

Comparing (3.43) with (3.51) we obtain the general equation of heat transfer:

$$\rho T\left(\frac{\partial s}{\partial t} + \mathbf{v} \cdot \text{grad s}\right) = \varepsilon_{ij}\frac{\partial v_i}{\partial x_j} + \text{div}\,(K\,\text{grad}\,T) \tag{3.52}$$

In the absence of viscosity and heat conduction, the entropy conservation condition is found: $ds/dt = 0$. The quantity $\rho T(ds/dt)$ represents the heat gained by the volume unit of the fluid; it is the sum of two terms: one is the energy dissipated in heat by the viscosity, the second is the heat transported by conduction within the considered volume.

3.7 Integration of Equations for a Real Fluid

Since the dynamic and thermal state of a fluid is characterized by the knowledge of the velocity field $\mathbf{v}(x, y, z, t)$, of the pressure $p(x, y, z, t)$, of the density $\rho(x, y, z, t)$ (the temperature $T(x, y, z, t)$ can be obtained from the equation of state), the study of a fluid dynamics problem requires the solution of a system of five scalar differential equations. In the case of ideal fluids, the system consists of the continuity equation, the Euler equation and the energy conservation equation which can also be expressed by the constancy of entropy. When the fluid is not ideal, the Euler equation must be replaced by the Navier–Stokes equation, if the fluid is incompressible, or the more general equation (3.32) and the condition of constancy of the entropy the equation of its balance given, for incompressible fluids, by (3.52) or energy conservation (3.43), including the effects of energy dissipation due to viscosity and thermal conduction.

Let us now consider the boundary conditions to be applied in the regions where the fluid is in contact with a solid; in the case of an ideal fluid, the condition is imposed that the normal component of the velocity vanishes, but no conditions can be placed on the tangential component since there are no tangential forces. In the case of real fluids, on the other hand, due to the presence of internal friction, we can impose that the tangential component of the velocity is also zero at the boundary (no-slip condition).

This system of equations is very complicated and can only be solved in particular situations. The main reason for the difficulty is due to the presence of the nonlinear term $(\mathbf{v} \cdot \mathrm{grad})\mathbf{v}$ in the acceleration expression. This term corresponds to a variation of velocity due to the transport of the fluid element considered and, by spreading the vorticity, is responsible for the amplification and distortion of the initial velocity field in an irregular configuration of fluctuations that span a wide range of scales (these correspond, in the Fourier analysis of the velocity field, to the wavelengths of the harmonic components necessary to describe the field).

Both inertia and viscous forces transfer energy and momentum but in a completely different way: viscous forces exchange energy and momentum on a scale equal to the mean free path of the particles while the forces of inertia have no preferential scales . The absence of a preferential scale and consequently the large number of scales involved makes the motion complicated and irregular and makes the resolution of the equations difficult, or even impossible.

The equations of motion, on the other hand, are easy to solve in those cases where, due to the particular physical conditions, this non-linear term is zero. Three examples of such motions are described in Appendix A. The assumptions relating to the characteristics of the motions assume that the viscosity is high enough, otherwise the hypotheses made are no longer valid.

An experience performed by Reynolds allows us to further clarify the situation. Let's consider the device in Fig. 3.1; it consists of a jar A containing water and a jar B filled with coloured liquid. By means of the valve V the flow rate can be continuously adjusted. When this is small, a thread of coloured fluid detaches from the vase B and moves regularly in the water without mixing with it. A regime of motion of this type

Fig. 3.1 Jars device to study
the effect of viscosity

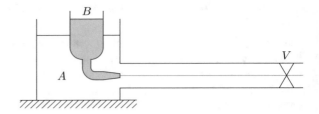

is called laminar: the trajectories of the macroscopic fluid particles are parallel and
the speed varies continuously both in space and in time.

As the flow rate increases, the motion of the colored liquid brush changes, initially
presenting an oscillating trend, and subsequently a disordered motion develops in
which the colored liquid mixes rapidly with the water. This regime is called turbulent
and is characterized by irregular trajectories, while the velocity presents fluctuations
both in space and in time.

Other examples are

(a) the motion of a liquid in a pipe with any section: when the velocity is small the
    motion is laminar but when it is large enough, the irregular character appears;
    the speed at which the transition occurs depends not only on the flow rate but
    also on the roughness conditions of the duct walls, e.g. if these are smooth the
    transition speed is greater;
(b) the Couette-Taylor motion: a fluid is contained in the interspace between two
    coaxial cylinders whose interior rotates with angular velocity N; also in this case
    if the angular velocity is progressively increased we see that there is a transition
    from laminar motion to an irregular turbulent motion.

The complex of experiments on these and other types of motion leads to the
identification of a parameter, the Reynolds number, defined by:

$$Re = \frac{Lv}{\nu} \tag{3.53}$$

where L and v are a characteristic size and speed. The Re parameter allows to
characterise the transition from one type of motion to another; more precisely when
the Reynolds number is less than a critical value Re_{cr} the motion remains laminar
while when the critical value is exceeded turbulent motion is established. This critical
value is not a universal constant and depends on particular concrete situations, even
though it is included in a narrow range between 100 and 1000.

This behaviour receives a clear interpretation when comparing the relative importance, in the Navier–Stokes equation, of the forces of inertia, described by the non-
linear term $(\mathbf{v} \cdot \text{grad})\mathbf{v}$, and viscous forces. In fact, it results for orders of magnitude:

$$\frac{|\nu \nabla^2 \mathbf{v}|}{|(\mathbf{v} \cdot \text{grad})\mathbf{v}|} \approx \frac{\nu(v/d^2)}{v^2/d} = \frac{\nu}{dv} = \frac{1}{Re} \tag{3.54}$$

When $Re \ll 1$ the viscous forces prevail which tend to dampen the inhomogeneities and maintain a regular character to the motion: vice versa when $Re \gg 1$ the term of the inertia forces is much more important and makes the equation non-linear.

Let's go back to the considerations made in Chap. 1 on vorticity. Starting from the Navier–Stokes equation and considering an isentropic motion, we obtained an equation for ω in presence of viscosity:

$$\frac{\partial \omega}{\partial t} = \text{rot}\,(\mathbf{v} \times \omega) + \nu\nabla^2\omega \tag{3.55}$$

To analyze the meaning of the new term introduced, suppose that the viscosity is large enough to make the first term on the second member negligible with respect to the second: in this case the Eq. (3.55) is equal to the diffusion equation (3.1) where the diffusivity is in this case the kinematic viscosity ν. Hence this term represents the diffusion of vorticity through the fluid by viscosity. Vorticity tends to diffuse in the liquid in a time of the order of L^2/ν, the lower the greater the viscosity (this time scale can be derived from the equation of motion with considerations of orders of magnitude).

In general, the relative weight of the two terms in the second member is described by the Reynolds number:

$$\frac{|\nu\nabla^2\omega|}{|\,\text{rot}\,(\mathbf{v} \times \omega)|} = \frac{\nu}{Lv} = \frac{1}{Re} \tag{3.56}$$

3.8 Case of Fluid with Low Viscosity

Let us now consider a motion with $Re \gg 1$ in which the term containing the viscosity is much smaller than the non-linear term and suppose that this happens because the viscosity is small. It would seem possible to describe, in an approximate way, the behaviour of the fluid ignoring the term due to viscosity and therefore making use of Euler's equation. That is, the fluid would be considered an ideal fluid. In reality, the results obtained this way often contradict the experimental data.

To clarify this aspect we consider the stationary motion of an incompressible fluid ($\text{div}\,\mathbf{v} = 0$) around a cylinder of circular section and infinite height and suppose that far from the cylinder the fluid is in uniform motion with velocity v_0 perpendicular to the cylinder axis.

Let us first deal with the problem with the theory of ideal fluids. Since we can think that the fluid started from a state of rest (irrotational state) it will be $\text{rot}\,\mathbf{v} = 0$. It is therefore a potential motion in which the potential of the velocity satisfies the equation (1.90):

$$\nabla^2\psi = 0 \tag{3.57}$$

Fig. 3.2 Trend of
trajectories with low
viscosity

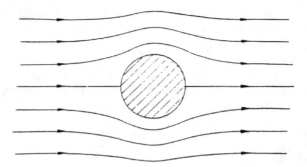

Fig. 3.2 Trend of trajectories with low viscosity

The solution of this equation must satisfy the boundary conditions: $v = v_0$ at infinite distance and $v_\perp = 0$ on the surface of the cylinder (since the fluid is ideal there is no reason to force the annulment of the tangential component). The trend of the trajectories is represented in Fig. 3.2.

Let us analyse the real behaviour of the fluid quantitatively. For very small values of Re (small velocities) the trend of the fluid trajectories is almost the same as that described by the theory of ideal fluids, but as Re increases, the situation changes: first vortices are formed behind the cylinder; subsequently these vortices detach and move away while new ones are formed. Furthermore, at first the vorticity spreads over a fairly large region but as the velocity increases further the diffusion time is too long and vorticity fills only a thin strip behind the cylinder: this region is characterised by a chaotic and irregular motion and is called the boundary layer. Outside this region, the motion can retain the regular character predicted by the theory of ideal fluids. The different situations are represented in Fig. 3.3.

We can therefore conclude that, in general, a real fluid with small but not zero viscosity does not behave like an ideal fluid with zero viscosity.

However, we cannot consider the equation of ideal fluids and impose the condition that $v = 0$ in the boundary: from the mathematical point of view, if we remove the term due to viscosity in the Navier–Stokes equation, the order is reduced by one. of the equation and therefore too many boundary conditions are applied.

The impossibility of describing the motion with the Euler equation depends on the circumstance that near a solid boundary the behavior of the ideal fluid is very different from what happens for a viscous fluid, even if it has a small viscosity, and this difference is well detected by experiments.

In the case of an ideal fluid, the tangential component of the velocity of the fluid in contact with the solid body is not zero, because there is no force acting in that direction. In a real fluid, on the other hand, the viscous forces also cancel the tangential component of the velocity of the fluid in contact with the solid: even if the viscosity coefficient is small, the velocity gradient is high so as to make the role of internal friction significant. As a consequence of this fact in a thin layer around the solid the viscosity cannot be neglected and this completely modifies the behavior of

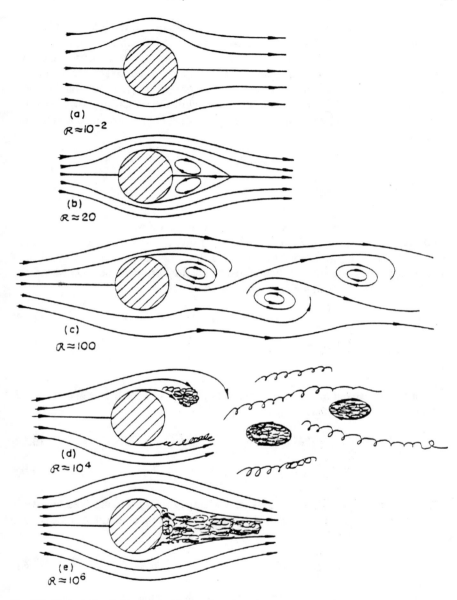

Fig. 3.3 Trajectories deformation at increasing viscosity

the fluid, as the vorticity spreads throughout the fluid due to the effect of viscosity but above all due to the action of the non-linear convective term.

It was found in Sect. 1.6 of Chap. 1 that in any motion of an ideal fluid starting from rest, vorticity cannot be generated and it has also been stated that this is in contrast with the observations. We can now solve this dilemma bearing in mind that in reality any fluid has a velocity gradient near the solid such that the viscosity, however small, is capable of producing and diffusing vorticity, which is then amplified by the term rot $(\mathbf{v} \times \boldsymbol{\omega})$.

3.9   Similar Motions

Let us consider the motion of an incompressible fluid and denote by d a characteristic length: it can represent for example the size of the region where the motion takes place or the size of an object moving in the fluid. Let us consider the motion of another incompressible fluid, characterized by the dimension d' and suppose that the equation is satisfied for all the corresponding dimensions:

$$d' = \alpha d \tag{3.58}$$

where α is a pure number. In such case the two motions are said to be geometrically similar.

We can study the first motion in a reference system (x, y, z) and the second in another system (x', y', z') such that:

$$x' = \alpha x \qquad y' = \alpha y \qquad z' = \alpha z \tag{3.59}$$

$$v'_x = \beta v_x \qquad v'_y = \beta v_y \qquad v'_z = \beta v_z \tag{3.60}$$

We indicate with v, ν, ρ, and p the speed, viscosity, density, and pressure relative to the first motion and with v', ν', ρ', and p' the corresponding quantities for the second motion. We will have:

$$v' = \beta v \qquad \nu' = \gamma \nu \qquad \rho' = \delta \rho \qquad p' = \epsilon p \tag{3.61}$$

The time scales t and t' and the forces F and F' are related by:

$$t' = \zeta t \qquad F' = \eta F \tag{3.62}$$

The quantities β, γ, δ, ϵ, ζ, and η are pure numbers.

The two motions are said to be similar or dynamically similar when the equations that describe them are equivalent. The equations of motion of the two fluids are respectively:

$$\frac{d\,\mathbf{v}}{d\,t} = \frac{\partial \mathbf{v}}{\partial t} + (\mathbf{v} \cdot \text{grad})\mathbf{v} = \mathbf{F} - \frac{1}{\rho}\,\text{grad}\,p + \nu\nabla^2\mathbf{v} \tag{3.63}$$

$$\frac{d\,\mathbf{v}'}{d\,t'} = \frac{\partial \mathbf{v}'}{\partial t'} + (\mathbf{v}' \cdot \text{grad}')\mathbf{v}' = \mathbf{F}' - \frac{1}{\rho'}\,\text{grad}'\,\mathrm{p}' + \nu'\nabla'^2\mathbf{v}' \tag{3.64}$$

In order to make the two equations equivalent, the coefficients of the homologous terms must be proportional; this implies that:

$$\frac{\beta}{\zeta} = \frac{\beta^2}{\alpha} = \eta = \frac{\eta}{\delta\alpha} = \frac{\gamma\beta}{\alpha^2}\,. \tag{3.65}$$

In particular, we obtain:

$$\frac{\alpha\beta}{\gamma} = 1\,, \tag{3.66}$$

from which, according to the definitions of α, β, and γ we have:

$$Re = \frac{dv}{\nu} = \frac{d'v'}{\nu'} = Re' \tag{3.67}$$

It is easy to verify that Re and Re' are dimensionless. The units of measure of the other quantities p, t, F can be adapted so that the other conditions (3.64) are also verified.

Equation (3.67) expresses the *law of similarity*: two geometrically similar motions are said to be similar or dynamically similar when they have the same Reynolds number. When viscosity is negligible (ideal fluids), Eq. (3.67) is no longer necessary and the condition of dynamic similarity is simply expressed by geometric similarity. These conclusions are exploited in the study of aircraft and ship design using models; from the theoretical point of view, as we will see below, they allow to find proportional relations between physical quantities through dimensional analysis.

3.10 Synthetic Summary

Thermodynamic equilibrium is characterised by the uniformity of the physical quantities (pressure, temperature, concentration of the different chemical components). Out of equilibrium these quantities are not uniform: phenomena intervene, called transport phenomena, through which uniformity is restored due to the effect of the motion of thermal agitation. Examples of these processes are the diffusion proper, the conduction of heat and the diffusion of the momentum due to viscosity. Regardless of the particular phenomenon, the equation they satisfy is the same.

Real fluids out of equilibrium are prone to transport phenomena. One of these is the conduction of heat: when a temperature gradient is created (and this is not too large) a heat flow occurs, described by the Fourier equation, which tends to uniform the temperature. The vehicle of this flux can be constituted by free electrons in a conducting metal and in a plasma, or by molecules and atoms in a fluid. The coefficient of thermal diffusion depends on thermal conductivity, density and specific heat. The usual description of conduction assumes that the mean free path of the particles is much smaller than the linear dimension over which there are variations in temperature. When this does not happen (as for example in the " hot medium") and the mean free path is comparable with the distance over which the temperature varies, the phenomenon of flow saturation occurs.

The transfer of momentum is caused by internal friction: this generates a force per unit area between two contiguous layers of fluid in relative motion; this force is called tangential stress, and it is equal to the flow of momentum through the unit of surface in the unit of time. This phenomenon is also described by a diffusion equation with the kinematic viscosity ν playing the role of diffusion coefficient .

In the case of a generic motion of an incompressible fluid, the general form of the internal friction is expressed by Eq. (3.27), or, dismissing the incompressibility restriction, by relation (3.30). In case of incompressible fluids the energy conservation equation is the Navier–Stokes equation, which is a second-order non-linear partial differential equation.

The fluids in which the transport phenomena cannot be ignored are called real fluids, as opposed to ideal fluids. Their energy balance can be obtained by taking into account that there is a flow of thermal energy due to conduction and a flow of energy produced by internal friction, which is expressed as the divergence of the vector $\mathbf{v} \cdot \boldsymbol{\varepsilon}$ with components $v_i \varepsilon_{ij}$.

By means of the energy balance equation and the Navier–Stokes equation we obtain the condition for the heat balance (general equation for heat transfer) in which the heat gain is expressed by the dissipation due to friction and by the heat flow due to conduction.

Laboratory experiments show the existence of two types of motions: laminar motions characterized by very regular fluid particle orbits and turbulent or vortical motions in which the trajectories are irregular and the speed varies significantly from point to point and, for a given point, over time. It turns out that the transition from laminar to turbulent motion occurs when the Reynolds number Re exceeds a critical value Re_{cr}. The Reynolds number allows us to give an interpretation of these facts in terms of the Navier–Stokes equation. In fact, the ratio between the orders of magnitude of the term containing the viscosity and of the non-linear term is equal to $1/Re$. Therefore when $Re \ll 1$ the viscous term prevails which gives the motion a regular character, when $Re \gg 1$ the non linear term prevails and the motion assumes a very complicated character.

It would seem that when $Re \gg 1$ and the viscous term is small it is possible to treat the fluid as ideal: that is, a fluid with very low viscosity can be approximated with the ideal fluid model. This is true except near the edges where the friction effect, even if small, cannot be ignored due to the high speed gradient.

Starting from the Navier–Stokes equation, the equation for vorticity is obtained: compared to that obtained for ideal fluids, a term appears that represents the diffusion of vorticity by internal friction; the diffusion coefficient is the kinematic viscosity.

Two motions are said to be geometrically similar when the linear dimensions differ only by a factor of scale; two geometrically similar motions are also dynamically similar when they have the same Reynolds number.

Appendix

Appendix A: Particular Solutions of the Navier–Stokes Equation

(1) (*Couette flux*) Let us consider a unidirectional stationary motion of an incompressible fluid, i.e. a motion in which the velocity has the same direction everywhere. Its modulus is independent of the distance in the direction of motion. Let, for example, be:

$$\mathbf{v} = (v, 0, 0) \quad \text{with } v = v(y, z) . \tag{3.68}$$

The nonlinear term in the equation of motion $(\mathbf{v} \cdot \text{grad})\mathbf{v}$ vanishes and the stationarity condition guarantees that the term $\partial \mathbf{v} / \partial t$ is also null. The Navier–Stokes equation projected on the x axis gives:

$$-\frac{\partial p}{\partial x} + \eta \left(\frac{\partial^2 v}{\partial y^2} + \frac{\partial^2 v}{\partial z^2} \right) = 0 , \tag{3.69}$$

while projecting onto the other axes we obtain:

$$\frac{\partial p}{\partial y} = 0 \qquad \frac{\partial p}{\partial z} = 0 . \tag{3.70}$$

Since the second term in (3.69) does not depend on x, it will be possible to write:

$$\frac{\partial p}{\partial x} = -\Delta , \tag{3.71}$$

where Δ, constant for the stationarity of the motion, represents the pressure gradient. We therefore have:

$$\frac{\partial^2 v}{\partial y^2} + \frac{\partial^2 v}{\partial z^2} = -\frac{\Delta}{\eta} . \tag{3.72}$$

The incompressible fluid is in motion under the action of a pressure gradient Δ between the two parallel planes of equations $y = 0$ and $y = d$, respectively, the first fixed and the second moving in the direction of the x axis with velocity v_0. In this case, the dependence on z is also broken and the (3.72) becomes:

$$\frac{\partial^2 v}{\partial y^2} = -\frac{\Delta}{\eta} .$$

(3.73)

Integrating with the conditions $v = 0$ for $y = 0$ and $v = v_0$ for $y = d$, we obtain:

$$v = \frac{\Delta}{2\eta} y(d - y) + \frac{v_0}{d} y .$$

(3.74)

When the two planes are in relative motion ($v_0 \neq 0$) and the pressure gradient is zero ($\Delta = 0$) the velocity profile is linear; if instead both planes are fixed the velocity profile is parabolic under the action of the pressure gradient Δ.

(2) (*Poiseuille flux*) A dynamically similar situation but different in part for the geometric situation is the motion of a fluid within a horizontal pipe with circular cross section of radius R. The problem can be studied in a two-dimensional polar reference system (there is no dependence on a coordinate parallel to the axis). The expression of the Laplacian in polar coordinates and in axial symmetry is:

$$\Delta^2 v = \frac{1}{2} \frac{\partial}{\partial t} \left(r \frac{\partial v}{\partial t} \right) .$$

(3.75)

The equation of motion is then:

$$\frac{1}{r} \frac{\partial}{\partial r} \left(r \frac{\partial v}{\partial r} \right) = \frac{\Delta}{\eta} ,$$

(3.76)

which must be solved with the boundary condition $v = 0$ on the circumference that defines the section of the pipe. The general solution is:

$$v = -\frac{\Delta}{2\eta} r^2 + a \log r + b ,$$

(3.77)

with constant a and b. It must be $a = 0$ otherwise the solution diverges in the center ($r = 0$). By imposing that $v = 0$ for $r = R$ we obtain:

$$v = \frac{\Delta}{2\eta} (R^2 - r^2) .$$

(3.78)

(3) (*Couette-Taylor flow*) Another case is the stationary motion of an incompressible fluid moving between two coaxial cylinders rotating in the same direction with different angular velocities. The current lines are circumferences perpendicular

to the axis. Velocity has only the non-zero tangential component v_θ which is a function of r :

$$\mathbf{v} = v_\theta \hat{\mathbf{e}}_\theta , \tag{3.79}$$

where $\hat{\mathbf{e}}_\theta$ is the vector perpendicular to the radial direction. Again, the nonlinear term $(\mathbf{v} \cdot \text{grad})\mathbf{v}$ is identically null. Furthermore, the pressure is only a function of r. Projecting the Navier–Stokes equation in the direction of r and in the perpendicular one obtains the equations

$$\frac{\rho v^2}{r} = \frac{d\,p}{d\,r} \tag{3.80}$$

and

$$\frac{\partial v}{\partial t} = \nu \left(\frac{\partial^2 v}{\partial r^2} + \frac{1}{r}\frac{\partial v}{\partial r} - \frac{v}{r^2} \right) \tag{3.81}$$

In stationarity conditions this equation becomes:

$$\frac{\partial^2 v}{\partial r^2} + \frac{1}{r}\frac{\partial v}{\partial r} - \frac{v}{r^2} = 0 . \tag{3.82}$$

The solution is of the type $v = A r^n$ (A and $n = $ cost.). Inserting this expression in the equation we obtain for n: $n \in [-1, 1]$, and hence:

$$v = ar + \frac{b}{r} . \tag{3.83}$$

The boundary conditions are:

$$r = R_1 \qquad v = \Omega_1 R_1 \tag{3.84}$$

and

$$r = R_2 \qquad v = \Omega_2 R_2 \tag{3.85}$$

By imposing these conditions we have:

$$a = \frac{\Omega_2 R_2^2 - \Omega_1 R_1^2}{R_2^2 - R_1^2} \tag{3.86}$$

$$b = \frac{(\Omega_1 \Omega_2) R_2^2 R_1^2}{R_2^2 - R_1^2} \tag{3.87}$$

We note two special cases:

(a) $\Omega_1 = \Omega_2 = \Omega$, $a = \Omega$, $b = 0$ and then $v = \Omega r$ (rigid rotation);
(b) $\Omega_2 = 0$ a $R_2 = \infty$, $a = 0$, $b = \Omega_1 R_1^2$ and therefore $v = \Omega_1 R_1^2 / r$.

References

1. Alonso M., Finn E.J.: Elements of Physics for the University, Chapter 24, Inter European Edition (1967)
2. Landau, L.D., Lifshitz, E.M.: Fluid Mechanics, Chapter 8. Pergamon Press, Oxford (1987)
3. Batchelor, G.K.: An Introduction to Fluid Dynamics, Chapter 5. Cambridge University Press, Cambridge (2012)
4. Feynman, R.P., Leighton, R.B., Sands, M.: The Feynman Lectures on Physics, Vol. II, Chapter 40 and 41 v41, Addison Wesley, Boston (1963)

Chapter 4
The Interstellar Medium

4.1 Thermal Balance of Interstellar Gas

Except in particular situations, the behaviour of ordinary gases is adiabatic. In fact, due to the low values of the viscosity coefficients and thermal conductivity there is no thermal contact with the surrounding medium and the dissipation of energy by viscosity is negligible. In general, viscosity and thermal conduction are also of little importance in interstellar gas. Consider the role of thermal conduction in a molecular cloud of size 1 pc, with temperature $T = 10$ K and density $n = 100$ cm^{-3}. Thermal conductivity is expressed by:

$$K = \frac{3}{2} n v_m l k_B \tag{4.1}$$

where v_m is the mean square velocity of the molecules and l their mean free path. It is:

$$v_m = \sqrt{\frac{3 k_B T}{m}} = 3.5 \cdot 10^4 \text{ cm s}^{-1} . \tag{4.2}$$

To evaluate l on the basis of $l = 1/n\sigma$ we adopt the geometric cross section for σ and $r = 1.1 \cdot 10^{-8}$ cm for the radius of the H_2 molecule. Ultimately we have:

$$K = 1.8 \cdot 10^4 \text{ erg cm}^{-1} \text{ s}^{-1} \text{ K}^{-1} . \tag{4.3}$$

The thermal diffusion coefficient is given by:

$$D_c = \frac{K}{\rho c_p} \tag{4.4}$$

where the density is $3.3 \cdot 10^{-22}$ g cm^{-3} and for the specific heat at constant pressure c_p from the theory of perfect diatomic gases we have $c_p = 1.4 \cdot 10^8$ erg g^{-1} K^{-1} and therefore $D_c = 3.9 \cdot 10^{17}$ cm^2 s^{-1}.

© The Author(s), under exclusive license to Springer Nature Switzerland AG 2021   65
G. Carraro, *Astrophysics of the Interstellar Medium*,
UNITEXT for Physics, https://doi.org/10.1007/978-3-030-75293-4_4

According to (3.7) the characteristic time for the diffusion of heat by conduction is given by

$$\tau_c = \frac{L^2}{D_c} = 7.8 \cdot 10^{11} \text{ years} .$$ (4.5)

Let's compare this time scale with the dynamic scale:

$$\tau_{din} = \frac{L}{c}$$ (4.6)

where c is the speed of sound, which for a molecular cloud is $4 \cdot 10^4$ cm s^{-1} so $\tau_{din} = 2.5 \cdot 10^6$ years. From this comparison it can be seen that the gas is not affected by the conduction phenomenon. Thermal conductivity is also negligible in HI and HII regions.

In the case of coronal gas, which has temperatures of the order of 10^6 K, the thermal conductivity is high; thermal conduction plays an important role in the transition region between this hot gas and the much colder and denser clouds. We repeat the previous evaluation also in this case, considering a coronal gas region of the size of 1 parsec, with a temperature of 10^6 K and density $n = 3 \cdot 10^{-3}$ g cm^{-3}. Using (3.16) we determine the value of K which is $6 \cdot 10^8$ erg cm^{-1} s^{-1} K^{-1}. The specific heat at constant pressure is $4.2 \cdot 10^8$ erg g^{-1} K^{-1} and the density $\rho = 2.5 \cdot 10^{-27}$ g cm^{-3}. The thermal diffusion coefficient is $D_c = 5.7 \cdot 10^{26}$ cm^2 s^{-1} and the time scale for the diffusion heat:

$$\tau_c = \frac{L^2}{D_c} = 520 \text{ years} .$$ (4.7)

The comparison of this time with the dynamic time $\tau_{din} = L/C \approx 6 \cdot 10^3$ years (with $c = 1.7 \cdot 10^7$ cm/sec) tells us that in this case the contribution of thermal conduction cannot be ignored.

Let us now consider the role of viscosity referring to a region HI with a temperature of 100 K. The expression of viscosity is given by (3.25) hence the diffusion coefficient of the momentum, given by $D_p = \eta/\rho$ is $D_p = l v_m/3$, while the viscosity diffusion time τ_ν, which is obtained from (3.24), is:

$$\tau_\nu = \frac{3L^2}{l v_m} .$$ (4.8)

The relationship between τ_ν and the dynamic time τ_{din} then holds:

$$\frac{\tau_\nu}{\tau_{dyn}} = 3\frac{L}{l}\frac{c}{v_m} \approx \frac{L}{l}$$ (4.9)

being $c \approx v_m$ (see Sect. 2.1). Therefore as long as $L \gg l$, as is usually the case, the viscosity is negligible; however, when significant changes in the dynamic variables take place on a L scale comparable to the average free path of the particles, then viscosity becomes important. This is the situation that occurs in shock waves (see Chap. 5).

With the exception of these particular situations, in general the effects of viscosity and thermal conduction can be neglected in the interstellar medium. However, interstellar gas does not behave adiabatically, as there is another cause of deviation from adiabaticity: interstellar space is permeated by electromagnetic radiation whose energy density is comparable to the kinetic energy density of the gas and is much larger of its thermal energy. In such conditions the interaction of matter with radiation gives rise to processes that increase or decrease the thermal energy of the gas.

Let us go back to the general equation of heat transfer (3.52), in which we take into account only the term due to thermal conduction. We can now introduce two further terms that represent the contribution of the processes of energy gain and loss due to the interaction between matter and radiation. We denote by L (loss) the quantity of energy radiated by the gas per unit of volume and in the unit of time (cooling) and similarly with G (gain) the quantity of energy gained per unit of volume and per unit of time (heating). The transfer equation then becomes:

$$\rho T \frac{ds}{dt} = \mathrm{div}(K\,\mathrm{grad}\,T) + (G - L) \,. \tag{4.10}$$

When the effects of heat conduction can be neglected, as is the case, the heat balance equation becomes:

$$T \frac{ds}{dt} = \frac{G - L}{\rho} \,. \tag{4.11}$$

In general, G and L are functions of temperature, density and chemical composition and are specified by the knowledge of the various mechanisms of interaction between gas and radiation. Let us explain (4.11) in the case of perfect gas. The first member of this equation represents the heat gained by the unit of mass in the unit of time. Using the first law of thermodynamics we can write:

$$T \frac{ds}{dt} = \frac{dQ}{dt} = \frac{d\epsilon}{dt} + p \frac{dV}{dt} \tag{4.12}$$

where ϵ is the internal energy of the unit of mass and V the specific volume. You can write:

$$\epsilon = \frac{3}{2} \frac{k_B T}{\mu m_H} \tag{4.13}$$

where μ is the average molecular weight and m_H the mass of the hydrogen atom. It will therefore be:

$$T \frac{ds}{dt} = \frac{d}{dt} \left(\frac{3}{2} \frac{k_B T}{\mu m_H} \right) - \frac{p}{\rho^2} \frac{d\rho}{dt} \,. \tag{4.14}$$

Remembering that:

$$\rho = N \mu m_H \qquad p = \frac{k_B T}{\mu m_H} \rho \tag{4.15}$$

we have:

$$T \frac{ds}{dt} = \frac{1}{\mu m_H} \frac{d}{dt} \left(\frac{3}{2} k_B T \right) - \frac{k_B}{\mu m_H} \frac{T}{\rho} \mu m_H \frac{dn}{dt} \tag{4.16}$$

hence the balance equation becomes:

$$n \frac{d}{dt} \left(\frac{3}{2} k_B T \right) - k_B T \frac{dn}{dt} = G - L . \tag{4.17}$$

Suppose n is constant (this could happen when the dynamic aspect is not relevant). A state of thermal equilibrium is defined by the condition:

$$G - L = 0 \tag{4.18}$$

this is an equation in T and, for a given density value, gives an equilibrium temperature T_{eq}. If this state of equilibrium is stable (the stability analysis is developed in Sect. 4.5), the system, displaced a bit from the equilibrium position, tends to return to it. For $|T - T_{eq}| \ll 1$ we can develop $G - L$ in series by neglecting the higher order terms than the first:

$$G - L = a_0 + a_1 (T - T_{eq}) + ... \tag{4.19}$$

where $a_0 = 0$ being T_{eq} equilibrium temperature: moreover, stability requires that $a_1 < 0$, in fact if $T < T_{eq}$ we have $G - L < 0$ and the system heats up, while if $T > T_{eq}$ you have $G - L > 0$ and the system cools down. Substituting in (4.17) we have:

$$\frac{d}{dt} \left(\frac{3}{2} n k_B T \right) = a_1 (T - T_{eq}) . \tag{4.20}$$

This equation can be transformed by substituting the variable

$$\frac{d}{dt} (\Delta E) = - \frac{\Delta E}{\tau_{cool}} \tag{4.21}$$

with:

$$\tau_{cool} = - \frac{3 n k_B}{2 a_1} = \text{cost.} \tag{4.22}$$

which has for solution:

$$\Delta E = (\Delta E)_0 \exp \left(- \frac{t}{\tau_{cool}} \right) . \tag{4.23}$$

The system then relaxes towards the stable equilibrium configuration with a time scale τ_{cool} which is called cooling time. The cooling time can also be defined starting from (4.17):

$$\tau_{cool} = \frac{|\Delta(3nk_B T/2)|}{|G - L|} \tag{4.24}$$

that is, the cooling time is equal to the ratio between the excess of the internal energy density with respect to the equilibrium value, and the net energy gain per unit of time and per unit of volume.

4.2 Thermodynamic State of Interstellar Gas

As has already been pointed out, in the case of the interstellar medium, the interaction between matter and radiation makes the motion non-adiabatic and the energy balance must take into account the energy absorbed and produced by the interaction processes.

If thermal conduction and viscosity can be neglected, the heat balance equation reduces to (4.11), which in the case of an ideal gas can be put in the form (4.17). Many times it happens that the dynamic time scale, that is the one on which significant variations of motion occur, is much greater than the time scale in which the heating and cooling processes operate effectively. In this case the first member of (4.11) is negligible with respect to each of the terms G and L and the thermal state of the gas is defined by the equation:

$$G - L = 0 . \tag{4.25}$$

In this case the gas temperature is determined by the interaction processes between matter and radiation and does not depend on the state of motion. Obviously it is a state of quasi-equilibrium because over a time interval comparable with dynamic time the density changes and therefore the (4.25) gives a new solution.

It is important to consider, in the context of the study of the thermal behaviour of the interstellar gas, how much it deviates from the condition of thermodynamic equilibrium. Let us consider a rigid and thermally insulated container, which contains a mixture of gas and radiation initially not in equilibrium. After a certain time (mechanical relaxation time) the system reaches mechanical equilibrium, after a longer time (thermal relaxation time) the thermal equilibrium is reached and after an even longer time (chemical relaxation time) it is established the chemical balance. This situation is called thermodynamic equilibrium. At thermodynamic equilibrium the gas temperature is uniform and there is no flow of thermal energy. The radiation obeys the Kirchhoff principle and the energy distribution is the Planckian one. The gas verifies the Maxwellian distribution of velocities and the Boltzmann and Saha equations hold.

There is no thermodynamic equilibrium in the interstellar medium. Let's see how matter behaves. In thermodynamic equilibrium the distribution of velocities is reached after many elastic collisions which distribute the energy in a manner defined by statistical mechanics. In the interstellar gas the inelastic collisions prevent the achievement of equilibrium: e.g. collisional excitations or ionizations transfer energy from matter to the radiation field. However, inelastic collisions are relatively rare and the velocity distribution is close to the Maxwellian one. The reason for this is that the gas is made up predominantly of hydrogen and helium and collisions between these elements and their ions and electrons, when the relative kinetic energy of collision is less than about 10 eV, are essentially elastic; excitations require more energy. OII, OIII and NII have smaller excitation energies but are too little abundant to affect distribution. This allows the establishment of a distribution close to the Maxwellian one.

4.3   Cooling Processes

The primary mechanism of each cooling process is the excitation by collision of molecules, atoms and ions by electrons or atoms of H:

$$A + B \rightarrow A + B^* \tag{4.26}$$

followed by the decay of the excited system by emission of radiation:

$$B^* \rightarrow B + h\nu \ . \tag{4.27}$$

The radiation escapes from the region if the density of the gas is low enough and the mean free path of the photons is greater than the size of the region. Generally this is the situation, only in the densest regions of molecular clouds the mean free path is shorter and in this case the problem must be addressed by solving the equation of radiative transport. Ultimately there is a decrease in the thermal energy of the gas.

For such a process to be effective, the following three conditions must be met.

(1) The excitation energy must be comparable to or less than the thermal energy. Therefore, if we indicate with ΔE the excitation energy must be:

$$\Delta E \leq \frac{3}{2} k_B T = 1.3 \cdot 10^{-4} T \ eV \tag{4.28}$$

otherwise the number of particles capable of producing the excitation would be too small, being limited to the most energetic particles of the Maxwellian distribution. The temperature T_{ex} so:

$$\frac{3}{2} k_B T_{ex} = \Delta E \tag{4.29}$$

it is called excitation temperature. For the transition to be energetically possible it must be:

$$T \geq T_{\text{ex}} = \frac{2\Delta E}{3k_B} = 4.8 \cdot 10^{15} \Delta E \text{ erg} = 7.7 \cdot 10^3 \Delta E \text{ eV} . \qquad (4.30)$$

(2) Collisions must be frequent and this requires that the abundances of the particles involved in the process be sufficiently large.
(3) During the collision the excitation probability must be high. To assess whether this condition is satisfied, the cross section must be estimated; in general if the transition is allowed, the cross section is large enough and in this case it can be considered that the condition is satisfied.

In collisions between atoms or ions and electrons, the number of excitations from the jth level to the kth level for cm^3 and per second will be $n_e n_j \gamma_{jk}$ where γ_{jk} is the transition probability per unit of time and n_j the numerical density of the jth level. The kinetic energy lost by the colliding electrons is the difference $E_{jk} = E_k - E_j$, but this is not the net loss because in part the de-excitation can take place by collision and in this case the energy E_{jk} is acquired by an electron. Denoting by γ_{kj} the probability of shock de-excitation the net loss per cm^3 per second we have:

$$L = n_e \sum_{j<k} E_{jk}(n_j \gamma_{jk} - n_k \gamma_{kj}) . \qquad (4.31)$$

In many cases of concrete interest, the contribution of only one excited level is important, furthermore the collisional de-excitation is negligible and the starting level is the fundamental one. In this case the previous relationship is reduced to:

$$L = n_e n_1 \gamma_{1k} E_{1k} \qquad (4.32)$$

where n_1 is the numerical abundance of the ion in the ground level.

The probability that an ion in the ground state, as a consequence of the collision with an electron, undergoes a transition in the unit of time up to the k state, and subsequently spontaneously decays down to the ground level, is expressed by:

$$P_{jk} = n_j v \sigma_{jk}(v) \qquad (4.33)$$

where v is the relative velocity of the two particles and $\sigma_{jk}(v)$ is the cross section. The coefficient γ_{jk} is the average of the previous expression, averaged over the Maxwellian distribution of velocities:

$$\gamma_{jk} = \frac{4}{\sqrt{\pi}} \left(\frac{\mu}{2k_B T} \right)^{\frac{3}{2}} \int_0^\infty v^3 \sigma_{1k}(v) \exp\left(-\frac{\mu v^2}{2k_B T} \right) dv \qquad (4.34)$$

where μ is the reduced mass.

Let us now examine the most important cooling processes in the different regions of the interstellar medium, characterised by a different temperature.

(A) **Molecular clouds** ($T \approx 10$ K).

At the temperature of 10 K the thermal energy is of the order of 0.001 eV which is comparable with the excitation energy of the rotational levels of the lighter molecules. Cooling then takes place through the excitation of these levels. Let us consider the case of the most abundant molecule H_2: condition (2) is obviously satisfied. Electric dipole transitions between rotational states are not allowed as the dipole moment is zero: therefore these transitions are quadrupole ones, and also must satisfy the selection rule:

$$\Delta J = \pm 2 \tag{4.35}$$

where J is the rotational quantum number that is associated with energy levels:

$$E_J = J(J+1)\frac{h^2}{8\pi^2 I} \tag{4.36}$$

where I is the moment of inertia of the molecule. Therefore the probability of excitation is not very large, but above all condition (1) is not verified as the least energetic rotational transition of the H_2 molecule is the transition $J = 0 \rightarrow J = 2$ which corresponds to an excitation energy equivalent to 510 K. Since the temperature of the molecular clouds is about $10 - 20$ K, the transition is not energetically possible. Moreover, even if this condition were satisfied, since the level $J = 2$ has a high average life, further excitation is more likely than decay. In molecular clouds, on the other hand, the cooling produced by the rotational transitions of the CO molecule is important, which, after that of H_2, is the most abundant molecule, its fractional abundance being 10^{-5}. The CO molecule has an electric dipole moment and therefore dipole transitions between rotational states are allowed. The excitation energy related to the transition from the $J = 0$ level to the $J = 1$ level, whose inverse process involves the emission of radiation with $\lambda = 0.26$ cm, corresponds to an excitation temperature of 5.5 K, lower than that typical of molecular clouds. CO is an effective cooling agent in these objects, unless the density is so high that the CO reabsorbs some or all of the photons emitted; in this case, for a correct evaluation of the phenomenon it is necessary to take into account the opacity of the material.

(B) **HI regions** ($T \approx 100$ K).

In the HI regions, which have temperatures of the order of 10^2 K, atoms or ions can be excited by collisions if the excitation energy is:

$$\Delta E \leq \frac{3}{2}k_B T = 0.03 \text{ eV} . \tag{4.37}$$

The transitions of CI, CII, OI, SiII are particularly important ($\Delta E = 10^{-2} - 3.6 \cdot 10^{-2}$ eV). C can exist in the ionised state when H is predominantly in the neutral state because its ionisation potential is lower (11.3 eV); it is ionised by cosmic rays and excited by collisions with electrons but above all with hydrogen atoms. The ground state of CII separates into two levels of fine structure with a separation corresponding to an excitation temperature of 92 K. The number of excitations for cm^3 and per second is given by an analogous formula to (4.32) where the density of the hydrogen atoms is substituted for the electron density. Using (4.34) with the appropriate expression of the cross section we have for the cooling of the CII:

$$L = 7.9 \cdot 10^{-27} n(H)n(CII)T^{-\frac{1}{2}} \exp\left(-\frac{92}{T}\right) \text{ erg cm}^{-3} \text{ s}^{-1} . \qquad (4.38)$$

(C) **Regions with intermediate temperatures** ($10^2 < T < 10^4$ K).
At these temperatures, ions such as FeII and SiII can be excited by collisions in metastable states with a relatively long average life; there is a fairly high probability that these, before decaying, are still excited at a higher level.

(D) **Ionized regions** ($T = 10^4$ K).
In the HII regions, whose temperatures are about 10^4 K, the cooling is determined by the excitation of ions of relatively abundant elements such as C, N, O, Ne which have excited levels that can be populated by collisions with electrons at the temperatures in question ($\Delta E < 1$ eV).

The most important processes are those related to the excitation of the metastable levels of OI, NII, OIII, NeIII whose excitation energy goes from thousandths of eV up to eV. Figure 4.1 (from Spitzer [1]), where L is denoted by λ and G with Γ, presents the values of $L/n_{and}n(H)$ for low density HII regions ($n_e < 10^2$ cm^{-3}) as a function of temperature. The abundances used are the cosmic abundances:

$$n(O) : n(Ne) : n(N) : n(H) = 6 \cdot 10^{-4} : 4 \cdot 10^{-5} : 10^{-4} : 1 . \qquad (4.39)$$

Figure 4.1 also shows the trend of the heating function (dashed curve) and the intersection of the two curves corresponds to the solution of the equation (4.25) which provides the equilibrium temperature. The cross section of the collisional excitation is very high, about 10^5 times greater than that due to the radiative capture, and only their relatively low abundance means that the gas temperature is not much lower.

As will be seen in the detailed study of HII regions, recombination of H can be a source of cooling. The corresponding energy loss is:

$$L = \frac{3}{2}k_B T N_{ric} \text{ erg cm}^{-3} \text{ s}^{-1} \qquad (4.40)$$

where $3k_B T/2$ is the thermal energy of the recombining electron which is released in the form of electromagnetic energy and N_{ric} is the recombination rate whose expression is :

$$N_{ric} = n_e^2 \beta(T) \qquad (4.41)$$

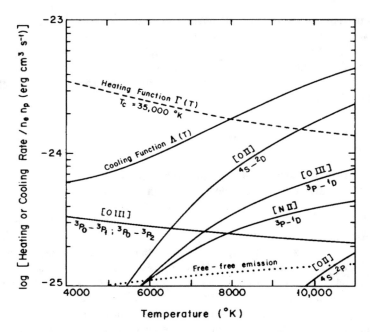

Fig. 4.1 Cooling rate of HII region dominant coolants: OI, OIII, NII and Ne III. Heating (Γ) and Cooling (Λ) temperature trends

β being the global recombination coefficient of H. However, this mechanism is less effective than the collisional excitation processes of ions.

(E) **Regions with temperatures in the range** $10^5 < T < 10^7$ K.

Ions of the iron group still retain bound electrons which can be excited by collision by cooling the gas in the radiative de-excitation. An approximate expression of the cooling rate in this temperature range is given by:

$$L = 1.33 \cdot 10^{-19} n(\mathrm{H}) n_e T^{-\frac{1}{2}} \ \mathrm{erg \ cm^{-3} \ s^{-1}} \,. \tag{4.42}$$

(F) **High temperature regions** ($T \approx 10^7$ K).

At these temperatures an effective cooling process is bremsstrahlung radiation (free-free processes) produced by free electron decelerations in the field of a positive ion. The emission takes place mainly at radio frequencies.

The general trend of the cooling function is represented in Fig. 4.2 (from Sutherland and Dopita [2]). The sharp rise for $T \geq 10^4$ K is due to the onset of collisional excitation of H and in part to the rapid increase of n_e following the ionization of H.

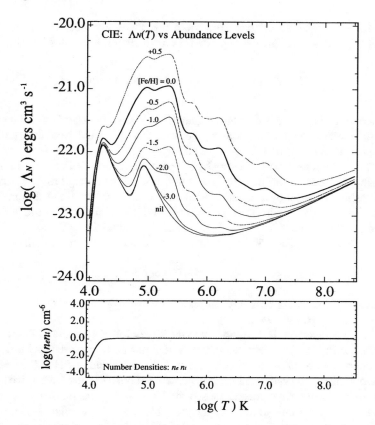

Fig. 4.2 Cooling functions for different abundance values. From Sutherland and Dopita [2]

4.4 Heating Processes

Regardless of the agent that produces it, the main mechanism of heating processes is the ionization of neutral atoms. We list below the main agents of our interest.

(A) **Heating by the stellar radiation field.**

In a radiation field, sufficiently energetic photons can ionize an atom. Let $E_2 = h\nu - I = h(\nu - \nu_0)$ be the energy of the ejected electron; not all this energy will be used to heat the gas by means of elastic collisions that distribute the excess energy of the electron, because in part it will be used in some inelastic collision. In particular, it can be used in the recombination of an electron, whose kinetic energy is E_1. We can write $E_1 = h\delta$ and then the energy gain from the gas will be:

$$E_2 - E_1 = h(\nu - \nu_0) - h\delta = h(\nu - \nu_0') \qquad (4.43)$$

where ν_0' defines an effective threshold that takes into account losses due to inelastic collisions. The heating rate for cm^3 is then for the ith species:

$$G = n_i \int_{\nu_0'}^{\infty} h(\nu - \nu_0')\sigma(\nu)\Phi(\nu)d\nu \text{ erg cm}^{-3} \text{ s}^{-1} \qquad (4.44)$$

where $\sigma(\nu)$ is the cross section for photoionization and $\Phi(\nu)$ is the photon flux of the ionizing field with frequency between ν and $\nu + d\nu$. Note that, unlike L, G linearly depends on the density of the gas.

In the HI regions, heating occurs by the photoionization of C, Si, Fe; in the HII regions, the heating is due to the ionization of H. In the HI regions, which remain so despite being hit by a radiation field, the average energy of the photons is less than 13.6 eV (otherwise they would ionize H) and is higher than the ionization potential of C which is 11.3 eV ; the energy introduced into the gas per photon is therefore less than 2.3 eV: in reality it is about one eV per photon. We have:

$$G = 2 \cdot 10^{-22} n(C) \text{ erg cm}^{-3} \text{ s}^{-1} \qquad (4.45)$$

where $n(C)$ is the number of C atoms per unit of volume.

(B) **Heating by cosmic rays.**

Taking into account the heating due to the ionization of C by the stellar radiation field, the Eq. (4.25) gives a temperature of about 20 K, considerably lower than the temperature of 80 K, observed in the diffuse clouds of HI . Evidently, to interpret this observation data, it is necessary to consider a more effective heating process. Cosmic rays consist mostly of high-energy protons and represent a potentially important source of energy supply to the interstellar medium. Their total energy density is estimated at 0.8 eV cm^{-3}. The relatively low energy protons, around $2 - 10$ MeV, are most effective in ionizing and heating the gas. However, low-energy cosmic rays are the ones that interact more easily with interstellar magnetic fields and therefore it seems that they do not propagate in a diffuse and uniform way in the interstellar medium. Furthermore, the Earth's estimate of the flux of cosmic rays is flawed for the same reason. The process involved in the heating by cosmic rays is the ionization of H. If $n(HI)$ is its numerical density, the free electron production rate for cm^3 and per second is $\zeta_{CR} n(HI)$ where ζ_{RC} is the ionization rate per unit of time. The heating rate is given by:

$$G = n(HI)\zeta_{CR}\langle E \rangle \qquad (4.46)$$

where $\langle E \rangle$ is the average value of the energy supplied to the gas. There are two levels of difficulty in estimating G; one is related to the evaluation of ζ_{CR}. The most recent measurements give a value:

$$\zeta_{CR} = 4 \cdot 10^{-17} \text{ s}^{-1} \qquad (4.47)$$

but there is probably an uncertainty of an order of magnitude in this estimate. The other difficulty is related to the estimation of $\langle E \rangle$ which must take into account the fact that each cosmic particle gives rise to a sequence of processes. The efficiency of heating is sensitive to the composition and density of the gas and its degree of ionization. In fact, cosmic protons ionize the atoms of H:

$$p + H \rightarrow p' + H^+ + e \tag{4.48}$$

for a proton of 2 MeV, the electrons produced by this reaction have a wide energy distribution around an average value of 30 eV. However, not all of this energy will appear as thermal energy, unless the medium is almost completely ionized, in which case the energy is distributed by elastic collisions. If the medium is predominantly neutral, the electron can produce excitations and further ionizations, e.g.

$$e + H(1s) \rightarrow e + H(2p) \rightarrow e + H(1s) + h\nu \tag{4.49}$$

or

$$e + H(1s) \rightarrow e + HII + e . \tag{4.50}$$

When the electron energy is below 13.6 eV, ionizations of H are no longer possible and when it is below 10.2 eV, its excitation is not possible either. Taking these processes into account, an average energy of 3.4 eV is injected into the medium for each electron produced. Ultimately we have from (4.46) for a medium with low fractional ionization:

$$G = 8.75 \cdot 10^{-12} \zeta_{CR} n(HI) \text{ erg cm}^{-3} \text{ s}^{-1} . \tag{4.51}$$

In a highly ionized medium, the energy of the secondary electrons goes mainly into thermal energy, because elastic collisions with other electrons are more likely than other processes.

(C) **Heating by X-rays.**
X-rays emitted by the "hot" gas, produced by supernova explosions, with energy of the order of 0.1 keV can be a source of ionization. In this case, however, it is the He, present with a number of atoms equal to $1/10$ of those of H, which plays the most important role due to the high absorption cross section of the X-rays. The electrons secondary can thermalize with the electrons of the environment or can produce further ionizations, as in the previous case. Detailed calculations show that for each photon X with energy of 50 eV only 6 eV are transformed into thermal energy in the limiting case of zero fractional ionization.

Mechanisms other than ionizations can also produce heating of the interstellar medium. The most important processes are:

(D) **Hydrodynamic heating.**
Even macroscopic motions can supply a considerable amount of heat to the interstellar gas. The effects of ordered motions in shock waves can be evaluated by the dynamic characteristics of the processes that produce them (supernova

remnants, stellar winds, HII regions and density waves of the spiral arms). An analysis of some of these processes will be found later. It is more difficult to evaluate the contribution to the warming of the interstellar medium due to the decay of turbulent motions, since the detailed character of the turbulence of the interstellar medium is not yet well known, which is partly different from the turbulence of ordinary incompressible fluids, of which we only possess a relatively in-depth knowledge (see Chap. 6). An expression of the thermal energy produced by the dissipation of turbulent energy is given by:

$$G_{\text{turb}} = 3.5 \cdot 10^{-28} v_{\text{turb}}^3 \frac{n(\text{H})}{R} \text{ erg cm}^{-3} \text{ s}^{-1} \qquad (4.52)$$

where v_{turb} is the turbulent velocity on the R scale equal to the size of the region concerned, R is measured in pc and v_{turb} in km / sec.

A cloud in non-free fall (i.e. when the force of gravity is partially opposed by another force) can be heated by the release of gravitational energy; for this to happen it is necessary that the radiant energy eventually obtained does not escape from the system but is absorbed again and transformed into thermal energy. For a cloud with temperature T and pressure ρ the energy released in the unit of volume in the unit of time is:

$$G_{\text{coll}} = \left| p\rho \frac{d}{dt} \frac{1}{\rho} \right| = \frac{p}{\rho} \frac{d\rho}{dt} . \qquad (4.53)$$

Thinking that ρ is the average density, it can be expressed by the mass and the radius of the collapsing cloud, hence:

$$G_{\text{coll}} = \left| \frac{k_B T}{\mu m_H} \frac{d}{dt} \left(\frac{3M}{4\pi R^3} \right) \right| = 3 \frac{k_B T}{\mu m_H} \rho \frac{d \ln R}{dt} . \qquad (4.54)$$

The quantity $dt / d \ln R$ is the time scale of the collapse, but since the collapse is not in free fall, otherwise there would be no heating, 2.49 cannot be used.

4.5   Thermal Instability

The equation of the interstellar gas heat balance (4.11) can be put in the form:

$$T \frac{ds}{dt} = -\lambda \qquad (4.55)$$

with

$$\lambda(\rho, T) = \frac{L(\rho, T) - G(\rho, T)}{\rho} . \qquad (4.56)$$

Let us consider a gas at rest and uniform, with density ρ_0 and temperature T_0, such that:

$$\lambda(\rho_0, T_0) = 0 . \tag{4.57}$$

The state characterised by the values (ρ_0, T_0) is a state of thermal equilibrium, therefore the continuity and momentum conservation equations are automatically satisfied. We determine under what conditions this state of equilibrium is stable. We perturb the system by setting:

$$\rho = \rho_0 + \delta\rho \qquad T = T_0 + \delta T . \tag{4.58}$$

Suppose that the perturbation occurs in such a way that a thermodynamic quantity A remains constant and we set:

$$s = s_0 + \delta s \qquad \lambda = \lambda_0 + \delta\lambda . \tag{4.59}$$

These quantities will obey the energy balance equation:

$$T\frac{ds}{dt} = -\lambda . \tag{4.60}$$

Taking into account that the unperturbed state satisfies (4.56) and that s is uniform, linearising the equation we obtain:

$$T_0\frac{d\delta s}{dt} = -\delta\lambda . \tag{4.61}$$

We are looking for a solution such as:

$$\delta s = (\delta s)_0 \exp(\sigma t) \tag{4.62}$$

which expression, substituted in (4.61), gives us:

$$\sigma = -\frac{1}{T_0}\left(\frac{\partial\lambda}{\partial s}\right)_A . \tag{4.63}$$

If $\sigma < 0$ the solution decreases exponentially and the equilibrium configuration is stable. If $\sigma > 0$ there is instability, so the instability condition is:

$$\left(\frac{\partial\lambda}{\partial s}\right)_A < 0 . \tag{4.64}$$

A realistic choice for A cannot be ρ because the variations of p consequent to an isochoric transformation give rise to motions that destroy the constancy of density; it can instead be p because the absence of a variation of p guarantees the conservation of mechanical equilibrium and preserves the density field. For an isobaric transformation $T\delta s = c_p \delta T$, so (4.64) is equivalent to:

$$\left(\frac{\partial \lambda}{\partial T}\right)_p < 0 . \tag{4.65}$$

consider

$$\left(\frac{\partial \lambda}{\partial T}\right)_p = \left(\frac{\partial \lambda}{\partial \rho}\right)_T \left(\frac{\partial \rho}{\partial T}\right)_p + \left(\frac{\partial \lambda}{\partial T}\right)_\rho . \tag{4.66}$$

Since for the isobaric transformation of perfect gas we have

$$\rho = \rho_0 \frac{T0}{T} \tag{4.67}$$

which gives us

$$\left(\frac{\partial \rho}{\partial T}\right)_p = -\rho_0 \frac{T_0}{T^2} . \tag{4.68}$$

Since $(\partial \rho / \partial T)_p$ is evaluated in (ρ_0, T_0) we have

$$\left(\frac{\partial \rho}{\partial T}\right)_p = -\frac{\rho_0}{T_0} \tag{4.69}$$

so the (4.66) holds

$$\begin{aligned}
\left(\frac{\partial \lambda}{\partial T}\right)_p &= \left(\frac{\partial \lambda}{\partial T}\right)_\rho - \frac{\rho_0}{T_0}\left(\frac{\partial \lambda}{\partial \rho}\right)_T \\
&= \left(\frac{\partial \lambda}{\partial T}\right)_\rho \left[1 - \frac{\rho_0}{T_0}\frac{(\partial \lambda / \partial \rho)_T}{(\partial \lambda / \partial T)_\rho}\right] .
\end{aligned} \tag{4.70}$$

The expression

$$\frac{(\partial \lambda / \partial \rho)_T}{(\partial \lambda / \partial T)_\rho} \tag{4.71}$$

can be evaluated by calculating λ in an equilibrium point (ρ, T) by means of a series development with an equilibrium starting point (ρ_0, T_0) as well: neglecting higher order terms we have:

$$\lambda(\rho, T) = \lambda(\rho_0, T_0) + \left(\frac{\partial \lambda}{\partial \rho}\right)_T d\rho + \left(\frac{\partial \lambda}{\partial T}\right)_\rho dT \tag{4.72}$$

where it is:

$$\rho = \rho_0 + d\rho \qquad T = T_0 + dT . \tag{4.73}$$

Taking into account that both (ρ_0, T_0) and (ρ, T) are equilibrium configurations, it follows that

$$\left(\frac{\partial \lambda}{\partial \rho}\right)_T d\rho + \left(\frac{\partial \lambda}{\partial T}\right)_\rho dT = 0 \qquad (4.74)$$

therefore

$$\frac{(\partial \lambda / \partial \rho)_T}{(\partial \lambda / \partial T)_\rho} = -\left(\frac{dT}{d\rho}\right)_0 \qquad (4.75)$$

with which the (4.70) becomes

$$\left(\frac{\partial \lambda}{\partial T}\right)_p = \left(\frac{\partial \lambda}{\partial T}\right)_\rho \left[1 + \frac{d \ln T}{d \ln \rho}\right] . \qquad (4.76)$$

From the equation of state of the perfect gas we deduce:

$$\frac{d \ln p}{d \ln \rho} = 1 + \frac{d \ln T}{d \ln \rho} \qquad (4.77)$$

so finally the condition of instability becomes:

$$\left(\frac{\partial \lambda}{\partial T}\right)_p = \left(\frac{\partial \lambda}{\partial T}\right)_\rho \frac{d \ln p}{d \ln \rho} < 0 . \qquad (4.78)$$

We distinguish two situations:

(a) if $(\partial \lambda / \partial T)_\rho > 0$ the instability condition holds

$$\frac{dp}{d\rho} < 0 \qquad (4.79)$$

that is, there is instability, and therefore growth of the perturbations, along the descending sections of the equilibrium curve $p(\rho)$;

(b) if $(\partial \lambda / \partial T)_\rho < 0$ the instability condition holds

$$\frac{dp}{d\rho} > 0 \qquad (4.80)$$

and the instability occurs along the ascending branches of the equilibrium curve $p(\rho)$.

If we limit the consideration to temperatures below 10^5 K and if we assume that the warming is produced by cosmic rays, we have:

$$L(\rho, T) = n^2 \mathcal{L} \qquad \text{with} \qquad \mathcal{L} = \mathcal{L}(T) \qquad (4.81)$$

$$G(\rho, T) = n \mathcal{H} \qquad \text{with} \qquad \mathcal{H} = \text{cost.} . \qquad (4.82)$$

It is:

$$\lambda = \frac{\rho}{\mu^2 m_H^2} \mathcal{L} - \frac{1}{\mu m_H} \mathcal{H} \tag{4.83}$$

and therefore

$$\left(\frac{\partial \lambda}{\partial T}\right)_\rho = \frac{\rho}{\mu^2 m_H^2} \left(\frac{\partial \mathcal{L}}{\partial T}\right)_\rho \tag{4.84}$$

therefore, according to Fig. 4.2, condition (a) is verified in the temperature interval considered.

Figure 4.3 shows the equilibrium configurations of the interstellar medium accord-ing to Field theory [3]. These are determined by solving the Eq. (4.25) using the expressions of the cosmic ray heating process and the collisional excitation cooling process of C, O, Si. The equilibrium configuration curve consists of three branches: the AB branch and CD branch are stable according to the (4.79) criterion while BC branch is unstable and therefore not significant. Based on these results, two phases can coexist in pressure equilibrium: a dense phase ($n = 1 - 10$ cm^{-3}) and cold ($T = 100$ K), e.g. clouds, and a rarefied ($n = 0.1$ cm^{-3}) and warm ($T = 10^4$ K) phase, e.g. the diffuse intra-cloud medium. The phase of the molecular clouds does not appear, which, being self-gravitating, is not in pressure equilibrium with the other phases. However, this representation of the interstellar medium will be put into crisis in the 70 s and replaced with a much more complex picture.

The existence of two phases in pressure equilibrium is the direct consequence of the nature of the cooling processes which takes the form of the cooling function as a function of temperature, expressed by the graph in Fig. 4.3. Let us consider the instability condition expressed by (4.65) with the (4.70) and calculate the partial derivatives using the expression of λ given by (4.83). They are:

$$\left(\frac{\partial \lambda}{\partial T}\right)_\rho = \frac{\rho}{\mu^2 m_H^2} \frac{d\mathcal{L}}{dT} \tag{4.85}$$

is

$$\left(\frac{\partial \lambda}{\partial \rho}\right)_T = \frac{\mathcal{L}}{\mu^2 m_H^2} \tag{4.86}$$

for which we have:

$$\left(\frac{\partial \lambda}{\partial T}\right)_p = \frac{\rho}{\mu^2 m_H^2} \frac{d\mathcal{L}}{dT} - \frac{\rho}{T} \frac{\mathcal{L}}{\mu^2 m_H^2} = \frac{\rho}{\mu^2 m_H^2} \left[\frac{d\mathcal{L}}{dT} - \frac{\mathcal{L}}{T}\right]. \tag{4.87}$$

Ultimately, the condition of instability is expressed through:

$$\frac{d \ln \mathcal{L}}{d \ln T} < 1 \tag{4.88}$$

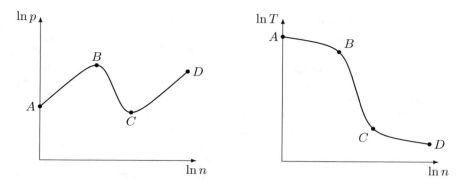

Fig. 4.3 Example of $p = p(\rho)$ and $T = T(\rho)$ for a case in which heating is caused by cosmic rays and cooling by collisional excitation of C, O, and Si. See text for details

therefore there is instability when $\ln \mathcal{L}$ grows as T grows slower than $\ln T$, while stability occurs when $\ln \mathcal{L}$ grows faster than $\ln T$.

The considerations made on thermal instability presuppose dynamic equilibrium, if this consideration is not verified the analysis is no longer correct. However, the results obtained are approximately valid when the dynamic time scale is considerably greater than the cooling time: in this case the thermal equilibria evolve slowly (on the dynamic time scale) due to the dynamic evolution.

It may happen that an unstable phase can be observed as the time of its disappearance, which is the cooling time, is sufficiently large.

4.6 Synthetic Summary

A comparison of the characteristic times for the diffusion of heat and of the momentum, deduced from the diffusion equation, with the dynamic times leads to the conclusion that: (a) the conduction of heat is negligible in all regions of the interstellar gas with the exception hot gas; (b) viscosity can be ignored when $L \geq l$, L being the distance along which there are significant variations of the physical quantities and l the mean free path of the particles. Except in particular situations (e.g. shock waves, turbulence at the scales of the order of the mean free path) the inequality $L \geq l$ is verified and therefore the viscosity is negligible.

Nevertheless, the behavior of the interstellar gas is not adiabatic because phenomena of interaction between matter and radiation can occur which determine a gain or loss of thermal energy by the gas. The energy balance is expressed by (4.11) where G and L represent respectively the energy gained and the energy lost per unit of volume and per unit of time.

Using the balance equation in the form applicable to an ideal gas, the cooling time is determined, which measures the time scale on which there are significant variations in the gas temperature due to these interactions. When the cooling time is less than the dynamic time over a time interval equal to the cooling time, the quantities such as pressure and density vary very little and the first-member term in the (4.17) is approximately zero: the temperature is determined by the balance between heating and cooling processes. When the cooling time is comparable to or greater than the dynamic time this approximation is invalid.

Generally the cooling processes are a consequence of collisional excitations when the photons produced by the de-excitation have a mean free path greater than the size of the region. For the process to be effective, three requirements must be met: (1) the excitation energy must be lower than the average energy of the thermal agitation motion; (2) the collisions must be numerous, which occurs when the colliding particles are numerous; (3) the cross section for excitation must be large, e.g. it is sufficient that the transition is allowed. The type of process depends on the gas temperature and consequently the rate of energy loss varies.

The heating processes are instead determined by ionization phenomena: the ionizing agents can be sufficiently energetic photons of the stellar radiation field, low-energy cosmic protons and soft X-rays. In evaluating the energy gain it is necessary to consider the possibility that the freed electron undergoes inelastic collisions with further ionizations or excitations: these will reduce the energy deposited in the gaseous medium.

The temperature obtained from the equilibrium condition does not necessarily represent a state that actually exists in reality: for this to happen, the solution must be stable, that is, small perturbations of the equilibrium state are extinguished. The condition for interstellar gas instability when the temperature is below 10 K is that the pressure decreases as the density increases. On the basis of this result, the nature of the interstellar medium in the past decades was studied on the assumption that this gas was in dynamic equilibrium, which was subsequently denied by observations. In those hypotheses it was found that two phases coexisted in pressure equilibrium: HI clouds with temperatures around 100 K and a rarefied diffuse medium with $T \approx 10^4$ K.

The stable or unstable character, and therefore the existence of several phases in equilibrium, is determined by the particular dependence of L on the temperature.

References

1. Spitzer, L.: Physical Processes in the Interstellar Medium, Chapter 6. Wiley, New York (1978)
2. Sutherland R.S., Dopita M.A.: Cooling functions for low-density astrophysical plasmas. In: ApJS, vol. 88, p. 253 (1993)
3. Field G.B.: Thermal instability. In: ApJ, vol. 142, p. 531 (1965)

Chapter 5
Shock Waves

5.1 Supersonic Motion and Discontinuity

Sound waves have been seen to propagate in the medium with speed

$$c = \sqrt{\left(\frac{\mathrm{d}\,p}{\mathrm{d}\,\rho}\right)_{\mathrm{ad}}}.$$ (5.1)

The speed of sound is a critical speed (remember what was said in Sect. 2.1 about the time required to restore dynamic equilibrium). In relation to c the fluid motions are classified into subsonic ($v < c$) and supersonic ($v > c$). We use to introduce the Mach number $Ma = v/c$, so the motions will be subsonic or supersonic depending on whether $Ma < 1$ or $Ma > 1$. The behavior of fluids is different in the subsonic case and in the supersonic case in two respects.

(1) **The mode of propagation of small disturbances.** Let us consider a source of acoustic waves integral with the fluid moving with uniform speed **v** with respect to a fixed observer O. Compared to O the speed of propagation of a perturbation within the fluid depends on two factors: the perturbation is dragged by the fluid with speed **v**, and in the fluid it propagates with speed $c\hat{\mathbf{n}}$, where $\hat{\mathbf{n}}$ is the vector of the direction of propagation of the perturbation with respect to the fluid. Therefore, with respect to the fixed observer O, the speed of propagation of the perturbation is $\mathbf{v} + c\hat{\mathbf{n}}$. Distinguishing between the subsonic and the supersonic cases we have that

    (a) in the case $v < c$ the possible speed values can be obtained by constructing the resultant of **v** and $c\hat{\mathbf{n}}$ at varying $\hat{\mathbf{n}}$; the extreme of the resultant distributes on a spherical surface (see Fig. 5.1) so the vector $\mathbf{v} + c\hat{\mathbf{n}}$ can have any direction and the propagation can reach each point of the gas;

    (b) in the case $v > c$ the direction of $\mathbf{v} + c\hat{\mathbf{n}}$ can only lie within an opening cone defined by

G. Carraro, *Astrophysics of the Interstellar Medium*,
UNITEXT for Physics, https://doi.org/10.1007/978-3-030-75293-4_5

Fig. 5.1 Regions that can be
reached by an acoustic
perturbation depending on
the Mach number

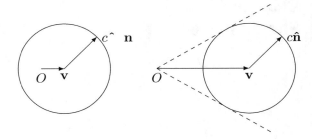

$$\tan \alpha = \frac{c}{v} = \frac{1}{Ma} \qquad (5.2)$$

and therefore the perturbation can only propagate within the cone; the disturbance on the other hand has no influence outside the cone.

(2) *The different degree of compressibility.* We have seen that, in the approximation in which infinitesimals of higher order can be neglected, it results:

$$\frac{\delta \rho}{\rho} \approx \frac{v}{c} = Ma \qquad (5.3)$$

so if $v/c \ll 1$ it will be $\Delta \rho \ll \rho$, while if $v \geq c$ the compression is relevant and can give rise to phenomena not considered so far. Supersonic motions occur in gases and the study of the phenomena associated with these motions is part of gasdynamics. To get an idea of what happens when the effects of compression become relevant, consider the case of an explosion in the atmosphere. The surrounding air is hit by a compression front that moves in the middle with a speed much greater than the speed of sound. We can schematize the situation by considering the action, on a gas enclosed in a tube, of a piston that moves with accelerated motion until it reaches supersonic speeds (Fig. 5.2). Suppose that the variations produced by the piston occur so rapidly that they are considered adiabatic.

When the piston starts at rest with a small displacement a signal propagates in the gas at the speed of sound and the gas is slightly compressed. The piston continues to accelerate with respect to the gas and another signal moves into the already compressed gas and compresses it further; this signal propagates with the characteristic speed of sound of the already compressed gas. Since, by definition, $c \sim \rho^{1/3}$, the sound is now a little faster. So if we arbitrarily divide the gas in front of the piston into zones 1, 2, 3, etc. the density will gradually decrease, i.e. $\rho(1) > \rho(2) > \rho(3) > ...$, as well as the speed of sound, i.e. $c(1) > c(2) > c(3) > ...$.

Thus the piston signals move in the gas and the gas far enough away remains unmodified until the signal arrives. The denser areas, on the left, accumulate on the less dense areas on the right; the compression of the areas closest to the piston grows more and more and the density profile steepens as shown in Fig. 5.3. In addition to

Fig. 5.2 Sketch of a tube filled with air and compressed by a piston

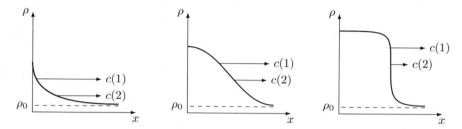

Fig. 5.3 Density gradient evolution for the case of the air compression in a tube

the density, the other quantities that describe the dynamic and thermodynamic state also undergo sudden variations. The steepening of the density profile cannot continue indefinitely: when the density variation scale

$$L_\rho = \frac{\rho}{|\,\text{grad}\,\rho\,|} \tag{5.4}$$

becomes of the order of the average free path l of the particles, in the transition region the dissipative effects due to viscosity become important (see Sect. 4.1) and the kinetic energy of the ordered motion is transformed into thermal energy.

We can therefore evaluate the thickness of the impact front by expressing the average free path of the particles:

$$l = \frac{1}{\sqrt{2}n_a\sigma} \tag{5.5}$$

where σ is the collision cross section, n_a is the density of the atoms in front of the front. Taking $\sigma = 10^{-15}$ cm^2 and $n_a = 10$ cm^{-3} we have $l = 7 \cdot 10^{13}$ cm $\approx 2.4 \cdot 10^{-5}$ pc.

In fully ionized gases the shock width is much larger: an estimate can be obtained from

$$l = 3.9 \cdot 10^{-5}\frac{m_H^2 v^4}{e^4 n_i} \tag{5.6}$$

where n_i is the density of the ions. This width depends greatly on the speed of the shock. If $v = 10^3$ km/s and $n_i = 10$ cm^{-3}, the width of the shock is $l = 0.006$ pc, but if the velocity is $5 \cdot 10^3$ km/s the width of the front becomes 4 pc. Since the gas is never completely ionized, the presence of neutral atoms makes the front thinner as in the case of (5.5). A magnetic field produces a significant reduction in the thickness of the impact front. In this case the width can be equal to the radius of the orbit of a charged particle in a magnetic field, which for non-relativistic protons in interstellar space is $3 \cdot 10^{11}$ cm $= 10^{-7}$ pc.

With the exclusion of the layer where the abrupt variations of the quantities that characterize the behavior of the gas take place, the viscosity is negligible and the gas behaves adiabatically. We ignore for now the possible interaction phenomena of matter with radiation that are important in interstellar gas.

The precise treatment of the conditions and processes in the abrupt variation layer is complicated and, moreover, unnecessary: the thickness of the layer where the effect of viscosity is felt is small (as mentioned, it is of the order of the average free path of atoms or molecules) with respect to the scales of variation of the quantities both in the compressed gas and in the unperturbed one: therefore this area can be assimilated to a mathematical surface through which the variables have discontinuities. Therefore, with the exclusion of what happens through this discontinuity surface, the fluid can be considered as non-viscous and the motion as adiabatic.

Nevertheless, it will be necessary to remember that, due to the presence of viscosity in the abrupt change layer, there is a production of entropy and therefore there will also be a discontinuity of entropy across the surface in question. In a strictly non-viscous fluid, surfaces of discontinuity cannot be created; when we treat the process by introducing a discontinuity we mask the role of viscosity. In the motion of a fluid in the absence of discontinuity the usual conservation laws (mass, momentum, energy) are expressed in differential form; in the presence of a surface of discontinuity they must be formulated in finite terms and represent boundary conditions which the equations of motion must satisfy. From the numerical point of view, situations like these are called Riemann problems, and the corresponding numerical solution techniques are called Riemann solvers.

5.2 Jump Conditions

Let us consider the motion of a gas that gives rise to a Σ discontinuity surface. In a non-stationary motion the surfaces of discontinuity are not fixed and the speed with which they move is different from the speed of the gas motion (in particular the gas can be at rest and the surface can move). Let O be a fixed reference system with respect to which the gas is in motion and with respect to which Σ is also in motion (see Fig. 5.4). Let us consider an element of Σ and a coordinate system O' with respect to which this element is at rest. The x axis of this reference is also oriented as the surface normal. Suppose that with respect to O the surface is now to the right. We denote with the index 2 the characteristic quantities of the gas to the left of the discontinuity and with the index 1 those to the right. The 2 gas has already been compressed, the 1 gas is unperturbed.

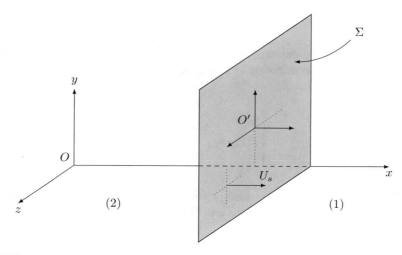

Fig. 5.4 Definition of shock integral reference frame and reference frame at rest

We express the conditions of connection to the discontinuity surface (jump conditions). They express the conservation laws of mass, momentum and energy through Σ.

(1) *Conservation of mass.* The mass entering from one side must equal the mass exiting the other side. The flow of matter through the Σ unit of surface is ρv_x, where v_x is the component of the fluid velocity perpendicular to the surface. Preserving mass requires:

$$\rho_1 v_{x,1} = \rho_2 v_{x,2} . \tag{5.7}$$

(2) *Conservation of momentum.* In Sect. 1.4 we have seen that, using the vector $p\hat{\mathbf{n}} + \rho\mathbf{v}(\mathbf{v} \cdot \hat{\mathbf{n}})$ where $\hat{\mathbf{n}}$ is the vector of a given direction, the flow of momentum can be expressed. Identifying $\hat{\mathbf{n}}$ with the $\hat{\mathbf{i}}$ versor of the x axis we have:

$$p\hat{\mathbf{i}} + \rho\mathbf{v}(\mathbf{v} \cdot \hat{\mathbf{i}}) = p\hat{\mathbf{i}} + \rho v_x\mathbf{v} . \tag{5.8}$$

The flow per unit area normal to the x axis of the component along the x axis of the momentum is expressed by:

$$p\hat{\mathbf{i}} \cdot \hat{\mathbf{i}} + \rho v_x\mathbf{v} \cdot \hat{\mathbf{n}} = p + \rho v_x^2 . \tag{5.9}$$

Similarly, the flow, through the unit of surface perpendicular to the axis x, of the components of the momentum along the axis y and along the axis z are respectively:

$$p\hat{\mathbf{i}} \cdot \hat{\mathbf{j}} + \rho v_x\mathbf{v} \cdot \hat{\mathbf{j}} = \rho v_x v_y \tag{5.10}$$

$$p\hat{\mathbf{i}} \cdot \hat{\mathbf{k}} + \rho v_x \mathbf{v} \cdot \hat{\mathbf{k}} = \rho v_x v_z \tag{5.11}$$

where $\hat{\mathbf{j}}$ and $\hat{\mathbf{k}}$ are the versors of the y and z axis respectively. By imposing the equality of the incoming and outgoing flux of the momentum through the discontinuity we have:

$$\begin{cases} p_1 + \rho_1 v_{x,1}^2 = p_2 + \rho_2 v_{x,2}^2 & (5.12) \\ \rho_1 v_{x,1} v_{y,1} = \rho_2 v_{x,2} v_{y,2} & (5.13) \\ \rho_1 v_{x,1} v_{z,1} = \rho_2 v_{x,2} v_{z,2} \,. & (5.14) \end{cases}$$

(3) *Conservation of energy.* The energy flowing through the unit of surface perpendicular to the vector unit $\hat{\mathbf{n}}$ in the unit of time is given by

$$\rho \mathbf{v} \left(\frac{1}{2} v^2 + w \right) \cdot \hat{\mathbf{n}} \tag{5.15}$$

(see (1.66)). Considering the direction perpendicular to Σ ($\hat{\mathbf{n}} = \hat{\mathbf{i}}$), the conservation of energy is expressed by:

$$\rho_1 v_{x,1} \left(\frac{1}{2} v_1^2 + w_1 \right) = \rho_2 v_{x,2} \left(\frac{1}{2} v_2^2 + w_2 \right) \,. \tag{5.16}$$

The relations (5.7), (5.12), (5.13), (5.14) and (5.16) represent the boundary conditions at the discontinuity surface. From them emerges the possibility of two types of discontinuity.

(a) In the first type there is no passage of matter through the surface of discontinuity. According to (5.7), this should be $v_{x,1} = v_{x,2} = 0$. The (5.16) condition is automatically satisfied. From the Eq. (5.12) we have $p_1 = p_2$, while from (5.13) and (5.14) it follows that the components v_y and v_z can show discontinuities. A discontinuity of this type is called *tangential discontinuity* or also *contact discontinuity*. In conditions similar to those found in the earth's atmosphere, the contact discontinuities are unstable and transform into a turbulent region; in the interstellar medium can be stabilized by strong magnetic fields.

(b) In the second type of discontinuity, *shock waves* or *shock*, there is flow of matter through the discontinuity. From the (5.7) being $\rho v_x \neq 0$ it results that the normal component of the velocity is not zero and is discontinuous. From (5.13) and (5.14) with (5.7) we have:

$$v_{y,1} = v_{y,2} \qquad v_{z,1} = v_{z,2} \tag{5.17}$$

that is, the tangential component of the velocity is continuous. The density, pressure and other thermodynamic quantities are instead discontinuous.

In shock waves the component of the velocity normal to the discontinuity surface and the thermodynamic parameters vary abruptly. We can distinguish two cases:

- the shock wave is normal when the gas flow velocity is normal to the discontinuity surface Σ; in this case the phenomenon can be studied in a reference system integral with the element of Σ considered, with respect to which the gas has zero the tangential component of the velocity;
- the gas velocity is not perpendicular to the discontinuity surface; the shock wave is said to be oblique: since, as we have seen, the tangential component of the velocity is continuous, it can be reduced to the case of a normal shock with an appropriate choice of the reference system in which the tangential component is zero.

5.3 Shock Waves

Let us consider a shock wave in a O reference in which the element of the discontinuity surface is at rest; the x axis also has the direction of the surface normal. The tangential component of the velocity, which is not discontinuous, is zero (normal shock wave). In this reference the gas moves in the direction of $1 \rightarrow 2$; the gas in 1 is that towards which the discontinuity surface is moving and the gas in 2 is what remains behind the shock.

We denote by v the velocity of the gas with respect to the discontinuity and we refer to a perfect monoatomic gas ($\gamma = 5/3$): the connection conditions can be given by the constancy, through the discontinuity surface, of the quantities:

$$\rho v = \text{cost.} = j \tag{5.18}$$

$$p + \rho v^2 = \text{cost.} = S \tag{5.19}$$

$$\rho v \left(\frac{1}{2} v^2 + w \right) = \text{cost.} \tag{5.20}$$

where the last equation, given the continuity of ρv, is equivalent to:

$$\frac{1}{2} v^2 + w = \text{cost.} = E . \tag{5.21}$$

These conditions of connection are called Rankine–Hugoniot (R–H) conditions.

The (5.21) can be further modified in the case of the perfect gas. Let f be the number of degrees of freedom, the internal energy of the unit of mass is:

$$\epsilon = \frac{f}{2} k_B T \frac{n}{\rho} = \frac{f}{2} \frac{k_B T}{\mu m_H} = \frac{f}{2} p V \tag{5.22}$$

where $V = 1/\rho$ is the specific volume; the enthalpy of the unit of mass is then:

$$w = \epsilon + \frac{p}{\rho} = \epsilon + pV = \frac{f}{2} pV + pV = \frac{f+2}{2} pV . \tag{5.23}$$

Specific heats are defined by f:

$$c_v = \frac{f}{2}\mathcal{R} \qquad c_p = c_v + \mathcal{R} = \frac{f+2}{2}\mathcal{R} \qquad \gamma = \frac{c_p}{c_V} = \frac{f+2}{f} \tag{5.24}$$

and through these relations f can be expressed as a function of γ

$$f = \frac{2}{\gamma - 1} \tag{5.25}$$

so, introducing this in the expression of w, we obtain:

$$w = \frac{\gamma}{\gamma - 1} pV . \tag{5.26}$$

The (5.21) then becomes:

$$\frac{1}{2}v^2 + \frac{\gamma}{\gamma - 1}\frac{p}{\rho} = \text{cost.} \tag{5.27}$$

and in the case of monoatomic gas we have:

$$\frac{1}{2}v^2 + \frac{5}{2}\frac{p}{\rho} = E = \text{cost.} . \tag{5.28}$$

It is interesting to know the effects of the shock on the dynamic and thermodynamic state of the gas; the direction of the transformation will be determined by the condition that the entropy increases due to the dissipative processes that take place within the shock. For the detailed derivation, see Appendix A while we briefly list the main conclusions here:

- the motion is initially supersonic but the velocity of the gas behind the shock becomes subsonic,
- the density behind the shock is greater than the density of the unperturbed gas,
- the pressure behind the shock front is greater than the pressure of the unperturbed gas,
- the temperature increases due to the passage of the shock wave.

It has been seen that the shock wave leads to an increase in entropy due to the dissipation of energy in the shock layer. The extent of this dissipation is determined by the conservation laws expressed by the Rankine–Hugoniot conditions and the width of the layer is such as to produce the increase in entropy required by the conservation laws.

5.4 Rankine–Hugoniot Conditions for an Ideal Gas

We keep the conventions on the reference system of the previous paragraph and apply the Rankine–Hugoniot conditions:

$$
\begin{cases}
\rho_1 v_1 = \rho_2 v_2 = j & (5.30) \\
p_1 + \rho_1 v_1^2 = p_2 + \rho_2 v_2^2 & (5.31) \\
\dfrac{1}{2} v_1^2 + w_1 = \dfrac{1}{2} v_2^2 + w_2 . & (5.32)
\end{cases}
$$

The (5.30) gives:

$$
v_1 = j V_1 \qquad v_2 = j V_2 \tag{5.32}
$$

where V is the specific volume. Substituting these in (5.31) we have:

$$
p_1 + j^2 V_1 = p_2 + j^2 V_2 \tag{5.33}
$$

from which:

$$
j^2 = \frac{p_2 - p_1}{V_1 - V_2} . \tag{5.34}
$$

These relationships relate the propagation velocity of the wave front with respect to the gas to the pressure and density of the gas on both sides of the discontinuity surface. From (5.32) we have:

$$
w_1 + \frac{1}{2} j^2 V_1^2 = w_2 + \frac{1}{2} j^2 V_2^2 \tag{5.35}
$$

and replacing j with (5.34)

$$
w_1 - w_2 + \frac{1}{2}(p_2 - p_1)(V_1 + V_2) = 0 . \tag{5.36}
$$

With the (5.26) expression of w the (5.36) becomes:

$$
\frac{\gamma}{\gamma - 1}(p_1 V_1 - p_2 V_2) + \frac{1}{2}(V_1 + V_2)(p_2 - p_1) = 0 \tag{5.37}
$$

hence the

$$
\frac{V_2}{V_1} = \frac{\rho_1}{\rho_2} = \frac{(\gamma + 1)p_1 + (\gamma - 1)p_2}{(\gamma - 1)p_1 + (\gamma + 1)p_2} . \tag{5.38}
$$

From the equation of state of the perfect gas we have:

$$\frac{T_2}{T_1} = \frac{p_2}{p_1}\frac{V_2}{V_1} = \frac{p_2}{p_1}\frac{(\gamma+1)p_1 + (\gamma-1)p_2}{(\gamma-1)p_1 + (\gamma+1)p_2}. \tag{5.39}$$

From (5.34) and (5.38) we get:

$$j^2 = \frac{p_2 - p_1}{V_1(1 - V_2/V_1)} = \frac{(\gamma-1)p_1 + (\gamma+1)p_2}{2V_1}. \tag{5.40}$$

From this we obtain the gas velocity with respect to the impact front:

$$v_1^2 = j^2 V_1^2 = \frac{1}{2}V_1[(\gamma-1)p_1 + (\gamma+1)p_2] \tag{5.41}$$

$$v_2^2 = j^2 V_2^2 = \frac{1}{2}V_2[(\gamma+1)p_1 + (\gamma-1)p_2]. \tag{5.42}$$

By knowing p_1, p_2 and V_1 in this way it is possible to obtain all the other quantities.

The parameters that define the state of the gas subjected to the shock action can also be expressed as a function of the Mach number Ma_1 of the gas 1. It is:

$$Ma_1^2 = \left(\frac{v_1}{c_1}\right)^2 = \frac{\rho_1 v_1^2}{\gamma p_1} \tag{5.43}$$

and using the first of the (5.41):

$$Ma_1^2 = \frac{\rho_1 V_1}{2\gamma p_1}[(\gamma-1)p_1 + (\gamma+1)p_2] = \frac{1}{2\gamma}\left[(\gamma-1) + (\gamma+1)\frac{p_2}{p_1}\right]. \tag{5.44}$$

We then have:

$$\frac{p_2}{p_1} = \frac{2\gamma Ma_1^2}{\gamma+1} - \frac{\gamma-1}{\gamma+1} \tag{5.45}$$

$$\frac{\rho_2}{\rho_1} = \frac{(\gamma+1)Ma_1^2}{2 + (\gamma-1)Ma_1^2} \tag{5.46}$$

$$\frac{T_2}{T_1} = \frac{[2\gamma Ma_1^2 - (\gamma-1)][2 + (\gamma-1)Ma_1^2]}{(\gamma+1)^2 Ma_1^2} \tag{5.47}$$

$$Ma_2^2 = \left(\frac{v_2}{c_2}\right)^2 = \frac{2 + (\gamma-1)Ma_1^2}{2\gamma Ma_1^2 - (\gamma-1)} \tag{5.48}$$

The limiting case of *strong shock waves* characterized by having $Ma \gg 1$ is interesting; from (5.45) it results $p_2/p_1 \gg 1$. From (5.38) or from (5.46) under the same conditions we obtain:

$$\frac{V_2}{V_1} = \frac{\rho_1}{\rho_2} = \frac{\gamma - 1}{\gamma + 1} . \tag{5.49}$$

If $\gamma = 5/3$ (i.e. $f = 3$) then $\rho_2/\rho_1 = 4$, while if $\gamma = 7/5$ (i.e. $f = 5$) then $\rho_2/\rho_1 = 6$. We also have:

$$\frac{T_2}{T_1} = \frac{p_2}{p_1} \frac{\gamma - 1}{\gamma + 1} \tag{5.50}$$

and from (5.48):

$$Ma_2^2 = \frac{\gamma - 1}{2\gamma} \tag{5.51}$$

hence, for example, in the monatomic case the gas behind a very strong shock moves with a speed which is approximately equal to $1/2$ of the corresponding speed of sound.

Therefore in strong shocks the ratios p_2/p_1 and T_2/T_1 diverge while the ratio ρ_2/ρ_1 remains finite and assumes relatively low values.

In the following for the study of some astrophysical phenomena we will need the description of a shock wave with respect to a reference system with respect to which the shock is in motion. For example, when we consider the motions produced by the interaction of stars with the interstellar medium, it is more convenient to use a reference system integral with the star. Let us consider the O' system with respect to which we have studied the phenomenon in the previous paragraphs; with respect to O', the discontinuity surface is fixed and v_1 and v_2 are the velocities of the unperturbed gas and of the one that has undergone the shock action, respectively. Let us now consider a second system O in which U_s is the rate of advance of the shock and u_1 and u_2 are the velocities of the gas. Suppose the references are oriented so that the motion occurs in both along the x axis. The transformations between the two systems are expressed by:

$$v_1 = u_1 - U_s \qquad v_2 = u_2 - U_s . \tag{5.52}$$

Normally the astrophysical situations foresee that $U_s \gg u_1$, that is that the unperturbed gas is practically stationary with respect to O, while the speed of advancement of the shock is very high, so that the strong shock approximation can be adopted . It is therefore:

$$v_1 = -U_s . \tag{5.53}$$

On the basis of the relations for strong shock waves we obtain:

$$\frac{v_1}{v_2} = \frac{\rho_2}{\rho_1} = 4 \tag{5.54}$$

and ultimately:

$$u_2 = \frac{3}{4} U_s . \tag{5.55}$$

Thus, the gas behind the shock moves in the same direction as the shock front with a velocity equal to 3/4 of its velocity.

Substituting the expressions of v_1 and v_2 given by the (5.52) into the Eq. (5.31) where p_1 i is neglected given the strong character of the shock, we have:

$$p_2 = \rho_1 v_1^2 - \rho_2 v_1^2 \left(\frac{\rho_1}{\rho_2}\right)^2 = \rho_1 v_1^2 - 4\rho_1 v_1^2 \frac{1}{16} = \frac{3}{4}\rho_1 U_s^2 . \tag{5.56}$$

The temperature behind the shock is given, according to the equation of state, by:

$$T_2 = \frac{\mu m_H}{k_B}\frac{p_2}{\rho_2} = \frac{3}{16}\frac{\mu m_H}{k_B}U_s^2 . \tag{5.57}$$

The internal energy and the kinetic energy of the unit of mass, for the gas behind the shock, are given respectively by:

$$\epsilon_2 = \frac{3}{2}\frac{p_2}{\rho_2} = \frac{9}{32}U_s^2 \tag{5.58}$$

$$e_{cin,2} = \frac{1}{2}u_2^2 = \frac{9}{32}U_s^2 . \tag{5.59}$$

5.5 Shock Waves in the Interstellar Medium

Quite recently, around the 1970s, it has emerged that the interstellar medium is dominated by the action of shock waves and that its dynamic evolution is profoundly influenced by them. The astrophysical phenomena in which shock waves are produced are:

– supernova explosions and supernova remnants (SNR),
– stellar winds (OB stars, Wolf-Rayet stars, red giants),
– collisions between clouds,
– HII regions in expansion,
– molecular flows, jets,
– merger of the disks of two spiral galaxies,
– processes of accretion on protostars and proto-stellar discs.

The environment in which such shocks are formed ranges from diffuse clouds, which contain atomic gas with density $n(H) = 0.1 - 10^2$ cm^{-3} up to the molecular clouds with density of $10^3 - 10^8$ cm^{-3}.

The shocks heat, compress and irreversibly modify the entropy of the gas they pass through and can give rise to gravitational or thermal instability.

In the conditions in which shock waves occur in the interstellar medium, energy conservation is generally not verified in the form expressed by (5.16): this is because

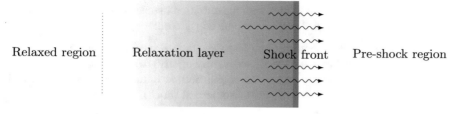

Fig. 5.5 Detaile structure of an ISM region which undergoes a shock

there may be a variation in the degree of ionization or because the gas compressed by the wave can radiate. In this case the shock structure can be divided into three regions (see Fig. 5.5): the *precursor layer*, i.e. the region where the ambient gas has not yet interacted with the wave shock; the *shock front*, i.e. the thin layer where the gas is compressed and heated by the dissipative processes: this layer is substantially the same as that described in the case of ordinary gases; the post-shock *relaxation layer* in which processes of inelastic collisions (ionizations, dissociations, excitations) cool the gas and relax it to a state of thermal equilibrium.

A shock of this type is called radiative: the condition for a radiative shock to occur is that the cooling time (4.20) is less than or equal to the dynamic time:

$$\tau_{din} = \frac{s}{v} \tag{5.60}$$

where s is the space covered by the shock wave since its formation and v the average speed of advance.

Let's consider the example of an HII region. To evaluate the importance of cooling processes, we compare the time scale of dynamic processes with the cooling time. The cooling rate L is expressed by:

$$L = 10^{-24} n_e n(H) \text{ erg cm}^{-3} \text{ s}^{-1} . \tag{5.61}$$

Putting $n(H) = n_e = 10^2 \text{ cm}^{-3}$ we have $L = 10^{-20} \text{ erg cm}^{-3} \text{ s}^{-1}$. At a temperature of $T = 10^4$ K, the internal energy of the volume unit is $U = 3nk_B T/2 = 2 \cdot 10^{-10} \text{ erg cm}^{-3}$. The cooling time is given by:

$$\tau_{cool} = \frac{U}{L} = \frac{2 \cdot 10^{-10}}{10^{-20}} = 2 \cdot 10^{10} \text{ s} = 700 \text{ years} . \tag{5.62}$$

Typical speeds in an HII region are a few times the speed of sound: $v = 30 \text{ km/s} = 3 \cdot 10^{-5} \text{ pc/year}$. The typical dimensions are $l = 1$ pc and therefore the time scale necessary for an appreciable expansion is $\tau_{din} = l/v = 3 \cdot 10^4$ years. Therefore the cooling time is much shorter than the dynamic time: this is a particular situation in which the phenomena due to the dynamics of the system are negligible compared to the interaction processes of matter with the radiation in determining the temperature of the gas. Two borderline situations can be considered:

(a) the radiation phenomena are completely negligible, the motion of the gas outside the shock is adiabatic and the shock is (improperly) called adiabatic;
(b) the gas radiates rapidly so that an equilibrium is immediately established and the temperature is almost constant (isothermal shock).

In the first case, in which the cooling phenomena are completely negligible, the behavior is that described in the previous paragraphs.

5.6   Isothermal Shock: Schematic Discussion

When the cooling time after the shock is less than the age of the shock itself, the processes of interaction with the radiation are important and affect the dynamics of the gas. The shock in this case is called radiative.

In radiative shocks, behind the impact front the temperature rises due to the dissipative processes as in the adiabatic case, but in this case, as the cooling processes are very effective, a state of equilibrium is rapidly achieved which, if there has not been a change in the state of ionization, is characterized by the same gas temperature in front of the shock.

The description of an isothermal shock is obtained using the jump (RH) conditions of the dynamical and thermodynamical variables between two control surfaces that include both the shock layer and the layer behind the shock front where the cooling processes take place. Of the Rankine–Hugoniot conditions, those relating to the conservation of mass and momentum remain valid:

$$\rho_1 v_1 = j = \rho_2 v_2 \tag{5.63}$$

$$p_1 + \rho_1 v_1^2 = p_2 + \rho_2 v_2^2 . \tag{5.64}$$

Instead of the energy jump condition we must consider the equations that determine the temperatures T_1 and T_2 from the thermal balance before the shock:

$$L(T_1, \rho_1) - G(T_1, \rho_1) = 0 \tag{5.65}$$

and beyond the relaxation layer:

$$L(T_2, \rho_2) - G(T_2, \rho_2) = 0 . \tag{5.66}$$

If the ionization state has not changed during the shock passage it must be $\mu_1 = \mu_2$, moreover it can be $T_1 = T_2$ if the solution temperatures of the equations (5.65) and (5.66) are equal : this happens when the cooling function is weakly dependent on density. In this case the shock is called isothermal. In the case of isothermal shock the result will be:

$$c_1^2 = \frac{p_1}{\rho_1} = c_2^2 = \frac{p_2}{\rho_2} .\tag{5.67}$$

If the shock wave is strong from the condition $v_1^2 \gg c_1^2$ we deduce:

$$\rho_1 v_1^2 \gg \rho_1 c_1^2 = p_1 \tag{5.68}$$

therefore in the junction condition (5.64) the gaseous pressure p_1 can be neglected and by using the (5.67) the (5.64) becomes:

$$\rho_1 v_1^2 = \rho_2 c_1^2 + \rho_2 v_2^2 \tag{5.69}$$

which with (5.63) becomes:

$$v_2^2 - v_1 v_2 + c_1^2 = 0 .\tag{5.70}$$

The solutions of this equation are:

$$v_2 = \frac{v_1}{2}\left(1 \pm \sqrt{1 - \frac{4c_1^2}{v_1^2}}\right) .\tag{5.71}$$

Still exploiting the hypothesis of strong shock we have:

$$\sqrt{1 - \frac{4c_1^2}{v_1^2}} \approx 1 - \frac{2c_1^2}{v_1^2} \tag{5.72}$$

and hence:

$$v_2 = \frac{v_1}{2}\left[1 \pm \left(1 - \frac{2c_1^2}{v_1^2}\right)\right] .\tag{5.73}$$

The solution $v_2 \approx v_1$ corresponds to the positive sign, which must be discarded as it does not represent a shock; the other solution is:

$$v_2 = \frac{c_1^2}{v_1} \tag{5.74}$$

and then it results:

$$\frac{v_1}{v_2} = \frac{\rho_2}{\rho_1} = \frac{v_1^2}{c_1^2} = Ma_1^2 .\tag{5.75}$$

The increase in density can be two orders of magnitude or even greater. The large density fluctuations observed in interstellar space can be explained by the theories of radiative shock waves. A compressed gas region with a density increase of several tens of times can form during the collision of two interstellar gas streams moving in an HI region with a relative velocity of the order of $8 - 10$ km/s.

Let us consider, as we have already done for the adiabatic shock wave, a fixed reference system with respect to which the shock front moves with velocity U_s (reference system O). The (5.52) and $U_s \gg u_1$ both hold.
    The pressure behind the shock according to the isothermal condition (5.67) is:

$$p_2 = \rho_2 c_1^2 . \tag{5.76}$$

By exploiting the first jump condition and the (5.74) we have:

$$p_2 = \rho_1 v_1^2 \tag{5.77}$$

while from (5.52) we have:

$$p_2 = \rho_1 (u_1 - U_s)^2 \approx \rho_1 U_s^2 \tag{5.78}$$

and therefore the pressure behind the shock depends on the speed of advancement of the shock with respect to the fixed reference system O.
    From (5.52) we have:

$$\frac{v_1}{v_2} = \frac{U_s}{U_s - u_2} \tag{5.79}$$

while on the other hand with (5.75) we have:

$$\frac{v_1}{v_2} = \frac{c}{v_1} c_1^2 / v_1 = Ma_1^2 \tag{5.80}$$

so by comparing (5.79) and (5.80) we get:

$$\frac{U_s}{U_s - u_2} = Ma_1^2 \tag{5.81}$$

from which we obtain:

$$u_2 = U_s \left(1 - \frac{1}{Ma_1^2}\right) \approx U_s \tag{5.82}$$

therefore the gas behind a strong isothermal shock moves in the same direction and with the same speed as the shock front.
    From (5.75) we have:

$$\frac{p_2}{\rho_1} = \frac{v_1^2}{c_1^2} = \frac{\mu_1 m_H}{k_B T_1} U_s^2 . \tag{5.83}$$

5.7   Structure of the Relaxation Layer

We have already mentioned the structure of a radiative shock: the entire region consisting of the shock front and the relaxation layer has a much greater thickness than non-radiative shock waves.

Let us now consider the structure of the relaxation zone of a strong shock. Immediately behind the shock front, the gas underwent the effect of the shock without the cooling processes having intervened. Hence its thermodynamic state can be identified from the relations of the adiabatic shock.

However, the cooling processes intervene to modify the state of the gas and their action progressively increases as it moves away from the impact front. We can describe the behavior of the relaxation layer by identifying each section parallel to the wave front with a x coordinate. For each value of x the first two R-H conditions give:

$$\rho v = j \tag{5.84}$$

$$p + \rho v^2 = S \tag{5.85}$$

with constant j and S. The second condition can be written in the form $p + j^2/\rho = S$: as the cooling proceeds the density increases and the role of the term $\rho v^2 = j^2/\rho$ (ram pressure) becomes less and less important. The initial value of the ratio $\rho v^2/p$ can be derived from the adiabatic shock formulas. From the definition of Ma_2 and from (5.48) we have:

$$Ma_2^2 = \frac{\rho_2 v_2^2}{\gamma p_2} \approx \frac{\gamma - 1}{2\gamma} \tag{5.86}$$

from which:

$$\frac{\rho_2 v_2^2}{p_2} = \frac{1}{2}(\gamma - 1) \tag{5.87}$$

where for $\gamma = 5/3$ the second member amounts to $1/3$. Since at the beginning of the relaxation process the ram pressure is about one third of the gaseous pressure and subsequently its importance gradually decreases, it is reasonable to neglect it in the (5.36): in this approximation the gaseous pressure is constant and from the equation status results:

$$T\rho = \frac{\mu m_H}{k_B} S = \text{cost. .} \tag{5.88}$$

At each point of the relaxation layer the equation of the heat balance in the form (4.17) holds, which can be written as:

$$\frac{3}{2} \frac{k_B}{\mu m_H} \rho \frac{dT}{dt} - \frac{p}{\rho} \frac{d\rho}{dt} = -L \tag{5.89}$$

where it is assumed that there is no heating process and for L we assume an expression like:

$$L = a\rho^2 T^l \tag{5.90}$$

with $a = const$ and l depends on the most relevant process, therefore ultimately on the temperature of the compressed gas. Taking into account the (5.88) we eliminate ρ in the (5.89) obtaining:

$$\frac{dT}{dt} = -\frac{2}{5}a\left(\frac{\mu m_H}{k_B}\right)^2 ST^{l-1}. \tag{5.91}$$

We can pass from a temporal description to a spatial description using the operator:

$$\frac{d}{dt} = v\frac{d}{dx} \tag{5.92}$$

for which we have:

$$v\frac{dT}{dx} = -\frac{2}{5}a\left(\frac{\mu m_H}{k_B}\right)2ST^{l-1} \tag{5.93}$$

where v can be dropped using (5.84) to get:

$$\frac{dT}{dx} = -\frac{2}{5}a\left(\frac{\mu m_H}{k_B}\right)^3 \frac{S^2}{j}T^{l-2}. \tag{5.94}$$

Integrating this equation with the condition $x = 0$ for $T = T_2$ where T_2 is the temperature behind the shock front, which can be obtained from the adiabatic shock formulas, we obtain the temperature profile $T(x)$ in the relaxation layer. The relation (5.88) allows successively to have the trend of the density, while the connection condition (5.84) allows to derive that of the velocity.

5.8   Observability of Radiative Shocks

The structure of the post-shock relaxation layer depends on the processes that take place there: ionizations by impact, photo-ionizations, recombinations, collisional excitations, dissociation of molecules. When the trend of density and temperature is known, the brightness and spectrum of the emitted radiation can be calculated.

The emission of light (lines and continuum) is dominated by the resonant and forbidden lines of H, He, He^+ and of ions of abundant elements (C, O, N, Ne). The " rates " of these processes depend on the temperature and therefore on the characteristics of the shock. For a shock with velocity $v = 100$ km/s the temperature behind the shock front is 10^5 K. The radiative cooling is produced by the excitation of the He^+ lines and the ions C^+ and O^+.

The temperature dependence of the cooling rate is:

$$L \sim T^l \tag{5.95}$$

where the value of l depends on the speed of the shock and the distance from the impact front.

The typical spectral characteristics of the emission produced by shock waves has been compared in the literature to the emission due to other causes, such as e.g. stellar radiation in the HII regions. These distinctions are important because they allow us to identify the source of excitation in many active galaxies.

Among the shocks, those produced in SNRs are characterized by optical emission of forbidden lines over a wide range of ionization states. The ratio of the [SII] line intensity to the Hα line is stronger in SNRs than in HII regions.

In the optical domain the main characteristics of the emission due to radiative shock waves are:

(a) strong forbidden emission lines, intense with respect to Hβ, of ions such as OI, NI, OII, SII,
(b) a high excitation temperature ($T > 2 \cdot 10^4$ K) measured by the ratio of the intensities of the lines of [OIII] 4363/(5007 + 4959),
(c) presence of different ionization stages, e.g. OI, OII, OIII, NeIII, NeV,
(d) the ratios [OI]/Hβ and [OII]/Hβ are more intense than the HII regions (see Fig. 5.6 from Fesen et al. [1], which shows data relating to SNR and HII regions of our galaxy).

Similar characteristics have also been highlighted in the IR and UV regions of the spectrum.

Fig. 5.6 Trend of temperature-sensitive lines ratios in SNR and HII regions of the Milky Way

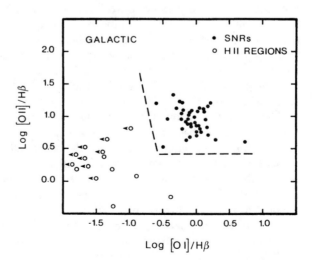

While every single spectral characteristic, both optical and UV, does not give a univocal response, the combination of lines corresponding to different ionization stages or different excitation temperatures allows, by means of models, to define the shock parameters. Once the source of the excitation has been recognized from the spectrum analysis, identifying it as a shock, the ratios of certain lines can be used to determine the shock velocity U_s, the pre-shock density n_0 and chemical abundances. Some lines are particularly suited for evaluating U_s, because their intensity varies significantly with U_s, e.g. [CIII], NV, [NeIII]. The temperature can be constrained by the intensity ratios of the lines of [0III] $4363/(5007 + 4959)$ and of [NeIII] $3342/3869$. The density can be inferred from other line ratios, such as the lines of $[OII]$ 3729/3726 and $[SIII]$ 6716/6731.

5.9   Synthetic Summary

The speed of sound represents a critical speed for the motion of fluids; there are two characteristics that distinguish subsonic and supersonic motions: the propagation of sound signals and compressibility. This last characteristic is decisive in the case of supersonic motions in which variations in physical quantities originate in regions with very small thicknesses, comparable with the average free path of molecules or atoms. In these conditions, as seen previously, energy is dissipated due to the viscosity.

We limit our considerations to a non-interstellar gas by assuming that thermal conduction is negligible and that viscosity should be taken into account only within the layer of abrupt variation of physical quantities.

The treatment of this phenomenon is done by assimilating the zone of abrupt variation of physical parameters with a discontinuity surface Σ and by imposing that through it the conservation of matter, momentum and energy are verified: these conditions will be expressed by equating the flux density on both faces of the surface. Obviously, through the discontinuity there will also be an increase in entropy. Outside the discontinuity the gas behaves adiabatically. the connection conditions are defined with respect to a reference system in which the discontinuity surface is fixed (it is advisable to indicate with the suffix 1 the gas not yet reached by the perturbation).

Two types of discontinuity are possible: (1) the tangential or contact discontinuity, in which there is no passage of matter through Σ, $v_1 = v_2 = 0$ and while the pressure field has no discontinuities, density and the tangential components of velocity can be discontinuous; (2) the shock wave in which there is passage of matter through Σ, in which pressure, density and the perpendicular component of the velocity are discontinuous while the tangential component is continuous. In this last case it is possible to study the motion with respect to a reference in which the velocity of the unperturbed gas has only the normal component different from zero. From the connection conditions, relations are derived by which the physical conditions of the gas behind the shock wave can be determined. In these relations the Mach number of the unperturbed gas can be used. The effects of the shock wave can also be studied with respect to a fixed reference system with respect to which the discontinuity moves.

By means of the concepts developed starting from the junction conditions it can be shown that in a shock wave the density, pressure and temperature increase and the gas behind the shock wave is subsonic.

In the case of the interstellar medium, it is necessary to verify whether the cooling and heating processes must be taken into consideration outside the discontinuity. This is evaluated by comparing the cooling time with the dynamic time of the shock wave defined as the ratio between the space covered by the formation of the shock wave and the speed. When $\tau_{cool} > \tau_{din}$ the effect of these processes can be ignored: the analysis made up to this point is valid and the shock wave is said to be adiabatic (with the convention that this terminology ignores what happens " inside " the surface of discontinuity). On the other hand, when $\tau_{cool} < \tau_{din}$ the cooling processes behind the discontinuity are important and cause the gas to relax towards an equilibrium situation in a time of the order of τ_{cool}. In this case the shock wave is called radiative and we can take two paths.

(A) We are not interested in what happens during the relaxation phase and we directly link the conditions of the gas before the passage of the shock wave with those of the gas that has undergone relaxation due to the cooling processes: in this case we can apply the first two connection conditions, while the third is replaced with the equation $G - L = 0$ which with the density ρ_2 defines the temperature T_2. In particular, if $T_2 = T_1$ the shock wave is called isothermal: this can happen if in the interval $[\rho_1, \rho_2]$ the function $T(\rho)$ is constant.

Even in the case of radiative shock waves it is possible to study the shock wave in a fixed reference system; this is of interest for the study of the interaction of the interstellar medium with stellar winds or supernova explosions.

(B) We want to know the trend of the physical conditions within the relaxation layer behind the shock wave: this can be used to study the spectrum of the radiation emitted by the layer, e.g. wave action uses stellar grains. In this case, the usual conditions of connection of the adiabatic shock waves allow to fix the initial conditions of relaxation. Taking a plane parallel to Σ within the relaxation layer, it is possible to determine for its points, by means of the first two connection conditions of ρv and of $p + \rho v^2$ and therefore, starting from the equation of the thermal balance, an equation in T is deduced which, integrated, gives the trend of the temperature within the relaxation layer. Since $p + \rho v^2 \approx p = \text{cost.}$, From this and from the equation of state we get the trend of ρ and, through the $\rho v = \text{cost.}$, we determine v.

Appendix

Appendix A: Effects of the Passage of a Shock Wave

We intend to study the effects of the shock on the dynamic and thermodynamic state of the gas, bearing in mind that the direction of the transformation will be determined by the condition that the entropy increases due to the dissipative processes that take place within the shock.

We introduce a reference speed defined by:

$$v^* = \frac{S}{j} . \tag{5.96}$$

The Eq. (5.19) can be written in a different form, dividing by ρ and remembering the definition of speed of sound; we have:

$$v^2 + \frac{3}{5}c^2 - \frac{S}{\rho} = 0 \tag{5.97}$$

and with (5.96) we obtain:

$$v^2 - v^*v + \frac{3}{5}c^2 = 0 . \tag{5.98}$$

The entropy of the unit of mass of a monatomic perfect gas is:

$$s = A\ln\frac{p}{\rho^{5/3}} \tag{5.99}$$

with $A = 3k_B/2\mu m_H$. It is:

$$s' = \exp\left(\frac{s}{A}\right) = \frac{p}{\rho^{5/3}} \tag{5.100}$$

where s' is a monotone function of s and therefore indicates the direction in which s grows.

Let us also say:

$$\Sigma = \frac{s' j^{2/3}}{(v^*)^{8/3}} = \frac{p}{\rho^{5/3}} \frac{(\rho v)^{2/3}}{(v^*)^{8/3}} = \frac{v^* - v}{v^*} \frac{v^{5/3}}{(v^*)^{5/3}} \tag{5.101}$$

and we introduce a dimensionless variable η defined as the ratio between v and the reference speed v^*:

$$\eta = \frac{v}{v^*} . \tag{5.102}$$

Note that η is always less than 1, in fact from the definition of v^* we have:

$$\eta = \frac{vj}{S} = \frac{\rho v^2}{\rho v^2 + p} < 1 . \tag{5.103}$$

It turns out then:

$$\Sigma = \eta^{\frac{5}{3}}(1 - \eta) \tag{5.104}$$

so σ is a dimensionless quantity which is an increasing monotone function of s.

Using the (5.98) the expression of E given by the (5.28) becomes:

$$E = \frac{1}{2}v^2 + \frac{5}{2}v^*v - \frac{5}{2}v^2 = v\left(\frac{5}{2}v^* - 2v\right) \tag{5.105}$$

which allows us to introduce dimensionless quantity

$$e_T = \frac{E}{(v^*)^2} = \frac{v\left(\frac{5}{2}v^* - 2v\right)}{v^*} = \eta\left(\frac{5}{2} - 2\eta\right) . \tag{5.106}$$

Two points characterizing the initial and final state of the gas must lie on the curve e_T and since energy is conserved, these points represent the intersection of the curve e_T with a straight line parallel to the abscissa axis. This allows once you have chosen the η value for a point to determine the value of η for the corresponding point. Once can graphically determine the pairs η_A and η_B. Ma which of the two values is the initial one and which is the final one? This can be determined by identifying in a diagram $\sigma - \eta$ the values of σ that correspond to η_A and η_B. Since σ must increase during the evolution of the system B is the starting point, while A describes the final situation. To preserve the meaning of the symbols, we indicate η_B with η_1 and η_A with η_2.

Having identified the direction in which the gas transformation proceeds, we can study how the characteristic quantities vary.

(1) The Mach number is given by:

$$Ma^2 = \frac{v^2}{c^2} \tag{5.107}$$

and from (5.98) we get:

$$Ma^2 = \frac{3}{5}\frac{v}{v^* - v} = \frac{3}{5}\frac{\eta}{1 - \eta} . \tag{5.108}$$

From the behavior of Ma^2 as a function of η we see that the motion is supersonic when $\eta > 0.63$ and otherwise subsonic. The value of $\eta = 0.63$ is the value at which σ has the maximum. Thus *motion is initially supersonic but the velocity of the gas behind the shock becomes subsonic.*

(2) Based on the definitions of j and v^* we get:

$$\frac{S}{\rho} = \frac{S j}{j \rho} = v^* v \; . \tag{5.109}$$

Since S/ρ has the size of the square of a velocity, let's say:

$$\lambda = \frac{S}{\rho (v^*)^2} = \eta \tag{5.110}$$

where λ is an increasing function of η and for $\eta_1 > \eta_2$ it results:

$$\lambda(\eta_1) > \lambda(\eta_2) \tag{5.111}$$

or

$$\frac{S}{\rho(\eta_1)(v^*)^2} > \frac{S}{\rho(\eta_2)(v^*)^2} \tag{5.112}$$

from which:

$$\rho_1 = \rho(\eta_1) < \rho(\eta_2) = \rho_2 \tag{5.113}$$

i.e. *the density behind the shock is greater than the density of the unperturbed gas.*

(3) From (5.98) we get

$$\frac{p}{\rho} = v^* v - v^2 \tag{5.114}$$

and dividing by $(v^*)^2$:

$$\frac{p}{\rho(v^*)^2} = \frac{v^* v - v^2}{(v^*)^2} \tag{5.115}$$

waves:

$$p = \rho(v^*)^2 \eta (1 - \eta) \; . \tag{5.116}$$

Dividing by S and remembering (5.110) we get

$$Sp = (1 - \eta) \tag{5.117}$$

so from $\eta_1 > \eta_2$ descends $p_1 < p_2$, that is *the pressure behind the shock front is greater than the pressure of the unperturbed gas.*

(4) The internal energy of the unit of mass holds

$$\epsilon = \frac{3}{2} \frac{k_B T}{\mu m_H} = \frac{3}{2} \frac{p}{\rho} \tag{5.118}$$

so based on the definition of c and the (5.98) we have:

$$\epsilon = \frac{3}{2}v(v^* - v) \, . \tag{5.119}$$

Dividing by $(v^*)^2$ we have:

$$e_I = \frac{\epsilon}{(v^*)^2} = \frac{3}{2}\eta(1 - \eta) \tag{5.120}$$

which has a maximum of $\eta = 0.5$. From the graph in Figure ?? it results that for each pair of values η_1 and η_2 which represents the conditions in front of and behind the shock, i.e. that they correspond to the intersection of the curve $and_T(\eta)$ with a horizontal line and are ordered in such a way that $\sigma(\eta_1) < \sigma(\eta_2)$, we have:

$$e_I(\eta_1) < e_I(\eta_2) \tag{5.121}$$

and since e_I is proportional to the temperature, it can be concluded that *the temperature increases due to the passage of the shock wave.*

Reference

1. Fesen, R.A., Blair, W.P., Kirshner, R.P.: Optical emission-line properties of evolved galactic supernova remnants. In: ApJ, vol. 292, p. 29 (1985)

Chapter 6
Turbulence

Werner Heisenberg once said: "When I meet God, I am going to ask him two questions: Why relativity? And why turbulence? I really believe he will have an answer for the first."

6.1 Experimental Aspects of Turbulence

Let us consider the device of Fig. 3.1; it consists of a A jar containing water and a B jar filled with colored liquid. By means of the valve V the flow rate can be continuously adjusted. When this is small, a thread of colored fluid detaches from the vase B and moves regularly in the water without mixing with it. A regime of motion of this type is called laminar: the trajectories of the macroscopic fluid particles are parallel and the speed varies continuously both in space and in time.

As the flow rate increases, the motion of the colored liquid brush changes, initially presenting an oscillating trend, and subsequently a disordered motion develops in which the colored liquid mixes rapidly with the water. This regime is called turbulent and is characterized by irregular trajectories, while the velocity presents fluctuations both in space and in time. The fluid velocity can be thought of as the sum of an average velocity v_m on which irregular fluctuations are superimposed $v(t)$, the average velocity, when the average is made over a sufficient time interval large, it is regular: statistically therefore the turbulent motion presents aspects of regularity.

Turbulence also occurs in other cases, for example in the following.

– Motion of a liquid in a duct with any section: when the velocity is small the motion is laminar but when it is large enough the irregular character appears. The speed at which the transition occurs depends not only on the flow rate but also on the roughness conditions of the duct walls: if these are smooth the transition speed is greater (the discussion of the laminar case is made in Chap. 3, Appendix A).

© The Author(s), under exclusive license to Springer Nature Switzerland AG 2021
G. Carraro, *Astrophysics of the Interstellar Medium*,
UNITEXT for Physics, https://doi.org/10.1007/978-3-030-75293-4_6

– Couette-Taylor motion: a fluid is contained in the interspace between two coaxial
  cylinders whose interior rotates with Ω angular velocity. Also in this case if the
  angular velocity is progressively increased we see that there is a transition from
  laminar motion to irregular turbulent motion (the treatment of the laminar case is
  made in Appendix 3.10).

The complex of experiments on these and other types of motion leads to the
identification of a parameter, the Reynolds number, already defined in Chap. 3, by
means of which it is possible to characterize the transition from one type of motion
to another. More precisely, when the Reynolds number is less than a critical value
Re_{cr} the motion remains laminar while when the critical value is exceeded turbulent
motion is established. This critical value is not a universal constant and depends on
particular concrete situations, even though it falls within a narrow range between
100 and 1000.

The fact that the role of the Reynolds number is fundamental leads us to analyze
this phenomenon in the light of the equations of fluid dynamics. In fact, we have
already seen that the Reynolds number is a measure of the relative importance of
the forces of inertia, described by the nonlinear term $(\mathbf{v} \cdot \mathrm{grad})\mathbf{v}$, and of the viscous
forces . When Re is small ($Re < Re_{cr}$) viscous forces prevail which tend to dampen
inhomogeneities and maintain a regular character to motion; the non-linear term is
negligible and in stationary conditions the Navier–Stokes equation becomes:

$$\mathrm{grad}\, p = \eta \nabla^2 \mathbf{v} \tag{6.1}$$

which is solvable in a very large number of cases.

As Re increases, the term of the inertia forces becomes increasingly important,
making the equation non-linear. It is this term that by spreading the vorticity in the
fluid, amplifies and destroys any initial motion in an irregular and unpredictable
configuration (but not in a statistical sense). This term involves a very large number
of scales (linear dimensions) that interact with each other. Both inertia and viscous
forces transfer energy and momentum but in a completely different way: viscous
forces exchange energy and momentum on a scale equal to the mean free path of
the particles while the forces of inertia have no preferential scales . The absence of
a preferential scale and consequently the large number of scales involved makes the
motion complicated and irregular and makes the resolution of the equations difficult,
or even impossible.

6.2 Stability of Stationary Motion and Turbulence Onset

For every problem with stationary external conditions the equations of fluid dynamics give a solution. However, not all the solutions of these equations correspond to
a motion observed in nature: the observed motions, in addition to obeying the fundamental equations, must possess the requisite of stability: this means that small

perturbations, which inevitably occur in the system, decrease in the time and eventually disappear. If, on the other hand, some of these perturbations increase over time, the motion is unstable and cannot exist as it is rapidly destroyed by the perturbations.

The stability analysis is done with the method of small perturbations. Let us consider a stationary motion of an incompressible fluid characterized by a velocity field $v_0(x, y, z)$, by the pressure $p_0(x, y, z)$ and by the density $\rho_0(x, y, z)$. These quantities satisfy the Navier–Stokes equation and the continuity equation in stationary form:

$$(\mathbf{v}_0 \cdot \text{grad})\mathbf{v}_0 = -\frac{1}{\rho_0} \text{grad } p_0 + \nu \nabla^2 \mathbf{v}_0 \tag{6.2}$$

$$\text{div } \mathbf{v}_0 = 0 . \tag{6.3}$$

Let $\mathbf{v}_1(x, y, z, t)$ denote a small non-stationary perturbation of the velocity and let $\mathbf{v}$ be the resulting velocity:

$$\mathbf{v} = \mathbf{v}_0 + \mathbf{v}_1 \tag{6.4}$$

then $\mathbf{v}$ will obey the equations:

$$\frac{\partial \mathbf{v}}{\partial t} + (\mathbf{v} \cdot \text{grad})\mathbf{v} = -\frac{1}{\rho} \text{grad } p + \nu \nabla^2 \mathbf{v} \tag{6.5}$$

$$\text{div } \mathbf{v} = 0 \tag{6.6}$$

where $p = p_0 + p_1$, p_1 being the pressure perturbation, and $\rho = \rho_0$ due to the incompressibility hypothesis. By substituting, taking into account that the unperturbed motion satisfies the (6.2) and (6.3) terms and neglecting the higher order terms than the first with respect to the perturbations, we arrive at:

$$\frac{\partial \mathbf{v}_1}{\partial t} + (\mathbf{v}_0 \cdot \text{grad})\mathbf{v}_1 + (\mathbf{v}_1 \cdot \text{grad})\mathbf{v}_0 = -\frac{1}{\rho_0} \text{grad } p_1 + \nu \nabla^2 \mathbf{v}_1 \tag{6.7}$$

$$\text{div } \mathbf{v}_1 = 0 . \tag{6.8}$$

The (6.7) and (6.8) constitute a linear system whose coefficients are functions of the coordinates x, y, z only and not of time. The most general solution can be represented as a sum of particular solutions of the type:

$$\mathbf{v}_1 \sim \exp(-i\delta t) . \tag{6.9}$$

The δ are determined by solving the system obtained from the Eqs. (6.7) and (6.8) after entering the expression of $\mathbf{v}$. In general the δ will be complex:

$$\delta = \omega + i\gamma \tag{6.10}$$

with ω and γ real

$$\mathbf{v}_1 \sim \exp(-i\omega t)\exp(\gamma t) \,. \tag{6.11}$$

If all frequencies δ have a negative imaginary part, the motion is stable; if at least one has a positive imaginary part, the perturbation increases over time and the motion is unstable. Let us now analyze the non-stationary motion that is established as a consequence of the instability in a stationary motion when $Re > Re_{cr}$. As long as $Re < Re_{cr}$ the imaginary part of all δ is negative. When $Re = Re_{cr}$ the imaginary part of a frequency $\delta_1 = \omega_1 + i\gamma_1$ vanishes and becomes positive when $Re > Re_{cr}$, causing instability. The $\mathbf{v}_1$ function looks like this:

$$\mathbf{v}_1 = \mathbf{A}(x, y, z) f(t) \tag{6.12}$$

with

$$f(t) = \cos t \exp(\gamma t)\exp(-i\omega t) \,. \tag{6.13}$$

This expression of $\mathbf{v}_1$ is valid only during a short time interval starting from the beginning of the instability; subsequently, due to the $\exp(\gamma t)$ factor, the perturbation grows and the approximation of neglecting the nonlinear terms will fail, so the analysis loses validity.

Instability determines the appearance of a periodic non-stationary motion: the distinction between unperturbed motion and perturbation is no longer significant and global motion is more a periodic motion with period $2\pi/\omega_1$; the time dependence can also be changed into the phase dependence $\phi_1 = \omega_1 t + \beta_1$ (β_1 being the initial phase) and the velocity is a periodic function of ϕ_1 with period 2π. It is not a harmonic function and in general it will be expressed by a Fourier series expansion

$$\mathbf{v} = \sum_p \mathbf{A}_p(x, y, z)\exp(-i\phi_1 p) \tag{6.14}$$

where the summation is extended to all integers p positive and negative and therefore includes ϕ_1 and its harmonics. The Eqs. (6.7) and (6.8) fix the coefficient γ_1 but not the initial phase β_1: the latter remains indeterminate because it depends on the particular conditions of the fluid at the instant in which instability manifests itself and it will be said that the periodic non-stationary motion has a degree of freedom.

What happens if the Reynolds number grows further? At a certain point, the non-stationary periodic motion becomes unstable in turn. The analysis of its stability, at least in principle, can be done in a similar way to that already adopted to analyze the instability of stationary motion; only the unperturbed motion $\mathbf{v}_0$ is now a function of time:

$$\mathbf{v}_0 = \mathbf{v}_0(x, y, z, t) \,. \tag{6.15}$$

The solution is

$$\mathbf{v} = \mathbf{v}_0 + \mathbf{v}_2 \tag{6.16}$$

where v_2 represents a small perturbation. By linearizing the system of equations and eliminating all the unknowns except one we obtain a linear equation, whose coefficients are now functions both of the coordinates x, y, z and of the time t, more precisely they are periodic functions of t with period $2\pi/\omega_1$. The solution of this equation has the form:

$$v_2 \sim \mathbf{B}(x, y, z, t) \exp(\gamma_2 t) \exp(-i\omega_2 t) \tag{6.17}$$

where $\mathbf{B}$ is a periodic function of t with period $2\pi/\omega_1$. Instability occurs when the imaginary part of another frequency $\delta_2 = \omega_2 + i\gamma_2$ becomes positive; the real part determines the new frequency that appears.

The global motion that emerges from the new instability is a motion characterized by two different periods: $2\pi/\omega_1$ and $2\pi/\omega_2$, correspondingly there are two arbitrary quantities, the initial phases β_1 and β_2 and we will say that there are two degrees of freedom.

According to the analysis formulated by Landau in 1944 as the Reynolds number increases, further instabilities appear, each of which is characterized by a new frequency and to which a degree of freedom is associated. The linear scales corresponding to the new modes are progressively decreasing: the order of magnitude of the distances with respect to which the speed changes appreciably is always smaller. According to this analysis, when Re has become very large, and a very large number of instabilities have arisen, each of which is associated with a new frequency and an additional degree of freedom, we are in the presence of turbulent motion. The general form of the velocity field can be expressed as a function of n different frequencies ω_j:

$$\mathbf{v}(x, y, z, t) = \sum_{p_1, p_2, \dots, p_n} \mathbf{A}_{p_1, p_2, \dots, p_n}(x, y, z) \exp\left(-i \sum_{p_j} \phi_j p_j\right). \tag{6.18}$$

These conclusions, however, have recently been invalidated in the sense that they are not of a general nature: it may be that in some situations the turbulent motion is triggered by a very large sequence of instabilities in the way just described, even if it seems that the most probable way in which it transits from laminar motion to turbulent motion is different.

In this regard, the analysis of the onset of turbulence in the Couette-Taylor motion is interesting (Fig. 6.1 from Fenstermacher et al. [1]). In the experiment, the rotation speed of the inner cylinder is progressively increased, while the speed of the liquid is measured at some points in successive instants, generating a time series $v(t_k)$ ($k = 1, 2, \dots, N$). This series is analyzed in Fourier series to study its spectrum. In the beginning, when the rotation speed is low, the liquid is animated by a rotary motion and the motion, due to the effect of viscosity, has a laminar character. If the rotation speed increases to the previous motion, a further pendicular component is added to the first characterized in which the fluid performs a circular motion in the vertical plane: donuts appear one on top of the other (Taylor vortex regime).

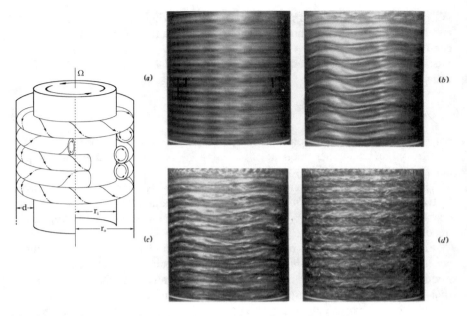

Fig. 6.1   Visual evolution of turbulence in a experiment. See text for details

This is still an ordered motion and the velocity of the fluid is constant over time (as long as the rotation speed is not varied). If you continue to increase the speed of rotation, the motion changes again: the separating lines of the donuts deform and take on an oscillatory character. The speed of the fluid is no longer constant but is a periodic function of time (not necessarily harmonic). If we do the Fourier analysis we find that the spectrum is made up of a frequency ω_1 and its harmonics (Fig. 6.2a). Increasing the rotation speed again the appearance of the motion becomes confused, but Fourier's analysis reveals that another frequency ω_2 has appeared in the spectrum, not commensurable with the first, and all frequencies $\omega = l\omega_1 + m\omega_2$, where l and m are integers (Fig. 6.2b). The appearance of another frequency corresponds to the presence of another mode; it was not easy to identify it but in the end it was identified as an oscillation in the width of the layers that separate the donuts. Finally, if the rotation speed is increased further, there is further instability and a continuous band appears in the frequency spectrum, with increasing extension and amplitude. At the same time and progressively the lines that previously characterized the spectrum disappear (Fig. 6.2c). The interpretation of this phenomenon is realized within the modern theories of chaos and is associated with the appearance in phase space of a strange attractor.

Other triggering mechanisms of turbulence have been studied and, although a definitive conclusion has not yet been reached, it seems that there is no single trigger mechanism even if the characteristics of fully developed turbulence are independent of the way in which it is triggered.

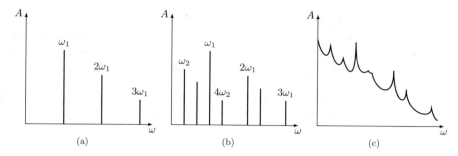

Fig. 6.2 Development of turbulence. More and more modes appears as time passes

6.3 Notes on the Statistical Description of Turbulence

Turbulent motion is characterized by poor predictability: the measurement of the instantaneous velocity in a point does not allow to deduce the velocity in the same point a few moments later nor that of the neighboring points at the same instant. Instead, the average values are significant; for example, in the turbulent motion of a liquid in a horizontal tube, points placed at equal distance from the axis of the tube have equal average velocities.

A statistical description of turbulence therefore appears preferable. The ideal aim in formulating a theory of turbulent motion, in the context of statistical mechanics, would be to develop a theory analogous to the kinetic theory of gases. The latter is based on an equation of motion for particles that interact only through elastic collisions; turbulence theory should instead be based on the Navier–Stokes equation and, due to its complexity, the problem has not yet been solved satisfactorily.

Much of our knowledge of turbulent motions comes from experiments, and many of the observable quantities are quantities that can be treated statistically. Each of the quantities relevant to dynamics (components of speed, pressure, density and temperature), if considered in a point of space, can be decomposed into an average value and a fluctuating part as a function of time. Exemplifying the discussion with the velocity component $v = v_i$ according to an axis we have:

$$v = \langle U \rangle + u \tag{6.19}$$

where $\langle U \rangle$ is the average value:

$$\langle U \rangle = \frac{1}{T} \int_0^T v \, dt \tag{6.20}$$

averaged over a sufficiently long time interval; furthermore it results:

$$\int_0^T u \, dt = 0 \tag{6.21}$$

that is, the average value of the fluctuation is zero. Analogous quantities can be defined by calculating the averages on a neighborhood of a point at a defined instant.

Often the average quantities are expressed by the probability distribution functions Pr. The quantity $\Pr(u)\,du$ gives the probability that the velocity is in the range $(u, u + du)$.

Information on the structure of the velocity fluctuations is given by other average quantities: one of these is the mean value of the square of the fluctuations $\langle u_i^2 \rangle$, where u_i is the ith component of the fluctuation of speed. The square root of this quantity is the magnitude

$$\sqrt{\langle u_i^2 \rangle} \tag{6.22}$$

is called the intensity of the ith component of the velocity and the quantity:

$$\sqrt{\langle q^2 \rangle} = \sqrt{\langle u_1^2 \rangle + \langle u_2^2 \rangle + \langle u_3^2 \rangle} \tag{6.23}$$

it is called the intensity of the turbulence speed or more simply the turbulence speed. It is related to the kinetic energy of the volume unit associated with speed fluctuations (turbulent kinetic energy):

$$\Sigma = \frac{1}{2}\rho\langle q^2 \rangle . \tag{6.24}$$

Further information on the turbulent velocity field at different points or times is obtained by correlation measurements. The correlation between the velocity fluctuations $u_{(1)}$ and $u_{(2)}$ at two points is defined as the mean value of the product of $u_{(1)}$ and $u_{(2)}$:

$$\langle u_{(1)}u_{(2)} \rangle \tag{6.25}$$

and the correlation coefficient is:

$$R = \frac{\langle u_{(1)}u_{(2)} \rangle}{\sqrt{\langle u_{(1)}^2 u_{(2)}^2 \rangle}} \tag{6.26}$$

If the fluctuations $u_{(1)}$ and $u_{(2)}$ are independent, the correlation is zero. However, turbulent motion is governed by the equations of fluid dynamics and therefore the fluctuations are not necessarily independent.

The correlation between two points generally depends between the distance r of the points and the orientation of their joining. When the two points coincide ($r = 0$) we have $R = 1$, while asymptotically R tends to zero. The trend of R as a function of r will have a qualitative trend of two types: R is always decreasing and remains positive or it can have a minimum with a negative value; the value of r corresponding to the minimum in this second case or the distance at which R has the value $1/e$ in the first case define the distance within which two points influence each other in motion.

The concept of correlation can also be formulated in higher orders, in which three, four, etc. are considered. points. For example, the correlation between the three-point velocities $u_{(1)}$, $u_{(2)}$, $u_{(3)}$ is $u_{(1)} u_{(2)} u_{(3)}$.

Complete determination of the turbulent velocity field would require knowledge of the correlations of each order. In practice it is possible to use the correlations of the second and third order.

The equations of motion can be rewritten by highlighting the mean value and the fluctuation for each quantity. For the continuity equation we have:

$$\frac{\partial}{\partial x_i}(U_i + u_i) = 0 \ . \tag{6.27}$$

Taking the average and keeping in mind that summation and spatial derivatives can be exchanged between them, we have:

$$\int \frac{\partial}{\partial x_i}(U_i + u_i)\,\mathrm{d}t = \frac{\partial}{\partial x_i}\int (U_i + u_i)\,\mathrm{d}t \tag{6.28}$$

$$\int \frac{\partial}{\partial x_i}(U_i + u_i)\,\mathrm{d}t = \frac{\partial U_i}{\partial x_i} = 0 \ . \tag{6.29}$$

Furthermore, it is possible to deduce from the starting equation

$$\frac{\partial u_i}{\partial x_i} = 0 \ . \tag{6.30}$$

We write the Navier–Stokes equation in steady state:

$$(U_j + u_j)\frac{\partial}{\partial x_i}(U_i + u_i) = -\frac{1}{\rho}\frac{\partial}{\partial x_i}(P_i + p_i) + v\frac{\partial^2}{\partial x_j \partial x_i}(U_i + u_i) \ . \tag{6.31}$$

Regarding the calculation of the average of the first member, we have:

$$\int U_j \frac{\partial U_i}{\partial x_j}\,\mathrm{d}t = U_j \frac{\partial U_i}{\partial x_j} \tag{6.32}$$

$$\int U_j \frac{\partial u_i}{\partial x_j}\,\mathrm{d}t = U_j \frac{\partial}{\partial x_j}\int u_i\,\mathrm{d}t = 0 \tag{6.33}$$

$$\int u_j \frac{\partial U_i}{\partial x_j}\,\mathrm{d}t = \frac{\partial U_i}{\partial x_j}\int u_j\,\mathrm{d}t = 0 \tag{6.34}$$

$$\int u_j \frac{\partial u_i}{\partial x_j}\,\mathrm{d}t = \langle u_j \frac{\partial u_i}{\partial x_j}\rangle \ . \tag{6.35}$$

This last result can be modified by remembering the equation of continuity (6.30):

$$\int u_j \frac{\partial u_i}{\partial x_j}\, \mathrm{d}t = \int \frac{\partial u_i u_j}{\partial x_j}\, \mathrm{d}t = \frac{\partial}{\partial x_j}(\langle u_i u_j\rangle) \tag{6.36}$$

for the second member of (6.31) it results:

$$\int \frac{1}{\rho}\frac{\partial}{\partial x_i}(P_i + p_i)\, \mathrm{d}t = \frac{1}{\rho}\frac{\partial P_i}{\partial x_i} \tag{6.37}$$

$$\int \nu \frac{\partial^2}{\partial x_i \partial x_j}(U_i + u_i)\, \mathrm{d}t = \nu \frac{\partial^2}{\partial x_i \partial x_j}U_i \tag{6.38}$$

and putting all these results together we finally have:

$$U_j \frac{\partial U_i}{\partial x_j} = -\frac{1}{\rho}\frac{\partial P_i}{\partial x_i} + \nu \frac{\partial^2 U_i}{\partial x_i \partial x_j} - \frac{\partial}{\partial x_j}(\langle u_i u_j\rangle)\,. \tag{6.39}$$

A relevant aspect of this equation is that the mean value of the velocity is influenced by the fluctuations through the last term deriving from the non-linear term of the equations: one cannot calculate the mean motion without having knowledge of the fluctuations.

An equation for fluctuating components can be derived by subtracting the mean motion equation from the Navier–Stokes equation: the third order correlation $\langle u_{(i)}u_{(j)}u_{(k)}\rangle$ appears in the resulting equation. One can proceed indefinitely by introducing successive equations in which correlation terms of increasing order appear. If the procedure is truncated to a given order, an unclosed system is obtained, that is, it contains more unknowns than equations; the impossibility of obtaining a closed system is a consequence of the non-linear character of the fluid dynamics equations.

Therefore the turbulence problem leads to a system with an infinite number of equations that clearly cannot be solved. We must necessarily limit ourselves to an approximation by considering a finite number of equations and addressing the *problem of closure* of the system by introducing a further suitably chosen condition.

In principle, the infinite system of equations we have considered allows us to determine the correlation functions of all orders and is equivalent to determining the probability distribution function. When we limit ourselves to a finite number of equations, closing the system appropriately, we can determine the correlation functions up to a given order and therefore we can only obtain some characteristics of the probability distribution function of the turbulent field.

From the experimental point of view, the study of turbulence is done using two techniques: (a) second order correlations, (b) Fourier analysis of the velocity field. There is an interesting relationship between the two points of view: the Fourier components are the Fourier transforms of the second order correlations of the velocities.

6.4   General Characteristics of Turbulence

It has been seen that turbulent motion is characterized by the simultaneous presence of many components of different scales, among which there is a continuous exchange of energy: this result is the consequence of the presence in the Navier–Stokes equation of the non-linear term that describes the forces of inertia.

As early as 1922 Richardson presented a picture of turbulence described as a hierarchy of turbulent elements (ETs), called eddies, of different scales. It can also be said that ETs are coherent structures in which there are close correlations between the speeds of their points and this coherence persists for some time. The ETs of a given scale originate from the loss of stability of the ETs of the immediately larger scale from which they acquire energy; these persist for a certain time and then in turn decay giving rise to ETs of lower dimensions among which their energy is distributed. There is therefore a cascading process of creation and destruction of turbulent elements which involves a transfer of energy from the largest to the smallest scales. This process stops at the scales at which viscosity becomes important and there is a significant dissipation of mechanical energy into heat.

The concept of a turbulent element can be further specified: an ET is a partial volume of fluid that has a common motion for a given time interval. Such motion will generally have a vortex component, but vorticity is not the essential property. An ET is characterized by its size λ and its velocity v_λ; it is destroyed after traveling a mixing length which is of the same order of magnitude as its linear dimensions λ, the life time is given by $\tau = \lambda/v_\lambda$.

There are ETs of all sizes from those of the entire system up to a limit defined by the molecular viscosity. The smaller ETs are contained within the larger ones: this is what is meant by the hierarchy of ETs.

The largest ETs are those that are comparable in size to the entire system and receive their energy from the outside. These external scales depend on the particular nature of the system; if for example we consider a liquid that passes through a duct in which a grate is inserted, the characteristic dimensions are the diameter of the duct, the dimensions of each mesh of the grate and the thickness of the bars that make up the grate. These ETs get their energy from the environment outside the system.

Up to now, we have spoken of scales of different dimensions excited by the forces of inertia (i.e. from the non-linear term that appears in the Navier–Stokes equation), of modes that manifest themselves when instability appears, modes that are characterized by periods and by different scales, of turbulent elements of different linear scales. It can be understood that all these concepts express the same ideas in different ways but appear to be formulated in a qualitative and vague way. If we want to give them a precise mathematical meaning we will have to resort to the Fourier analysis of the velocity field, with which we can solve the motion in a certain number of components whose physical meaning is well defined. Since the wavelength, or the wavenumber which is inversely proportional to the wavelength, is the parameter that specifies the different components, Fourier analysis corresponds to the resolution of the motion in components of different linear dimensions.

Fourier analysis also gives meaning to the idea of the different degrees of freedom possessed by the fluid; small and large-scale components are not related to limited portions of the fluid in the way in which the different degrees of freedom of a gas are related to different molecules but nevertheless we can think of a turbulent motion as the superposition of a large number of component motions of different sizes that make additive contributions to the total energy and interact with each other in the way required by the non-linear term of the equation of motion. The precise form of motion associated with each component is not of great importance from a physical point of view but can be thought of as a transverse plane wave that does not propagate over time. Suppose that the velocity field is expressed by:

$$\mathbf{v}(\mathbf{r}, t) = \sum_k \mathbf{u}(\mathbf{k}) \exp(i\mathbf{k} \cdot \mathbf{r}) \tag{6.40}$$

where k is the wave number which is inversely proportional to the linear scale λ of the motion component. The previous relation itself does not provide a description of turbulence if the interaction between the different components is not specified: these relations are obtained from the fluid dynamics equations for an incompressible fluid.

By introducing the expression (6.40) in the continuity equation and in the Navier–Stokes equation, the following conclusions can be drawn:

– viscosity does not transfer energy from larger components to smaller ones, in fact there is no coupling in the term that contains viscosity;
– pressure does not appear, as it is not responsible for the transfer of energy but its role is to create a state of isotropy;
– the inertial terms are responsible for the coupling between the different components.

6.5   Kolmogorov's Theory

By exploiting the predominantly qualitative framework just described and formulating a similarity hypothesis for smaller ETs, Kolmogorov derived a theory describing the fundamental characteristics of the turbulence of incompressible fluids. Let's try to clarify on which assumptions Kolmogorov's theory is based.

It has been said previously that the modes associated with the largest scales receive energy from outside the system and therefore their behavior will depend on the modalities of this transfer. Furthermore, the rate of decrease in the kinetic energy of turbulence can be derived experimentally and appears to be:

$$\frac{\mathrm{d}\,u^2}{\mathrm{d}\,t} = -A\frac{u^3}{D} \tag{6.41}$$

with $A \approx 1$, where D is the system size and u is the speed on the D scale. From this relation we get the decay time:

$$\tau = \frac{1}{A}\frac{D}{u} \tag{6.42}$$

which is of the same order as the life time of the elements of dimension D. This circumstance means that the modes associated with the largest scales are not in statistical equilibrium. As will be seen further on, the life times of the ETs decrease as the scale decreases and, for sufficiently large wave numbers, we can assume that the corresponding degrees of freedom are approximately in statistical equilibrium: the interval of their dimensions is called the interval of equilibrium. These modes are excited by the transfer of energy by the forces of inertia. Thus while the modes corresponding to the size of the system are affected by external conditions, as we consider larger waven-umbers (i.e. smaller linear scales) the influence of external conditions is less. This circumstance is such that, although the motion associated with large-scale components may have preferential directions, the motion associated with sufficiently large wave numbers tends to be isotropic. We denote by L the maximum dimension of the modes in statistical equilibrium.

Let us consider the developed turbulent motion relative to a very large Reynolds number. The forces of inertia excite a very large number of wave numbers and only for very high wave numbers the forces of inertia are balanced by the viscous forces which dissipate the kinetic energy. Let k_0 be the wavelength of the mode by which dissipation becomes important and let $l = 2\pi/k_0$ be the corresponding wavelength. For each mode a Reynolds number can be defined:

$$Re_\lambda = \frac{\lambda v_\lambda}{\nu} . \tag{6.43}$$

The dimension l of the scale at which dissipation occurs can be defined on the basis of the condition $Re_l = 1$.

The subset (l, L) of the equilibrium interval is called the inertial subinterval and is characterized by the circumstance that the behaviour of the corresponding modes is defined only by the force of inertia without viscosity having any role in it. Obviously the motion associated with the inertial sub-interval is statistically independent of both the major scale modes and the modes responsible for the dissipation. We limit our considerations to the inertial sub-range in which viscosity is negligible. Let us recall what was said in Chap. 3 about similar motions, i.e. that geometrically similar motions that obey the same equations differ only for a variation of scale: this property, called scale invariance, is therefore applicable to the different components of the subinterval inertial and allows to derive relationships between the characteristic quantities of motion through dimensional analysis.

Kolmogorov's theory of the turbulence of incompressible fluids is based on these considerations. More explicitly, the fundamental principle assumes that the various quantities relating to a generic scale λ of the equilibrium interval depend only on the scale itself, on the energy transfer rate per unit of mass ε and on the viscosity ν while in the inertial sub-range they are independent of viscosity and therefore depend only on the scale itself and on ε.

We apply this principle to derive the velocity relative to the λ scale: v_λ will depend only on ε and λ. We determine a suitable product of powers of ε and of λ having the dimensions of a velocity.

Let's say:

$$v_\lambda \propto \lambda^a \varepsilon^b .$$ (6.44)

By imposing that the dimensions of the two members are equal, we obtain the system in the unknowns a and b:

$$a + 2b = 1 \qquad 3b = 1$$ (6.45)

which has as solution $a = b = 1/3$, hence:

$$v_\lambda \propto (\lambda \varepsilon)^{1/3} .$$ (6.46)

In steady state conditions the energy transfer rate must be independent of the scale, and also must be equal to the dissipation rate, therefore:

$$v_\lambda \propto \lambda^{1/3} .$$ (6.47)

This is the Kolmogorov–Obukhov relation according to which the velocity on the λ scale is proportional to the cubic root of the linear dimension. It should be remembered that this relation and the others obtained from the scale invariance with dimensional analysis are only relations of proportionality and the constant of proportionality cannot be derived in the context of this theory.

The previous relation can be written by introducing the values of the maximum scale to express ε, from which we obtain:

$$v_\lambda = v_L \left(\frac{\lambda}{L} \right)^{\frac{1}{3}} .$$ (6.48)

From the Kolmogorov–Obukhov relation it is possible to derive the spectrum of turbulence, that is the law with which the energy of motion is distributed at different scales. Let $\rho F(k) \, dk$ be the energy of the volume unit residing in the components having a wavenumber in the interval $(k, k + dk)$. If v is the average speed of turbulent motion it will be:

$$\frac{1}{2} \rho v^2 = \int_k^\infty \rho F(k) \, dk$$ (6.49)

where k is relative to the maximum scale. If we refer to the generic λ scale, we have:

$$\frac{1}{2} \rho v_\lambda^2 = \int_k^\infty \rho F(k) \, dk$$ (6.50)

with $k = 2\pi/\lambda$. Let's say:

$$F(k) \propto k^{-\alpha} \qquad (\alpha > 0) \tag{6.51}$$

yes you have

$$v_\lambda^2 \propto 2 \int_k^\infty k^{-\alpha} \, dk \propto k^{-\alpha+1} \tag{6.52}$$

that is

$$v_\lambda \propto \lambda^{\frac{\alpha-1}{2}} \tag{6.53}$$

from this and from (6.47) it follows that:

$$\alpha = \frac{5}{3} \, . \tag{6.54}$$

It has already been said that the lower extreme l of the inertial subinterval can be determined by the condition $Re_l = l v_l / v = 1$. Inserting the (6.48) into this one obtains:

$$l = \left(\frac{L^{\frac{1}{3}} v}{V_L} \right)^{\frac{3}{4}} = L Re^{-\frac{3}{4}} \, . \tag{6.55}$$

As the Reynolds number increases, the scale at which dissipation takes place decreases; the corresponding speed is:

$$v_l = v_L \left(\frac{l}{L} \right)^{\frac{1}{3}} = v_L Re^{-\frac{1}{4}} \, . \tag{6.56}$$

It is also possible to estimate the number of degrees of freedom of turbulent motion corresponding to a given value of the Reynolds number. The number of degrees of freedom per unit volume is a function of ε and the viscosity v. Let's say:

$$n \propto \varepsilon^a v^b \tag{6.57}$$

with a and b unknowns. By imposing the equality of the dimensions of the two members of this equation we find the system:

$$2a + 2b = -3 \qquad 3a + b = 0 \tag{6.58}$$

which has $a = 3/4$ and $b = -9/4$ as solutions. Therefore:

$$n \propto \varepsilon^{\frac{3}{4}} v^{-\frac{9}{4}} \, . \tag{6.59}$$

Using the (6.46) applied to the maximum scale we have:

$$n \propto \left(\frac{v_L^3}{L}\right)^{\frac{3}{4}} v^{-\frac{9}{4}} = L^{-3} Re^{\frac{9}{4}} . \tag{6.60}$$

The total number of degrees of freedom N is obtained by multiplying n by the volume L^3:

$$N \propto Re^{\frac{9}{4}} . \tag{6.61}$$

The constant of proportionality can be determined by remembering that at the onset of instability, that is, when $Re = Re_{cr}$, it is $N = 1$ for which we obtain:

$$N = \left(\frac{Re}{Re_{cr}}\right)^{\frac{9}{4}} . \tag{6.62}$$

6.6   Turbulence of the Interstellar Medium

Spectroscopic observations of molecular clouds show a broadening of the emission lines (e.g. of CO) greater than can be expected from thermal Doppler broadening.

In a motion of thermal agitation in thermodynamic equilibrium the number of particles having velocities between (v_x, v_y, v_z) and $(v_x + \mathrm{d}\,v_x, v_y + \mathrm{d}\,v_y, v_z + \mathrm{d}\,v_z)$ is given by:

$$\mathrm{d}\,N = \left(\frac{m}{2\pi k_\mathrm{B} T}\right)^{\frac{3}{2}} \exp\left[-\frac{m(v_x^2 + v_y^2 + v_z^2)}{2k_\mathrm{B} T}\right] \mathrm{d}\,v_x \,\mathrm{d}\,v_y \,\mathrm{d}\,v_z . \tag{6.63}$$

To find the distribution as a function of only the v_z component, it must be integrated on the other components:

$$\mathrm{d}\,N_z = \left(\frac{m}{2\pi k_\mathrm{B} T}\right)^{\frac{3}{2}} \exp\left(-\frac{m v_z^2}{2k_\mathrm{B} T}\right) \mathrm{d}\,v_z \int_{\mathbb{R}^2} \exp\left[-\frac{m(v_y^2 + v_z^2)}{2k_\mathrm{B} T}\right] \mathrm{d}\,v_y \,\mathrm{d}\,v_z \tag{6.64}$$

that is

$$\mathrm{d}\,N_z = \sqrt{\frac{m}{2\pi k_\mathrm{B} T}} \exp\left(-\frac{m v_z^2}{2k_\mathrm{B} T}\right) \mathrm{d}\,v_z . \tag{6.65}$$

If we consider a source that, with respect to an observer O, approaches or moves away with the velocity v_z, we have a variation Δv of the frequency with respect to that v_0 of a stationary source:

$$\frac{\Delta v}{v_0} = \frac{v_z}{c} . \tag{6.66}$$

If the emission comes from a gas in thermal agitation, to obtain the number of emitters at frequencies $(v, v + \mathrm{d}\,v)$ just calculate the number of sources with velocity

v_z included in the interval $(v_z, v_z + d v_z)$, v and v_z being linked by the relation (6.66). Therefore the number of such sources will be:

$$d n = \sqrt{\frac{m}{2\pi k_B T}} \exp\left[\frac{mc^2(v - v_0)^2}{2k_B T v_0^2}\right] \frac{c}{v_0} d v .$$ (6.67)

The intensity of the line will be:

$$I(v) = I_0 \exp\left[-\frac{(v - v_0)^2}{2\delta^2}\right]$$ (6.68)

with

$$\delta^2 = \frac{k_B T v_0^2}{mc^2} .$$ (6.69)

The half width of the line (FWHM: full width half maximum) is defined based on:

$$I(v) = \frac{1}{2}I_0$$ (6.70)

from which we obtain:

$$v_1 = v_0 - 1.18\delta$$ (6.71)

$$v_2 = v_0 + 1.18\delta$$ (6.72)

is

$$\Delta v = v_2 - v_1 = 2.35\frac{v_0}{c}\sqrt{\frac{k_B T}{m}}$$ (6.73)

or

$$\Delta v = \frac{2.35}{\lambda_0}\sqrt{\frac{k_B T}{m}} .$$ (6.74)

The broadening in wavelength is given by:

$$\Delta\lambda = \frac{2.35\lambda_0}{c\sqrt{3}}v_{th}$$ (6.75)

or in terms of the rate of thermal stirring:

$$\Delta\lambda = 4.5 \cdot 10^{-11}\lambda_0 v_{th} \text{ cm} .$$ (6.76)

We evaluate the thermal Doppler broadening of the CO row at 2.6 mm: at the temperature of 10 K, representative of the molecular environment, it results $v_{th} = 0.35$ km/s and the corresponding widening is $\Delta\lambda \approx 4 \cdot 10^{-7}$ (40 Å). In reality, a higher widening is measured: $\Delta\lambda \approx 10^{-5}$ which would correspond to temperatures much higher than 10 K if the widening were of thermal origin. Speeds of the order of $1 - 20$ km/s are deduced from the enlargements observed.

The interpretation of the nature of the motion responsible for this enlargement is not immediate: it could be an ordered motion or a disordered motion. In the first case we can think of a rotational motion, an expansion, a collapse or an oscillation. However, all these motions would presuppose a gradient that is not found. In the case of the collapse then an even more serious difficulty arises: the star formation rate would be much higher than observed observationally and the molecular gas would run out too quickly. Oscillation also appears to be ruled out, although there are no adequate models.

The possibility therefore remains that the motion has a disordered character. In this case there are essentially two interpretations: a turbulent motion or a disordered motion of a magnetic nature. We will now examine the first possibility reserving the analysis of the second for another chapter when the topic of magnetohydrodynamics will be addressed.

The identification of the motion of interstellar gas as a turbulent motion cannot be automatic and indeed there is a heated debate on this topic. The individual indications are not entirely convincing but the various arguments as a whole move in this direction. It is obvious that in order to have a sure answer it would be necessary to observe the characteristics of the velocity field, but for now the data and in particular their resolution do not allow this identification.

Let's look at the most important of the arguments in favor of turbulence.

(1) *Reynolds number.* Let's consider a molecular cloud (but our considerations can be made with similar results also for the other components of the interstellar medium) and evaluate its Reynolds number.

According to the diffusion theory, the dynamic viscosity η is given by:

$$\eta = \frac{1}{3}\rho v_{th} l \tag{6.77}$$

where l is the mean free path of the particles which, in turn, is expressed by:

$$l = \frac{1}{n\sigma} \tag{6.78}$$

where n is the numerical density of the particles and σ the collision cross section. Recalling that the kinematic viscosity v is obtained from η dividing by ρ from the previous relations we obtain:

$$v \approx \frac{v_{th}}{n\sigma} . \tag{6.79}$$

Assuming $\sigma = 10^{-15}$ cm^2, $v_{th} = 10^4$ cm s^{-1}, corresponding to temperatures of the order of 10 K and $n = 10^2$ cm^{-3} we have $v = 3 \cdot 10^{16}$ cm^2 s^{-1}.

In evaluating Re we take a characteristic speed of 10 km/s and a size of the order of parsecs. With these data we obtain $Re = 10^8$ which, being well above the critical value, suggests that the interstellar medium has a turbulent character.

(2) *Relations of scale.* By a statistical analysis of the data relating to molecular clouds (see Chap. 12), Larson found a correlation between the dispersion of velocity where the dimension L of the clouds over a large range of sizes. The most important aspect is that the relationship, obtained by interpolating the observational data:

$$\sigma_v \propto L^{0.38} \tag{6.80}$$

bears a remarkable resemblance to the Kolmogorov–Obukhov relation (6.47), relating to the turbulence of incompressible fluids. Subsequent works of the same type have produced similar reports of the type:

$$\sigma_v \propto L^q \tag{6.81}$$

with a different exponent but still contained in the interval $0.3 < q < 0.6$, the most reliable value being 0.5. This circumstance has led to believe that the character of the motion was turbulent, even though there are convincing indications that Kolmogorv's theory of incompressible turbulence cannot be applied to the interstellar medium (see later in this same paragraph).

(3) *Fractal nature of the cloud contour.* Figure 6.3 shows the contours of the clouds' complex in the vicinity of the Cassiopea Cas OB6 regions obtained by imaging it at various wavelengths to highlight three different scales: a superficial analysis of these two maps and analogous ones seem to suggest a character of self-similarity at different scales and fractal geometry can be used to give quantitative confirmation.

Let us consider a closed curve and let P be its perimeter and A the enclosed area; the relationship:

$$P \propto A^{\frac{D}{2}} . \tag{6.82}$$

defines the fractal dimension of the curve. An analogous definition can be given for a closed surface in three-dimensional space and for the volume enclosed by it. For a classic closed curve, such as a circle, $D = 1$ (for the spherical surface $D = 2$) and D tends to 2 for a very convoluted curve that fills the entire plane; in the analogous three-dimensional situation, D tends to 3. Using different tracers to define the contours of the clouds, fractal sizes ranging from 1.26 to 1.60 are found whether the measurements concern molecular or atomic gas: this indicates that there are no fundamental differences in the physics that controls the edge of the clouds. What is the relationship with turbulence? The fractal dimensions of

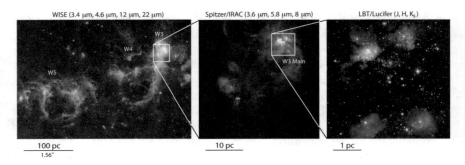

Fig. 6.3 Projected structure of the Cassiopea OB6 star forming region. From Gouliermis [2]

many surfaces in turbulent motions were measured by dissecting these surfaces with planes; for the curves obtained from the intersection we find a value of the fractal dimension of 1.35, close to those found for the motions of molecular clouds. However, it should be noted that in the case of molecular clouds what is obtained is the fractal dimension of the projection on the plane of the celestial vault and there is no proof that the two quantities must coincide.

(4) *Intermittence of the velocity field.* By analyzing the absorption lines of CaII and NaI of the interstellar medium, it was found that the velocity distribution of the absorbing atoms is better approximated by an exponential curve than by a Gaussian. The CO emission lines also show wings that are wider than the Gaussian predicted (see Fig. 6.4). If this phenomenon is truly representative of the velocity distribution, it reveals a greater number of large-amplitude velocity fluctuations than is predicted by the normal distribution. This phenomenon is known from laboratory experiments related to turbulence and goes by the name of intermittence. Essentially it highlights the non-uniform character of the turbulence and in particular the concentration of the energy dissipation in a fraction of the volume occupied by the fluid. Consequently, the spatial distributions of the velocity field are not Gaussian, since the velocity gradient has a higher than normal probability of assuming values far from the average. This phenomenon, if confirmed with a greater number of observations, could provide a valid element of confirmation of the turbulent nature of the interstellar gas.

One of the most substantial objections to the hypothesis of the turbulent nature of the motions of the molecular gas is constituted by the circumstance that the dissipation of the energy of these motions is relevant above all in consideration of the fact that the motions are, at the scales at which the measurements of the speed, supersonic and therefore presuppose the presence of shock waves and dissipation. In the absence of an adequate feeding mechanism, the turbulence would be destined to decay rapidly and this would lead, as will be seen later, to the instability of the clouds and therefore their collapse.

Mechanisms proposed to feed turbulence are, in clouds where the star formation process is already underway, the phenomena of energy input into the gas by stars

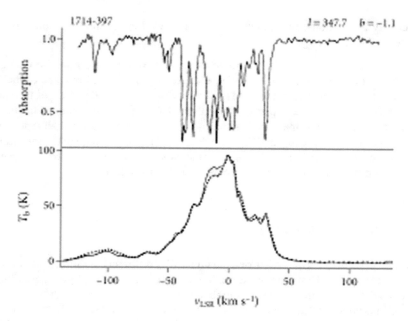

Fig. 6.4 A typical CO line in absorption and emission in HI clouds

(stellar winds, HII regions, supernova explosions) and, independently of presence of stars, the differential galactic rotation: this, through the sliding of two layers of gas with different speed, by friction phenomena converts the energy of the ordered motion into the energy of the disordered motion.

Accepting the turbulent nature of the motion of the molecular gas, however, we cannot think of applying our knowledge of the turbulence of the incompressible gas to this. In fact, the conditions prevailing in molecular clouds differ in several aspects from the hypotheses underlying the known theories of turbulence:

- in the interstellar medium the Mach numbers corresponding to the velocity fluctuations are large, i.e. these motions are supersonic and therefore compressibility is an important characteristic of the motion;
- it can be expected that there are shock waves at all scales and therefore energy dissipation is possible, through shocks, at all scales and not only at the smallest scale;
- there is evidence that, in the interstellar medium, energy is injected at all scales and not only at the larger scale: the mechanisms can be supernova explosions, stellar winds, expanding HII regions;
- the behavior of the interstellar medium is controlled by autogravitation, a characteristic completely absent in studies of the theory of turbulence; it is now known that autogravitation is at least comparable to other forces in clouds of any size and since the intensity of this force is sufficient to produce the observed density and

velocity fluctuations, autogravitation is probably a fundamental ingredient of any interstellar gas theory.

These considerations indicate, not so much that we must abandon the idea of a turbulent interstellar medium, but rather that the theory of the turbulence of incompressible fluids is inadequate and it is necessary to broaden the picture of turbulent phenomena to include the aspects just mentioned. Currently the theory of the turbulence of the interstellar medium is at a much more uncertain stage than that of the incompressible turbulence, but there is no doubt that our knowledge of the interstellar medium, of the formation of clouds and of stars will reach a satisfactory degree of depth when we have solved the problems related to turbulence.

6.7 Synthetic Summary

A whole series of experiments carried out under very different conditions indicate that there is a transition from laminar to turbulent motion when the physical conditions are changed so that the Reynolds number grows above a critical value. This fact can be interpreted in the light of the fluid dynamics equations: small values of Re correspond to the dominance of viscous forces that tend to dampen inhomogeneities, large values of Re indicate the domain of the non-linear term in the Navier equation-Stokes, responsible for the complexity of motion.

There seems to be no single trigger mechanism for turbulence: we thought of a sequence of instabilities that occur successively, corresponding to each of which a periodicity appears to which a degree of freedom is associated. On the other hand, there are situations, as in the Couette-Taylor motion, in which after the manifestation of two instabilities with the subsequent appearance of two periods a motion develops which has all the characteristics of a turbulent motion.

It is therefore believed that there are several ways in which the turbulent character manifests itself: once it is fully developed its characteristics are independent of the trigger mode.

Given the high number of degrees of freedom of turbulent motion, the deterministic approach is impracticable and one has to resort to the statistical method. The statistical treatment is based on the use of the correlations between the fluctuations of the quantities, in particular the velocity, in different points of the fluid. For example, the correlation of the second order is defined as the average value of the product of the velocities in two points of the fluid at the same instant and the definition easily extends to the correlations of the higher orders which involve a progressively increasing number of points.

The simplest of the correlations are: the mean values (first order), the variance of the velocity fluctuations describing the intensity of the turbulence, the correlation coefficients at two different points. Most of the work on turbulence moves in this direction, limiting, for the sake of simplicity, consideration to the lowest orders. In this case, however, it must be borne in mind that very rough information is obtained on the spatial structure of turbulence.

By introducing the mean values and the fluctuations of the different quantities into the continuity equation and the Navier–Stokes equation, we arrive at an equation for mean values that contains a second order correlation. The equation for the floating components contains the third order correlation and so on. Turbulence is then described by a system of infinite equations: if it is truncated to a certain order, there is the problem of the closure, that is, of the breakeven of the number of equations and the number of unknowns.

There are two experimental techniques for the study of turbulence: second order correlations and Fourier analysis of the velocity field, where the Fourier components are the Fourier transforms of the second order correlations.

It has been noted that the non-linear term that appears in the Navier–Stokes equation is responsible for the simultaneous presence of many components of different scales between which there is a continuous exchange of energy. Studying correlations indicates that correlations exist between points in regions of different sizes. These considerations lead to formulate a picture of turbulence which is described as a hierarchy of turbulent elements (ET, in English eddies) of different scales. These turbulent elements are coherent structures in which there are close correlations between the velocities of its points and this coherence persists for some time.

The ETs of a given scale originate from the loss of stability of the ETs of the immediately larger scale from which they acquire energy; in turn they decay giving rise to ETs of lower dimensions among which their energy is divided. This cascade process stops at the scales at which viscosity becomes important and there is a significant dissipation of mechanical energy into heat.

The theory of homogeneous turbulence of incompressible fluids formulated by Kolmogorov adds to this qualitative model a hypothesis of similarity for the turbulent elements of the different scales. Kolmogorov derived the relation $v_\lambda \propto \lambda^{\frac{1}{3}}$ between the turbulence velocity on the λ scale and the scale itself. From this we deduce that the turbulent energy spectrum is a power law with exponent $-5/3$.

The emission lines of the molecular gas, in particular the CO line at 0.26 cm, have a Doppler broadening much greater than that due to the motion of thermal agitation: while the thermal expansion corresponding to the temperature of 10 K is of the order of a few tens of Angstrom the one observed is over 10^3 Å. There are reasons to exclude that it is due to some overall motion (such as rotation, collapse or expansion) and therefore it appears to be characterized as a disordered motion. A possible interpretation is that it is a turbulent motion and in fact many elements are in favor of this hypothesis (high Reynolds number, relationship between velocity deduced from Doppler enlargement and size of the clouds that has close analogy with the Kolmogorov relationship- Obukhov, fractal nature of the outline of the clouds, intermittence) even if singly none appears decisive. It should be noted that this interpretation is contrasted by another according to which the velocity field responsible for the broadening of the lines is due to the propagation within the cloud of many magnetohydrodynamic waves.

It should be noted that, in the context of the hypothesis of the turbulent origin of molecular velocities, the Kolmogorov theory could hardly be valid since many of the premises underlying this theory are missing (incompressibility, injection of energy only at major scales and dissipation only at the smallest scales, absence of self gravitation).

References

1. Fenstermacher P., Swinney H.L., Gollub J.P.: Dynamical instabilities and the transition to chaotic Taylor vortex flow. J. Fluid Mech. **94**, 103 (1985)
2. Gouliermis, D.A.: Unbound young stellar systems: star formation on the loose, PASP, 130, 1 (2018)

Chapter 7
Electrodynamics and Magnetohydrodynamics

7.1 Plasmas. Debye Screening Length

Electrodynamics studies the behaviour of a system composed of electromagnetic fields and a conducting fluid (liquid or gaseous). The fields act on the electrons and on the ions producing currents. In turn currents cause alterations in the electromagnetic field: hence, we have to do with a complicated coupled system of matter and fields. We focus on conductor fluids, that is, on ionised gases. The fundamental difference between a non-conducting gas and an ionised gas is related to the nature of the interaction mechanism between particles. In the case of a neutral gas the interaction forces have a very limited range; the effect on the trajectories is felt only when two particles are very close: the binary collision model well represents the situation and the kinetic theory developed starting from these ideas is able to give a correct description of the gas behaviour.

In the following we will first consider a fully ionised gas, then we will deal with the situation in which the gas is only partially ionised.

In the case of an ionised gas with a fairly high density, the interactions cannot generally be described as binary collisions but the motion of each particle is determined by the average electric field produced by all the surrounding particles. In this case all the particles intervene in a collective way on the motion of each of them: this collective character is an essential characteristic of plasmas. If the density of the ionised gas is very low, the collective effects are negligible compared to binary collisions and the gas behaves like a simple system of charges in an electromagnetic field, not like a plasma.

In order to define what we mean with plasma it is therefore necessary to specify in a quantitative way the conditions under which the collective effects are relevant. This is achieved by introducing the Debye screening length, which is also indicated as Debye radius.

In conditions of thermodynamic equilibrium in a plasma the laws of Statistical Mechanics can be applied and basing on these the distribution of the Ze charges (we also include the case of the electron by setting $Z = -1$) in an electric field having

© The Author(s), under exclusive license to Springer Nature Switzerland AG 2021   135
G. Carraro, *Astrophysics of the Interstellar Medium*,
UNITEXT for Physics, https://doi.org/10.1007/978-3-030-75293-4_7

potential $\Phi(r)$ is given by:

$$n_Z = \langle n_Z \rangle \exp\left[\frac{-(Ze\Phi)}{k_B T}\right] \tag{7.1}$$

where n_Z is the number of such ions for cm^{-3} and $\langle n_Z \rangle$ is the average value at infinite distance from the charge generating the field. Suppose the potential energy is small compared to the thermal energy of the particles

$$Ze\Phi \ll k_B T . \tag{7.2}$$

We will then have, by developing (7.1) in series and neglecting higher order terms based on (7.2):

$$n_Z = \langle n_Z \rangle \left[1 - \frac{Ze\Phi}{k_B T}\right] . \tag{7.3}$$

For the sake of simplicity, we limit ourselves to considering the case of a gas chemically made up of hydrogen only, so that two species are present: electrons with medium density $\langle n_e \rangle$ and protons with medium density $\langle n_p \rangle$. The average densities must satisfy the condition of macroscopic neutrality of the charges:

$$\langle n_e \rangle = \langle n_p \rangle; \Sigma_Z Z < n_Z > e = 0 . \tag{7.4}$$

Furthermore, the densities n_e and n_p must satisfy the Poisson equation:

$$\nabla^2 \Phi = -\frac{\eta}{\epsilon_0} = -\frac{1}{\epsilon_0}\Sigma_Z Z n_Z e , \tag{7.5}$$

where η indicates the charge density and ϵ_0 is the electric constant or vacuum permittivity. Its value is $8.854 \times 10^{-12} F/m$.
By introducing (7.3) and taking into account the condition of neutrality we obtain:

$$\nabla^2 \Phi - k^2 \Phi = 0 \tag{7.6}$$

where we introduced k^2 as :

$$k^2 = \frac{e^2 \Sigma_Z < n_Z > Z^2}{\epsilon_0 k_B T} = \frac{2n_e e^2}{\epsilon_0 k_B T} . \tag{7.7}$$

The last expression comes from the fact that in a pure H plasma $\Sigma_Z Z^2 < n_Z >= n_e + n_p = 2n_e$. It is straightforward to show that k has the dimensions of the inverse of a length.
A spherically symmetrical solution (the electric field is a central field) of this equation is:

$$\Phi = \frac{A}{r} \exp(-kr) \tag{7.8}$$

where A is an integration constant. Considering in $r = 0$ a proton that produces the field, for $r \to 0$ must be:

$$\Phi = \frac{e}{4\pi \epsilon_0 r} . \tag{7.9}$$

This way we get:

$$A = \frac{e}{4\pi \epsilon_0} , \tag{7.10}$$

and therefore:

$$\Phi(r) = \frac{e}{4\pi \epsilon_0 r} \exp\left(-\frac{r}{\lambda_D}\right) \tag{7.11}$$

where we put

$$\lambda_D = \frac{1}{k} \tag{7.12}$$

which is called *Debye screen length*. Expanding the exponential in series we obtain:

$$\Phi(r) = \frac{e}{4\pi \epsilon_0 r} - \frac{e}{4\pi \epsilon_0 \lambda_D} . \tag{7.13}$$

We can now give the formulas (7.11) and (7.13) the following interpretation: the potential generated by the proton is partially shielded by a charge of opposite sign distributed within a sphere of radius λ_D. The Debye's screening length is expressed by:

$$\lambda_D = \left(\frac{\epsilon_0 k_B T}{2 n_e e^2}\right)^{\frac{1}{2}} . \tag{7.14}$$

The condition that the potential energy is less than the thermal energy gives:

$$\frac{1}{4\pi d} \ll \frac{\epsilon_0 k_B T}{2 e^2} \tag{7.15}$$

where d is the average distance between the charges. Since $d \approx n_e^{-\frac{1}{3}}$ we have $d \ll \lambda_D$, i.e. within the sphere of radius λ_D there are many charges (electrons and ions), however, always such that the overall charge is equal and opposite to that of the central ion. Introducing the values of the various physical constants it results that:

$$\lambda_D = 7.9 \left(\frac{T}{n_e}\right)^{\frac{1}{2}} \text{ cm} . \tag{7.16}$$

Let L be the characteristic linear dimension of the ionized gas; if $\lambda_D \geq L$ the shielding effect does not take place. From this relationship it follows that the density of the gas is small and the average distance between the particles is large: binary collisions count and there are no collective effects; the gas must be treated as a system of independent particles and we cannot speak of plasma. If, on the other hand, $\lambda_D \leq L$, the collective effects are important; the latter condition defines a plasma.

For instance, in a typical HII region we have:

- $L = 1$ pc $= 3 \cdot 10^{16}$ cm
- $n \leq 5$ cm^{-3}
- $n_e \leq 2.5$ cm^{-3}

so $\lambda_D = 500$ cm: the condition $\lambda_D \ll L$ is largely satisfied.

7.2   Plasma Oscillations

In a plasma macroscopically there is no separation of charges of opposite sign and the neutrality condition is verified. However, under the action of an external field, the charges have a tendency to separate, but this same phenomenon creates fields that oppose the separation.

Taking into account the collective phenomena, the electrons oscillate (the ions, having a greater inertia, are practically fixed) and the recall force is the Coulomb interaction between ions and electrons: the frequency of these oscillations is called plasma frequency.

We denote by n_i and n_e respectively the density of ions and electrons and suppose that initially the system is in equilibrium: it will be $n_i = n_e = n_0$. Suppose that the electrons are displaced from the equilibrium position, while the ions remain stationary due to the greater inertia. The motion takes place along the x axis. Let us consider a parallelepiped-shaped volume with base area S and height Δx in the direction of the x axis. We denote by $s(x)$ the shift undergone by the electrons in x; if with a and with b we denote the positions of the two bases of the parallelepiped, we can express $s(b)$ by developing s in series and ignoring terms of higher order than the first

$$s(b) = s(a) + \frac{d\,s}{d\,x}\Delta x \tag{7.17}$$

and therefore the elongation suffered by the parallelepiped will be:

$$\Delta s = \frac{d\,s}{d\,x}\Delta x \ . \tag{7.18}$$

If n_0 is the number of electrons for cm^3 in the new situation, it will be:

$$n_0 S\Delta x = nS\Delta x\left(1 + \frac{d\,s}{d\,x}\right) \tag{7.19}$$

from which we obtain:

$$n = \frac{n_0}{1 + \frac{ds}{dx}} \approx n_0 \left(1 - \frac{ds}{dx} \right). \tag{7.20}$$

The excess of electrons per unit of volume is:

$$n - n_0 = -n_0 \frac{ds}{dx} \tag{7.21}$$

and excess charge density:

$$\eta = -e(n - n_0) = en_0 \frac{ds}{dx}. \tag{7.22}$$

The electric field produced by this charge is defined by the

$$\mathrm{div}\mathbf{E} = \frac{\eta}{\epsilon_0} \tag{7.23}$$

and being due to a motion along the axis x it will have the direction of this. Therefore

$$\frac{dE}{dx} = \frac{en_0}{\epsilon_0} \frac{ds}{dx} \tag{7.24}$$

which, integrated with the condition $E = 0$ for $s = 0$, gives:

$$E = \frac{en_0}{\epsilon_0} s. \tag{7.25}$$

The electrostatic force acting on each electron is:

$$F = -eE = -\frac{e^2 n_0}{\epsilon_0} s \tag{7.26}$$

and the equation of motion is:

$$m_e \frac{d^2 s}{dt^2} = -\frac{e^2 n_0}{\epsilon_0} s. \tag{7.27}$$

placing:

$$\omega_p^2 = \frac{e^2 n_0}{m_e \epsilon_0}, \tag{7.28}$$

the previous equation becomes:

$$\frac{d^2 s}{dt^2} + \omega_p^2 s = 0. \tag{7.29}$$

This equation describes and oscillating motion with frequency

$$\nu_p = \frac{1}{2\pi} \left(\frac{e^2 n_0}{m_e \epsilon_0} \right)^{\frac{1}{2}} \tag{7.30}$$

said plasma frequency; introducing the constants' values we obtain:

$$\nu_p = 9 \cdot 10^2 n_0^{\frac{1}{2}} . \tag{7.31}$$

The electron oscillations are a consequence of the collective behaviour of the plasma. Their existence is subordinated to the absence of any mechanism that destroys their coherence and therefore to the circumstance that each electron of the same plane perpendicular to the displacement vibrates in phase. If the ionisation degree is small, the binary collisions, no longer negligible, are able to break the coherence because during a period of oscillation the number of collisions is then expected to be quite large.

It has been said that the plasma is macroscopically neutral, but that external fields can produce a partial separation of the charges of the opposite sign, to which the plasma reacts with oscillations of frequency ν_p. However, it may happen that the effects of this separation are not negligible. For example, consider a plasma subject to the action of a magnetic field **B**. A characteristic quantity defining the motion in these conditions is the cyclotron frequency

$$\nu_c = \frac{eB}{2\pi m_e} . \tag{7.32}$$

If the collision frequency ν_{coll} is smaller than the cyclotron frequency, the electrons and ions spiral several times before their motion is altered by collisions and therefore the separation of the charges produced by an external field is important. If, on the other hand, the collision frequency is larger than the cyclotron frequency, the collisions are more frequent and tend to prevent the separation of the charges.

The description of a plasma is more or less complicated depending on whether these effects are completely negligible or not. Whenever the effects of the separation of charges cannot be neglected, the study of plasmas must be done (a) with a model with two conducting fluids (ions and electrons), or (b) using the methods of statistical mechanics (Boltzmann equation): in this case we are in the realm of plasma physics.

When the separation of the charges does not produce sensible effects, the mechanical behaviour of the plasma can be described with a model of a single conducting fluid, similarly to the treatment of fluid dynamics; the variables density, pressure, speed, etc. must satisfy the known hydrodynamical equations into which the effects of electro-magnetic fields are introduced. These equations then must be completed with Maxwell's equations describing the electromagnetic field. This is the approximation routinely referred to with the term magnetohydrodynamics (MHD).

Electrodynamics is therefore divided into plasma physics and magnetohydrodynamics. The precise definition of the conditions in which the MHD approximation is valid would require a complicated analysis that goes beyond the tasks of this course: it is sufficient to point out that in the astrophysical situations that will deal with this approximation is valid. In only one case we will deal with, i.e. for the ambipolar diffusion, it will be necessary to take into account two fluids: the fluid of the charged particles and the fluid of the neutral particles.

7.3 A Note on Systems of Measurement Units

In the following we will make use of the MKSA system, which, except for some small differences, coincides with the International System officially adopted in Italy since 1979.

Since in the literature most of the works relating to electromagnetism and magnetohydrodynamics use older systems it is necessary to be able to pass from these systems to the International System.

We limit ourselves here to analyzing the Gauss system, which is the most common among the systems of units of measurement. It is known that there is a CGS-es (electrostatic) system which, starting from Coulomb's law:

$$F = \frac{kq_1q_2}{r^2} \tag{7.33}$$

and setting $k = 1$, it defines the unit of measurement of the electric charge, as that charge which, placed in a vacuum at a distance of 1 cm from an equal charge, repels it with the force of a dyn. There is also a CGS-em (electromagnetic) system in which the unit of measurement of the current is defined starting from the relationship that expresses the magnetic dipole moment of a coil though which current is passing:

$$m = HiS \tag{7.34}$$

where S is the area of the loop. The unit of measurement of the current is defined by setting $h = 1$ and assuming as unit the current which, flowing in a 1 cm^2 loop, determines a unit magnetic moment.

The two systems are connected and one electromagnetic unit of current is equal to $3 \cdot 10^{10}$ electrostatic units.

The reason for the existence of an electrostatic system and an electromagnetic system is due to the fact that in the first half of the last century the study of electrostatic phenomena and magnetism progressed separately and only later was their interdependence identified.

The Gauss system uses mixed units taken partly from the electrostatic system and partly from the electromagnetic system. Magnetic quantities are measured in electromagnetic units, electrical quantities in electrostatic units. Although there are

Table 7.1 Definition of electromagnetic quantities in different unit systems

Quantity	Gauss system	MKSA system
Speed of light	c	$(\epsilon_0 \mu_0)^{-\frac{1}{2}}$
Electric field	E	$(4\pi \epsilon_0)^{\frac{1}{2}} E$
Electric potential	Φ	$(4\pi \epsilon_0)^{\frac{1}{2}} \Phi$
Load	q	$(4\pi \epsilon_0)^{-\frac{1}{2}} q$
Charge density	η	$(4\pi \epsilon_0)^{-\frac{1}{2}} \eta$
Current	I	$(4\pi \epsilon_0)^{-\frac{1}{2}} I$
Current density	j	$(4\pi \epsilon_0)^{-\frac{1}{2}} j$
Magnetic induction	B	$(4\pi / \mu_0)^{\frac{1}{2}} B$
Electrical conductivity	σ	$\sigma / 4\pi \epsilon_0$
Resistance	R	$4\pi \epsilon_0 R$
Impedance	Z	$4\pi \epsilon_0 Z$
Inductance	L	$4\pi \epsilon_0 L$
Capacity	C	$C / 4\pi \epsilon_0$

differences in the use of the system, electric charge, electric field, potential and capacitance are generally measured in electrostatic units, while magnetic induction, resistance, current and magnitudes are measured in electromagnetic units. connected to it.

Table 7.1 provides the conversion coefficients from an equation expressed in Gaussian units to the corresponding equation expressed in the MKSA system.

It is found, for example, in the texts in which the Gauss system is used, that Maxwell's equation of the electric field is written in the form:

$$\text{rot } \mathbf{E} = -\frac{1}{c} \frac{\partial \mathbf{B}}{\partial t} . \tag{7.35}$$

Using the indications of the table we obtain the form of the same equation in the MKSA system:

$$\text{rot } (\sqrt{4\pi \epsilon_0} \mathbf{E}) = -\sqrt{\epsilon_0 \mu_0} \frac{\partial}{\partial t} \left(\sqrt{\frac{4\pi}{\mu_0}} \mathbf{B} \right) \tag{7.36}$$

which becomes

$$\text{rot } \mathbf{E} = -\frac{\partial \mathbf{B}}{\partial t} . \tag{7.37}$$

7.4 Maxwell's Equations

Referring to an inertial system, Maxwell's equations are expressed in the form:

$$\begin{cases} \text{div}\mathbf{E} = \dfrac{\eta}{\epsilon_0} & (7.38) \\[2mm] \text{div}\mathbf{B} = 0 & (7.39) \\[2mm] \text{rot }\mathbf{E} = -\dfrac{\partial \mathbf{B}}{\partial t} & (7.40) \\[2mm] \text{rot }\mathbf{B} = \mu_0 \mathbf{j} + \epsilon_0 \mu_0 \dfrac{\partial \mathbf{E}}{\partial t} & (7.41) \end{cases}$$

where η denotes the charge density, $\mathbf{j}$ the current density, and μ_0 the magnetic constant or magnetic vacuum permeability, whose value is $1.256 \times 10^{-6} H/m$. Equation (8.39) states that the magnetic field is solenoidal. Equation (8.40) is referred to as Faraday–Newman–Lenz equation, while equation (8.41) is referred to as Ampere-Maxwell equation. In this last equation the second term in the right hand side is called displacement current. Furthermore the charge density and the current density satisfy the conservation equation, which is a consequence of the previous equations:

$$\frac{\partial \eta}{\partial t} + \text{div}\mathbf{j} = 0 \ . \tag{7.42}$$

We remind also that the Ohm's law is expressed by:

$$\mathbf{j} = \sigma \mathbf{E} \ , \tag{7.43}$$

being σ the electric conductivity. In the ISM σ is very high, typical values being $10^{13} - 10^{14} \Omega^{-1} m^{-1}$.

Maxwell's equations are invariant with respect to a transformation from one reference system to another that moves with respect to the first of uniform rectilinear motion (Lorentz transformations).

Let us consider a O' system with respect to which a charge q is at rest and feels the field of another charge Q as well at rest. Let us indicate with quotes the quantities measured with respect to this reference. Let O be a reference frame with respect to which O' moves in a uniform rectilinear motion with velocity $v \ll c$. We indicate without quotes the quantities measured in this last reference system. We will have that O' sees q subjected to a force $\mathbf{F}' = q\mathbf{E}'$ while O sees that the charges are in motion and finds that q is subject to a force $\mathbf{F} = q\mathbf{E} + q\mathbf{v} \times \mathbf{B}$. Since according to the Lorentz transformation $\mathbf{F} = \mathbf{F}'$ will be:

$$\mathbf{E}' = \mathbf{E} + \mathbf{v} \times \mathbf{B} \ . \tag{7.44}$$

The **B** field transforms according to:

$$\mathbf{B}' = \mathbf{B} - \frac{1}{c^2}\mathbf{v} \times \mathbf{E} \qquad (7.45)$$

while it will be:

$$\mathbf{j}' = \mathbf{j} - \eta\mathbf{v} \qquad (7.46)$$

with

$$\eta' = \eta . \qquad (7.47)$$

The fields $\mathbf{E}$ and $\mathbf{B}$ do not have an independent existence as separate entities: the fundamental complex is the electromagnetic field, splitting it into the electric component and the magnetic component is an operation entirely related to the motion of the observer .

The magnetohydrodynamics of the interstellar medium can be developed entirely under the following conditions:

– plasma speeds are much lower than the speed of light: $v \ll c$;
– the size of the systems involved are very large;
– in the ISM the electric conductivity σ is very high .

Taking into account these conditions, the following relevant simplifications hold:

(A) from the Faraday–Neumann–Lenz relation (7.40), considering magnitude orders, we have $E/L \sim B/T$, where L and T are the characteristic length and time. We then have:

$$E \sim B(\frac{L}{T}) \sim Bv \qquad (7.48)$$

Hence we can replace E/B with v when considering order of magnitude relationships.

(B) If we compare the displacement current term with the term rot $\mathbf{B}$, we have:

$$\epsilon_0\mu_0\frac{\partial\mathbf{E}}{\partial t} \sim \frac{1}{c^2}(\frac{E}{T}) \qquad (7.49)$$

$$\text{rot } \mathbf{B} \sim \frac{B}{L} \qquad (7.50)$$

and therefore:

$$\frac{\epsilon_0\mu_0\partial\mathbf{E}/\partial t}{\text{rot } \mathbf{B}} \sim \frac{1}{c^2}(\frac{E}{T})(\frac{L}{B}) = (\frac{v}{c})^2 \ll 1 \qquad (7.51)$$

Hence (7.41) reduces to :

$$\text{rot } \mathbf{B} = \mu_0\mathbf{j} \qquad (7.52)$$

(C) Another way to look at it is that in Maxwell's equation (7.41), where the very same term term can be neglected because σ is very high in the ISM:

$$\left| \frac{(\epsilon_0\mu_0)\partial\mathbf{E}/\partial t}{\mu_0\mathbf{j}} \right| \approx \frac{\epsilon_0 E}{T\sigma E} \ll 1 \, . \tag{7.53}$$

The equation (7.41) then becomes:

$$\text{rot } \mathbf{B} = \mu_0\mathbf{j} \, . \tag{7.54}$$

(D) The electric force is much smaller than the magnetic one so that the former can be neglected compared to the latter (refer to the next chapter for the expression of the magnetic force). Indeed:

$$\frac{f_e}{f_m} = \left| \frac{\eta\mathbf{E}}{(1/\mu_0)(\text{rot } \mathbf{B}) \times \mathbf{B}} \right| \approx \left| \frac{\epsilon_0 E \text{div} \mathbf{E}}{(1/\mu_0)(\text{rot } \mathbf{B}) \times \mathbf{B}} \right| \approx \frac{E^2}{c^2 B^2} \ll 1 \, . \tag{7.55}$$

As for the force, the energy density of the electric field as well can be neglected compared to that of the magnetic field.

(E) Since σ is very large from the Ohm equation $\mathbf{j}/\sigma$ tends to zero and therefore

$$\sigma\mathbf{E} \sim \sigma(\mathbf{v} \times \mathbf{B}) \tag{7.56}$$

and we recover the approximation in item A. Besides, the plasma is then essentially neutral, with no free charges ($\mathbf{j} \sim 0$).

(F) As a consequence of the previous item, we can safety consider $\eta = 0$. In an HII region, for example, the ratio between the difference among electron and ions over ions $(n_e - n_i)/n_i$ is roughly $5 \cdot 10^{-13}$. Putting $\eta = 0$ doesn't necessarily implies putting $\text{div}\mathbf{E} = 0$, though.

(G) From this results, and also considering (7.46), one can write:

$$\mathbf{j}' = \mathbf{j} \tag{7.57}$$

Thnne, using Ohm's law and (7.45), we can conclude:

$$\sigma\mathbf{E}' = \sigma(\mathbf{E} + \mathbf{v} \times \mathbf{B}) \tag{7.58}$$

(H) Let's finally consider the right hand side of (7.45). We have that:

$$\frac{1}{c^2}\mathbf{v} \times \mathbf{E} \sim \frac{1}{c^2}v^2 B \sim (v/c)^2 B \ll B \tag{7.59}$$

with implies that in (7.45) we can neglect the second term and, in other words, conclude that the magnetic field $\mathbf{B}$ is invariant with respect to a reference frame change.

7.5   Magnetic Field Equation

We now derive an equation that describes in a very general way the behavior of the magnetic field. From (7.40) and (7.56) we obtain:

$$\frac{\partial \mathbf{B}}{\partial t} = - \mathrm{rot} \left(\frac{\mathbf{j}}{\sigma} \right) + \mathrm{rot} \, (\mathbf{v} \times \mathbf{B}) \, . \qquad (7.60)$$

With (7.59) we have:

$$\frac{\partial \mathbf{B}}{\partial t} = \mathrm{rot} \, (\mathbf{v} \times \mathbf{B}) - \mathrm{rot} \, (\eta_{\mathrm{m}} \, \mathrm{rot} \, \mathbf{B}) \qquad (7.61)$$

where is it

$$\eta_{\mathrm{m}} = \frac{1}{\sigma \mu_0} \qquad (7.62)$$

is the magnetic diffusivity, which in general will be a function of the spatial coordinates.

From (7.61) a further relationship can be derived which is valid only in the hypothesis that the electrical conductivity is uniform over the whole fluid. Remembering the vector identity

$$\mathrm{rot} \, \mathrm{rot} \, \mathbf{u} = \mathrm{grad} \, \mathrm{div} \mathbf{u} - \nabla^2 \mathbf{u} \qquad (7.63)$$

we get the:

$$\frac{\partial \mathbf{B}}{\partial t} = \mathrm{rot} \, (\mathbf{v} \times \mathbf{B}) + \eta_{\mathrm{m}} \nabla^2 \mathbf{B} \, . \qquad (7.64)$$

This equation defines the temporal evolution of the magnetic field.

7.6   Freezing of the Magnetic Field Lines of Force

The Eq. (7.64) allows to study the evolution of the magnetic field: the first term with the second member describes the process of generation of the magnetic field by the motion of the plasma; the second term describes the decay of the field due to its diffusion following ohmic losses.

In general, regardless of astrophysical situations, the solution of the equation (7.64) depends on the relative importance of the two terms in the second member. It is:

$$\left| \frac{\mathrm{rot} \, (\mathbf{v} \times \mathbf{B})}{\nu_{\mathrm{m}} \nabla^2 \mathbf{B}} \right| \approx \frac{\upsilon L}{\nu_{\mathrm{m}}} \qquad (7.65)$$

where, as usual, L and υ are a characteristic length and speed. We use here the term ν_m in place of η_m for the sake of a better comparison with previously introduced

Reynolds numbers. The quantity is defined as the magnetic Reynolds number:

$$Re_m = \frac{vL}{v_m} = \mu_0 \sigma vL .$$ (7.66)

When $Re_m \ll 1$ the effects of the second term prevail; the Eq. (7.64) reduces to:

$$\frac{\partial \mathbf{B}}{\partial t} = \eta_m \nabla^2 \mathbf{B}$$ (7.67)

which is a diffusion equation. From it it results, with considerations on orders of magnitude, that an initial configuration of the field decays, due to the ohmic dissipation, in a time:

$$\tau = \mu_0 \sigma L^2$$ (7.68)

called diffusion time, where L is a characteristic length.

When $Re_m \gg 1$ the effects of the first term prevail; the Eq. (7.64) reduces to:.

$$\frac{\partial \mathbf{B}}{\partial t} = \text{rot} \, (\mathbf{v} \times \mathbf{B}) .$$ (7.69)

The condition $Re_m \gg 1$ is never satisfied under laboratory conditions (for mercury it is $\eta_m = 0.8 \text{ m}^2 / \text{s}$). In the interstellar medium, on the other hand, this condition is easily verified due to the high electrical conductivity and large dimensions. Consequently, the decay time of the magnetic field is very long: for a HII region the diffusion time is 10^{14} years, higher than the Hubble time; in general, due to the typical conductivities of the interstellar medium and for large length scales, the magnetic field is constant over a period of time comparable to or greater than the age of the universe.

Therefore, due to the high conductivity and large size, the (7.64) equation for the interstellar medium takes the form (7.69). With this equation a very important result can be obtained.

Let us consider a material circuit l, tied to the particles of the fluid: it is closed and a direction of travel is fixed on it. We calculate the variation over time of the flow of $\mathbf{B}$ through l:

$$\Phi_B = \int_S \mathbf{B} \cdot d\sigma$$ (7.70)

where S is any surface that has the circuit as boundary (being $\mathbf{B}$ a solenoidal vector as shown by (7.39), the flow is the same whatever the surface having l as boundary). Let l_1 be the curve that is superimposed on the material line at instant t and S_1 a surface having l_1 for contour, the normal to S_1 is oriented counterclockwise with respect to the direction of travel of l_1; similarly let l_2 be the line superimposed on l at the instant $t + \delta t$ and S_2 a surface having l_2 for contour: the normal to it is oriented with the same criterion. Let S_3 be the surface defined by the trajectories of the particles of l

during the interval δt; its normal is oriented outside the volume bounded by S_1, S_2 and S_3.

Let it be:

$$\delta \Phi_B = \int_{S_2} \mathbf{B}(t + \delta t) \cdot d\boldsymbol{\sigma} - \int_{S_1} \mathbf{B}(t) \cdot d\boldsymbol{\sigma} . \qquad (7.71)$$

We expand $\mathbf{B}(t + \delta t)$ in series keeping only the first two terms of the development:

$$\mathbf{B}(t + \delta t) = \mathbf{B}(t) + \frac{\partial \mathbf{B}}{\partial t} \delta t \qquad (7.72)$$

hence we have:

$$\Delta \Phi_B = \int_{S_2} \mathbf{B}(t) \cdot d\boldsymbol{\sigma} - \int_{S_1} \mathbf{B}(t) \cdot d\boldsymbol{\sigma} + \delta t \int_{S_2} \frac{\partial \mathbf{B}}{\partial t} \cdot d\boldsymbol{\sigma} . \qquad (7.73)$$

Since $\mathbf{B}$ is solenoidal, the outgoing flow from the closed surface formed by S_1, S_2 and S_3 is zero. Then the net flow is:

$$\int_{S_2} \mathbf{B}(t) \cdot d\boldsymbol{\sigma} - \int_{S_1} \mathbf{B}(t) \cdot d\boldsymbol{\sigma} + \int_{S_3} \mathbf{B} \cdot d\boldsymbol{\sigma} = 0 . \qquad (7.74)$$

With this relation the (7.73) becomes:

$$\delta \Phi_B = - \int_{S_3} \mathbf{B}(t) \cdot d\boldsymbol{\sigma} + \delta t \int_{S_2} \frac{\partial \mathbf{B}}{\partial t} \cdot d\boldsymbol{\sigma} . \qquad (7.75)$$

Let us consider the flow through S_3: the infinitesimal element of this surface can be expressed in the form:

$$d\mathbf{s} \times \mathbf{v}\delta t \qquad (7.76)$$

where $d\mathbf{s}$ is an infinitesimal element of the line l_1 and the flow through this element is:

$$\mathbf{B} \cdot (d\mathbf{s} \times \mathbf{v})\delta t \qquad (7.77)$$

therefore:

$$\int_{S_3} \mathbf{B} \cdot d\boldsymbol{\sigma} = \delta t \oint_{l_1} (\mathbf{v} \times \mathbf{B}) \cdot d\boldsymbol{\sigma} . \qquad (7.78)$$

Using Stokes' theorem we obtain:

$$\oint_{l_1} (\mathbf{v} \times \mathbf{B}) \cdot d\boldsymbol{\sigma} = \int_{S_1} \mathrm{rot}\,(\mathbf{v} \times \mathbf{B}) \cdot d\boldsymbol{\sigma} \qquad (7.79)$$

and then from (7.75):

$$\delta\Phi_B = \delta t \left[\int_{S_2} \frac{\partial \mathbf{B}}{\partial t} \cdot d\boldsymbol{\sigma} - \int_{S_1} \text{rot}\,(\mathbf{v} \times \mathbf{B}) \cdot d\boldsymbol{\sigma} \right] . \tag{7.80}$$

The time derivative of the Φ_B stream is then:

$$\frac{d\Phi_B}{dt} = \lim_{\delta t \to 0} \frac{\delta\Phi_B}{\delta t} = \int_{S_1} \left[\frac{\partial \mathbf{B}}{\partial t} - \text{rot}\,(\mathbf{v} \times \mathbf{B}) \right] \cdot d\boldsymbol{\sigma} . \tag{7.81}$$

Recalling the (7.69) from the previous one we obtain:

$$\frac{d}{dt} \int_S \mathbf{B} \cdot d\boldsymbol{\sigma} = 0 . \tag{7.82}$$

(7.82) expresses Alfven's theorem: the conservation of the induction flux associated with a material line and the solenoid character of $\mathbf{B}$, by virtue of which the flux associated with a flow tube is constant lead to conclude that the lines of force are plasma frozen. The lines of force adhere to the particles of the fluid and in the case of transverse motion are dragged by the fluid.

We can give (7.69) a different form that allows us to derive a conclusion similar to the one we have just found. Let's consider the vector identity:

$$\text{rot}\,(\mathbf{a} \times \mathbf{b}) = \mathbf{a}\,\text{div}\mathbf{b} - \mathbf{b}\,\text{div}\mathbf{a} + (\mathbf{b} \cdot \text{grad}\,)\mathbf{a} - (\mathbf{a} \cdot \text{grad}\,)\mathbf{b} . \tag{7.83}$$

With this relation and with (7.69) we obtain:

$$\frac{d\mathbf{B}}{dt} = (\mathbf{B} \cdot \text{grad}\,)\mathbf{v} - \mathbf{B}\,\text{div}\mathbf{v} . \tag{7.84}$$

Using the continuity equation we obtain from the last relation the:

$$\frac{d}{dt}\left(\frac{\mathbf{B}}{\rho}\right) = \frac{1}{\rho}(\mathbf{B} \cdot \text{grad}\,)\mathbf{v} . \tag{7.85}$$

Let $\delta\mathbf{l}$ be an element of a material line and let $\mathbf{v}$ be the velocity at one extreme. The speed at the other extreme, at the same instant, will be:

$$\mathbf{v}' = \mathbf{v} + (\delta\mathbf{l} \cdot \text{grad}\,)\mathbf{v} . \tag{7.86}$$

In the interval dt the length of the line element will be varied by:

$$d\,\delta\mathbf{l} = (\mathbf{v}' - \mathbf{v})\,dt = (\delta\mathbf{l} \cdot \text{grad}\,)\mathbf{v}\,dt . \tag{7.87}$$

The variation in the unit of time will be:

$$\frac{d\,\delta\mathbf{l}}{dt} = (\delta\mathbf{l} \cdot \text{grad}\,)\mathbf{v} . \tag{7.88}$$

Comparing (7.88) with (7.85) we see that $\delta \mathbf{l}$ and $\mathbf{B}/\rho$ satisfy the same equation. Therefore, if at the initial instant they have the same direction they will always be parallel and their sizes will vary proportionally:

$$\frac{\mathbf{B}}{\rho} \propto \delta \mathbf{l} \,. \tag{7.89}$$

If two particles are on the same line of force they always remain there and their distance varies as $\mathbf{B}/\rho$. Furthermore, if $\rho = $ cost and $\delta \mathbf{l}$ increases, i.e. if the line of force undergoes a stretch, the $\mathbf{B}$ field increases.

These results are valid when the electrical conductivity is infinite, if there is some ohmic dissipation the lines of force will slide slowly with respect to the fluid.

7.7   The Galactic Magnetic Field: Origin and Methods of Measure

The standard model of the Big Bang does non include magnetic fields. They can be generated by plasma processes, but in this case they are of small entity. In fact, if the forces acting on ions and electrons are difference, separation of charges occur which generate an electric field $\mathbf{E}$. If this field is not potential, from the Faraday law a magnetic field $\mathbf{B}$ is generated, which however is very small when compared with observations. We need therefore to look for an amplifying mechanism. We know that a dynamo can play this role. Let's consider a conducting solid disc which rotates around an axis. An initial current can generate a magnetic field $\mathbf{B}$. The electrons in the disc are moving across the field and thus feel the Lorentz force $F = e\mathbf{v} \times \mathbf{B}$. This is perpendicular both to the direction of motions and to the field direction. It will then be directed along the disc radius. If the disc is rotating anti-clockwise and the field $\mathbf{B}$ is directed upwards a current moving towards the disc periphery originates. This current flows into the coil amplifying $\mathbf{B}$ which, in turn, amplifies the current. When considering a magnetohydrodynamical dynamo, electrons move across a fluid, and not a solid dynamo. The freezing of the field force lines implies that the field moves with the fluid and therefore the lines of force are distorted by the motion. If particles to which a line of force is tied move perpendicularly with different velocity the line of force gets stretched and, in turn, the magnetic field $\mathbf{B}$ increases. This way, the particles kinetic energy is transformed into magnetic energy. For this to work, however, we need, first, a pristine magnetic field and, second, the motion should be as described. The evolution of the magnetic field in time is described by Eq. (7.64). For any given velocity $\mathbf{v}$, the solution of the equation would have the form: $\mathbf{B} \sim exp(qt)$. In full generality, the eigenvalues will be complex; the case with $q = 0$ corresponds to a stationary field. The case $q = ip$ with p real describes and oscillating period field. In both case the dynamo is neutral since it does not generate or destroy the field. If the velocity $\mathbf{v}$ is small every solution q has a negative real part, which corresponds to a decaying field. At increasing $\mathbf{v}$, when the neutral value is overpassed, q has a positive real part, and the field would be exponentially increasing. This increase does not continue forever, because of the interaction of the field with the fluid motion. A periodic motion can in general sustain a magnetic field **B**. Example of such motions are the Galactic differential rotation or the convective motion in the external layer of a planet.

There are essentially four methods to obtain information on the interstellar magnetic field:

(1) *Optical polarization of stellar radiation*: the light coming from distant stars can intersect dust particle along the line of sight. It results then polarised. Polarisation is produced by dust grains, and it is proportional to reddening or extinction. As a consequence of collisions among them grains can rotate and the interaction with an existent magnetic field induces their alignment along it. Polarisation can give information on the direction of the magnetic field perpendicular to the line of sight.

(2) *Rotation measures of radio-sources (Faraday rotation)*: radiation polarised in a plane, like the synchrotron one, and which propagates across a magnetic field, undergoes a plane rotation proportional to $\lambda^2 RM$, where λ is the radiation wavelength, while RM is the rotation measure along the line of sight, defined as: $RM = 8.1 \cdot 10^5 \int n_e \mathbf{B} \, d\,l$. Normally one uses the pulsar light. The average value for **B** obtained this way is $\sim 2.2 \cdot 10^{-6}$ Gauss, and the direction of the field is parallel to the Galactic plane.

(3) *Zeeman effect*: a magnetic field splits the 21 cm transition upper level in three sub-levels, with a frequency value of 2.8 **B** Hz, being **B** measured in μG. The value obtained this way amounts to $7.0 \cdot 10^{-5}$ G

(4) *Synchrotron radiation*: electrons moving at relativistic speed can interact with the magnetic field emitting synchrotron radiation. To explain this radiation a field of 10^{-5} G is required adopting the solar vicinity value for the density of relativistic particles.

7.8   Synthetic Summary

This chapter and the following ones concern the study of the dynamics of systems made up of conducting fluids immersed in magnetic fields.

A plasma is an ionized gas with a sufficiently high density so that each particle is simultaneously affected by the action of a very large number of particles, i.e. the collective effects are important; if the density is very low each particle is affected only by the action of a few other particles.

The conditions in which collective effects are important can be quantified by the Debye screen length λ_D: s and $\lambda_D \geq L$, where L is the linear dimension characteristic of the system, the shielding effect does not take place and the system is only an ionized gas but not a plasma; if on the contrary $\lambda_D \leq L$ the collective effects are important and this last condition defines a plasma.

A plasma is macroscopically neutral: a possible separation of charges of opposite sign produced by an external field develops a field that opposes the separation: in this way a recall force develops which produces oscillations with a characteristic frequency (plasma frequency). These fluctuations are also a consequence of collective effects.

It may happen that even a minimal separation of the charges has an effect on the motion of the plasma: when this happens the study of plasmas must be approached either with a model with two conducting fluids (ions and electrons) or through the methods of statistical mechanics. When the separation of the charges does not produce sensible effects, the mechanical behavior of the plasma can be described with a model of a single conducting fluid, to which the equation of the fluids in which the effects of magnetic fields are introduced, which must be completed with Maxwell's equations that describe the electromagnetic field (magnetohydrodynamic treatment). In the case of interest of the interstellar medium, the conditions for applying the magnetohydrodynamic approximation holds.

In the study of interstellar plasmas, some approximations allowed by the particular nature of these systems can be introduced: (a) the velocities of the plasma are much lower than the speed of light, (b) the dimensions of the systems involved are very large, c) the conductivity electricity of interstellar plasmas is very large. Consequently, it can be shown that: (a) $\mathbf{B}$ is invariant, (b) a condition of macroscopic neutrality holds (c) in Maxwell's equation (7.41) the term of the displacement current can be neglected, (d) The electric force can be neglected compared to the magnetic one (the same goes for the corresponding energy densities).

From Maxwell's equation we deduce the equation for the temporal evolution of the magnetic field: in the conditions valid for the interstellar medium, the diffusion term linked to the ohmic dissipation can be neglected and the equation takes the form (7.64). By means of it we can prove Alfven's theorem: the flux of magnetic induction associated with a closed material line remains constant during motion. The fundamental consequence of this theorem is the freezing of the lines of force in the plasma in the sense that the motion of the plasma in a direction perpendicular to the lines of force is not allowed while the motion along the lines of force is allowed (the proof of this property is completed in next chapter).

Appendix

Appendix A: Energy Balance Equation for the Magnetic Field

Multiplying (7.64) scalarly by **B** yields:

$$\mathbf{B} \cdot \frac{\partial \mathbf{B}}{\partial t} = \mathbf{B} \cdot \text{rot} \, (\mathbf{v} \times \mathbf{B}) + \frac{1}{\mu_0 \sigma} \mathbf{B} \cdot \nabla^2 \mathbf{B} \, . \tag{7.90}$$

From the vectorial identities:

$$\text{div}(\mathbf{a} \times \mathbf{b}) = \mathbf{b} \cdot \text{rot} \, \mathbf{a} - \mathbf{a} \cdot \text{rot} \, \mathbf{b} \tag{7.91}$$

$$\text{rot rot} \, \mathbf{a} = \text{grad div} \mathbf{a} - \nabla^2 \mathbf{a} \tag{7.92}$$

we have

$$\begin{aligned} \text{div}(\mathbf{B} \times \text{rot} \, \mathbf{B}) &= (\text{rot} \, \mathbf{B})^2 - \mathbf{B} \cdot \text{rot rot} \, \mathbf{B} \\ &= (\text{rot} \, \mathbf{B})^2 + \mathbf{B} \cdot \nabla^2 \mathbf{B} \end{aligned} \tag{7.93}$$

from which

$$\mathbf{B} \cdot \nabla^2 \mathbf{B} = -(\text{rot} \, \mathbf{B})^2 + \text{div}(\mathbf{B} \times \text{rot} \, \mathbf{B}) \, . \tag{7.94}$$

By introducing (7.94) into (7.90) we get:

$$\frac{1}{2} \frac{\partial B^2}{\partial t} = \mathbf{B} \cdot \text{rot} \, (\mathbf{v} \times \mathbf{B}) + \frac{1}{\mu_0 \sigma} [-(\text{rot} \, \mathbf{B})^2 + \text{div}(\mathbf{B} \times \text{rot} \, \mathbf{B})] \, . \tag{7.95}$$

Considering the identity (7.91) and setting $\mathbf{a} = \mathbf{B}$ and $\mathbf{b} = \mathbf{v} \times \mathbf{B}$ we obtain

$$\text{div}[\mathbf{B} \times (\mathbf{v} \times \mathbf{B})] = (\mathbf{v} \times \mathbf{B}) \cdot \text{rot} \, \mathbf{B} - \mathbf{B} \cdot \text{rot} \, (\mathbf{v} \times \mathbf{B}) \tag{7.96}$$

or

$$\mathbf{B} \cdot \text{rot} \, (\mathbf{v} \times \mathbf{B}) = (\mathbf{v} \times \mathbf{B}) \cdot \text{rot} \, \mathbf{B} - \text{div}[\mathbf{B} \times (\mathbf{v} \times \mathbf{B})] \, . \tag{7.97}$$

Substituting (7.97) into (7.95) you get

$$\frac{1}{2} \frac{\partial B^2}{\partial t} = (\mathbf{v} \times \mathbf{B}) \cdot \text{rot} \, \mathbf{B} - \text{div}[\mathbf{B} \times (\mathbf{v} \times \mathbf{B})] + \frac{1}{\mu_0 \sigma} [-(\text{rot} \, \mathbf{B})^2 - \text{div}(\mathbf{B} \times \text{rot} \, \mathbf{B})] \tag{7.98}$$

or:

$$\frac{1}{2}\frac{\partial B^2}{\partial t} = \mathbf{v} \cdot (\mathbf{B} \times \mathrm{rot}\, \mathbf{B}) - \frac{1}{\mu_0 \sigma}(\mathrm{rot}\, \mathbf{B})^2 - \mathrm{div}[\mathbf{B} \times (\mathbf{v} \times \mathbf{B}) - \frac{1}{\mu_0 \sigma}(\mathbf{B} \times \mathrm{rot}\, \mathbf{B})] \,.$$

$$(7.99)$$

On the other hand, the rot $\mathbf{B}$ appearing in the second term at the second member can be eliminated by means of the Eq. (7.59) and we have:

$$\frac{1}{2}\frac{\partial B^2}{\partial t} = \mathbf{v} \cdot (\mathbf{B} \times \mathrm{rot}\, \mathbf{B}) - \frac{\mu_0}{\sigma}\mathrm{j}^2 - \mathrm{div}[\mathbf{B} \times (\mathbf{v} \times \mathbf{B}) - \frac{1}{\mu_0 \sigma}(\mathbf{B} \times \mathrm{rot}\, \mathbf{B})]$$

$$(7.100)$$

which represents the equation of the magnetic energy balance.

To get an insight on its physical meaning, we integrate the previous relation over a fixed volume V_0:

$$\frac{\mathrm{d}}{\mathrm{d}t}\int_{V_0}\frac{B^2}{2\mu_0}\,\mathrm{d}V = \frac{1}{\mu_0}\int_{V_0}\mathbf{v} \cdot (\mathbf{B} \times \mathrm{rot}\, \mathbf{B})\,\mathrm{d}V - \int_{V_0}\frac{1}{\sigma}\mathrm{j}^2\,\mathrm{d}V$$

$$- \frac{1}{\mu_0}\int_{S_0}[\mathbf{B} \times (\mathbf{v} \times \mathbf{B}) - \frac{1}{\mu_0 \sigma}\mathbf{B} \times \mathrm{rot}\, \mathbf{B}]\cdot \mathrm{d}\sigma$$

$$(7.101)$$

where S_0 is the outline of V_0. The energy density of the electric field $\epsilon_0 E^2/2$ is negligible compared to that of the magnetic field on the basis of the approximations introduced previously. Hence the first member of (7.103) represents, to less than infinitesimals, the variation in the unit of time of the energy of the electromagnetic field within V_0.

The first term in the second member is the work done by the magnetic force in the unit of time. In fact it is:

$$\mathbf{v} \cdot (\mathbf{j} \times \mathbf{B})\,\mathrm{d}V = -\mathbf{v} \cdot \left[\mathbf{B} \times \left(\frac{1}{\mu_0}\mathrm{rot}\, \mathbf{B}\right)\right]\mathrm{d}V\,.$$

$$(7.102)$$

The second term at the second member of (7.103) gives the power dissipated by the Joule effect and the third term the electromagnetic energy flow through S_0. In the absence of dissipative effects ($\sigma \to \infty$) the (7.90) becomes:

$$\frac{1}{2}\frac{\partial B^2}{\partial t} = -\mu_0\mathbf{v} \cdot (\mathbf{j} \times \mathbf{B}) - \mathrm{div}[\mathbf{B} \times (\mathbf{v} \times \mathbf{B})]\,.$$

$$(7.103)$$

The (7.57), in the same hypotheses, gives:

$$\mathbf{v} \times \mathbf{B} = -\mathbf{E}$$

$$(7.104)$$

hence (7.105) becomes:

$$\frac{1}{2}\frac{\partial B^2}{\partial t} = -\mu_0\mathbf{v} \cdot (\mathbf{j} \times \mathbf{B}) - \mathrm{div}(\mathbf{E} \times \mathbf{B})\,.$$

$$(7.105)$$

Recalling that the Poynting vector is defined by

$$\mathbf{P} = c^2 \epsilon_0 (\mathbf{E} \times \mathbf{B}) = \frac{1}{\mu_o} (\mathbf{E} \times \mathbf{B}) \qquad (7.106)$$

we have:

$$\mathrm{div}\mathbf{P} = \frac{1}{\mu_0} \mathrm{div}(\mathbf{E} \times \mathbf{B}) \ . \qquad (7.107)$$

It turns out then:

$$\frac{\partial}{\partial t} \left(\frac{B^2}{2\mu_0} \right) = -\mathbf{v} \cdot (\mathbf{j} \times \mathbf{B}) - \mathrm{div}\mathbf{P} \qquad (7.108)$$

and integrating on the volume V_0 we obtain:

$$\frac{\partial}{\partial t} \int_{V_0} \frac{B^2}{2\mu_0} \, \mathrm{d}V = \int_{V_0} \mathbf{v} \cdot (\mathbf{j} \times \mathbf{B}) \, \mathrm{d}V - \int_{S_0} \mathbf{P} \cdot \mathrm{d}\sigma \qquad (7.109)$$

which clarifies the meaning of the third term in (7.105) in the absence of dissipation.

Chapter 8
Motion of a Plasma in a Magnetic Field

8.1 Expressions of Magnetic Force

The magnetic stress tensor is defined by:

$$\sigma_{ij} = \frac{1}{\mu_0}\left(B_i B_j - \frac{B^2}{2}\delta_{ij}\right).$$ (8.1)

The ith component of the magnetic force acting on the surface unit oriented pendicular to the versor $\hat{\mathbf{n}} = (n_1, n_2, n_3)$ will be:

$$f_i = \sigma_{ij}n_j = \frac{1}{\mu_0}B_i(B_j n_j) - \frac{B^2}{2\mu_0}n_i.$$ (8.2)

We choose the x_3 axis in the direction of $\mathbf{B}$ and then $\mathbf{B} = (0, 0, B)$. We'll have:

$$f_1 = \frac{1}{\mu_0}B_1(B_j n_j) - \frac{B^2}{2\mu_0}n_1 = -\frac{B^2}{2\mu_0}n_1$$ (8.3)

$$f_2 = \frac{1}{\mu_0}B_2(B_j n_j) - \frac{B^2}{2\mu_0}n_2 = -\frac{B^2}{2\mu_0}n_2$$ (8.4)

$$f_3 = \frac{1}{\mu_0}B_3(B_j n_j) - \frac{B^2}{2\mu_0}n_3.$$ (8.5)

If the considered surface is perpendicular to the x_3 axis it will be $n_1 = n_2 = 0, n_3 = 1$
waves

$$f_1 = f_2 = 0 \qquad f_3 = \frac{B^2}{\mu_0} - \frac{B^2}{2\mu_0},$$ (8.6)

if it is perpendicular to the x_1 axis it will be $n_1 = 1, n_2 = n_3 = 0$ and

© The Author(s), under exclusive license to Springer Nature Switzerland AG 2021
G. Carraro, *Astrophysics of the Interstellar Medium*,
UNITEXT for Physics, https://doi.org/10.1007/978-3-030-75293-4_8

$$f_1 = -\frac{B^2}{2\mu_0} \qquad f_2 = f_3 = 0, \tag{8.7}$$

while if it is perpendicular to the x_2 axis you will have $n_1 = n_3 = 0$ and $n_2 = 1$ and it will be:

$$f_1 = 0 \qquad f_2 = -\frac{B^2}{2\mu_0} \qquad f_3 = 0. \tag{8.8}$$

Therefore the magnetic force is equivalent to a hydrostatic pressure

$$p_m = \frac{B^2}{2\mu_0} \tag{8.9}$$

and to a tension

$$\frac{B^2}{\mu_0} \tag{8.10}$$

directed in the direction of the field.

These results can be interpreted differently: parallel to **B** the magnetic force per unit area has component:

$$F_{\parallel} = \frac{B^2}{\mu_0} - \frac{B^2}{2\mu_0} = \frac{B^2}{2\mu_0} \tag{8.11}$$

while the perpendicular component is:

$$F_{\perp} = -\frac{B^2}{2\mu_0}. \tag{8.12}$$

Therefore the magnetic force can also be decomposed into a tension along the direction of the lines of force with intensity $B^2/2\mu_0$ and in a pressure equal to $B^2/2\mu_0$ perpendicular to the lines of force .

The magnetic force per unit of volume is given by:

$$F_i = \frac{\partial \sigma_{ij}}{\partial x_j} = \frac{1}{\mu_0} B_j \frac{\partial B_i}{\partial x_j} + \frac{1}{\mu_0} B_i \, \text{div} \, \mathbf{B} - \frac{\partial}{\partial x_i} \left(\frac{B^2}{2\mu_0} \right) \tag{8.13}$$

or

$$F_i = \frac{1}{\mu_0} B_j \frac{\partial B_i}{\partial x_j} - \frac{\partial}{\partial x_i} \left(\frac{B^2}{2\mu_0} \right) \tag{8.14}$$

and with vector notation:

$$\mathbf{F} = \frac{1}{\mu_0} (\mathbf{B} \cdot \text{grad}) \mathbf{B} - \text{grad} \left(\frac{B^2}{2\mu_0} \right). \tag{8.15}$$

Remembering the vector relation:

$$\text{grad}\,(\mathbf{a} \cdot \mathbf{b}) = (\mathbf{a} \cdot \text{grad})\mathbf{b} + (\mathbf{b} \cdot \text{grad})\mathbf{a} + \mathbf{a} \times \text{rot}\,\mathbf{b} + \mathbf{b} \times \text{rot}\,\mathbf{a} \qquad (8.16)$$

and putting $\mathbf{a} = \mathbf{b} = \mathbf{B}$ in it we obtain:

$$\text{grad}\,(\mathbf{B}^2) = 2(\mathbf{B} \cdot \text{grad})\mathbf{B} + 2\mathbf{B} \times \text{rot}\,\mathbf{B} \qquad (8.17)$$

from which

$$\frac{1}{\mu_0}(\mathbf{B} \cdot \text{grad})\mathbf{B} - \text{grad}\left(\frac{\mathbf{B}^2}{2\mu_0}\right) = -\frac{1}{\mu_0}\mathbf{B} \times \text{rot}\,\mathbf{B}\,. \qquad (8.18)$$

Keeping in mind the (8.15) and the (8.18) we find the:

$$\mathbf{F} = \mathbf{j} \times \mathbf{B}\,. \qquad (8.19)$$

8.2 Fundamental Equations of Magnetohydrodynamics

The equations of motion are obtained by generalizing those of fluid dynamics with the inclusion of magnetic terms.

(A) *Equation of continuity.* It does not require any changes:

$$\frac{\partial\rho}{\partial t} + \text{div}\,(\rho\mathbf{v}) = 0\,. \qquad (8.20)$$

(B) *Equation of momentum.* Obviously it is necessary to include the electromagnetic force, taking into account that in the approximation of the magnetohydrodynamics the force due to the electric field can be neglected with respect to the magnetic force. Therefore:

$$\rho\left[\frac{\partial\mathbf{v}}{\partial t} + (\mathbf{v} \cdot \text{grad})\mathbf{v}\right] = -\,\text{grad}\,p + \eta\nabla^2\mathbf{v} + \mathbf{j} \times \mathbf{B}\,. \qquad (8.21)$$

With the results of the previous paragraph, the magnetic force can be given the form deriving from (8.16) and the equation of motion becomes:

$$\rho\left[\frac{\partial\mathbf{v}}{\partial t} + (\mathbf{v} \cdot \text{grad})\mathbf{v}\right] = -\,\text{grad}\,p - \text{grad}\left(\frac{\mathbf{B}^2}{2\mu_0}\right) + \frac{1}{\mu_0}(\mathbf{B} \cdot \text{grad})\mathbf{B} + \eta\nabla^2\mathbf{v}\,.$$
$$(8.22)$$

As has already been said, the term $\text{grad}\,(\mathbf{B}^2/2\mu_0)$, by analogy with the first term with the second member, can be thought of as a magnetic pressure (in fact its dimensions are those of a pressure).

(C) *Equation of energy conservation.* A first version can be obtained from the first law of thermodynamics taking into account: (a) heat transferred by thermal conduction and (b) heat produced by the Joule effect, i.e.

$$\rho T \frac{ds}{dt} = \text{div} (K \, \text{grad} \, T) + \frac{j^2}{\sigma} . \tag{8.23}$$

As was done in the case of fluid dynamics (see Chap. 3, Sect. 3.7) in the Eq. (8.23), the contributions of the heating processes can be inserted, if necessary and cooling so the equation becomes:

$$\rho T \frac{ds}{dt} = \text{div} (K \, \text{grad} \, T) + \frac{j^2}{\sigma} + (G - L) . \tag{8.24}$$

When the time scale of these processes is smaller than the time scale of thermal conduction and electromagnetic energy dissipation, as well as the dynamic time scale, this equation is reduced to

$$G - L = 0 . \tag{8.25}$$

A second version can be obtained from the generalisation of:

$$\frac{\partial U}{\partial t} = - \text{div} \, \mathbf{g} \tag{8.26}$$

where U represents the energy of the unit of mass and $\mathbf{g}$ is the flux density of the energy. Taking into account that in the MHD approximation the energy density of the electric field is infinitesimal compared to the energy density of the magnetic field, the total energy of the unit of volume is:

$$U = \frac{1}{2} \rho v^2 + \rho \epsilon + \frac{B^2}{2\mu_0} . \tag{8.27}$$

With the procedure described in Appendix A we obtain the analogue of (8.26) where U is given by (8.27) and $\mathbf{g}$ is expressed by:

$$\mathbf{g} = \rho \mathbf{v} \left(\frac{1}{2} v^2 + w \right) - K \, \text{grad} \, T + \frac{1}{\mu_0} \left[\mathbf{B} \times (\mathbf{v} \times \mathbf{B}) - \frac{1}{\mu_0 \sigma} \mathbf{B} \times \text{rot} \, \mathbf{B} \right] . \tag{8.28}$$

The first term contains the flow of mechanical energy transferred with the movement of the fluid as well as the work of the pressure force (see Chap. 1, Sect. 1.5), the second the flow of thermal energy due to conduction and the third represents the flow of electromagnetic energy. Let's check the physical meaning of this last term in the case $\sigma \to \infty$, which therefore holds:

$$\frac{1}{\mu_0}[\mathbf{B} \times (\mathbf{v} \times \mathbf{B})] . \tag{8.29}$$

Again in the hypothesis $\sigma \to \infty$ from Joule's law we obtain:

$$j = -\mathbf{v} \times \mathbf{B} \tag{8.30}$$

and hence:

$$\frac{1}{\mu_0}[\mathbf{B} \times (\mathbf{v} \times \mathbf{B})] = \frac{1}{\mu_0}(\mathbf{E} \times \mathbf{B}) = \varepsilon_0 c^2 \mathbf{E} \times \mathbf{B} = \mathbf{P} \tag{8.31}$$

which is the Poynting vector. By integrating the divergence of this vector on a fixed volume V_0 we have:

$$\frac{1}{\mu_0} \int_{V_0} \text{div} [\mathbf{B} \times (\mathbf{v} \times \mathbf{B})] \, d\,V = \int_{V_0} \text{div} \, \mathbf{P} \, d\,V = \int_{S_0} \mathbf{P} \cdot d\,\boldsymbol{\sigma} \tag{8.32}$$

which represents the flow of electromagnetic energy across V_0.

To the previous equations we need to add:

(a) the equation describing the magnetic field (7.64),
(b) Maxwell's equation div $\mathbf{B} = 0$,
(c) the equation for the current

$$\mathbf{j} = \frac{1}{\mu_0} \text{rot} \, \mathbf{B} \tag{8.33}$$

(d) the equation for the electric field

$$\mathbf{E} = \frac{\mathbf{j}}{\sigma} - \mathbf{v} \times \mathbf{B} . \tag{8.34}$$

The resolution of this system of equations generally presents insurmountable difficulties and the approximate numerical resolution requires complicated programs and very long computation times. Sometimes an MHD problem can be transformed with a suitable procedure into a fluid dynamics problem; an example of this type is presented in Appendix B.

8.3 Motion of a Plasma with Respect to Magnetic Lines of Force

Let us once again place ourselves in the conditions in which the freezing of the magnetic lines of force in the plasma has occurred. Based on vector identity:

$$\mathbf{a} \times (\mathbf{b} \times \mathbf{c}) = (\mathbf{a} \cdot \mathbf{c})\mathbf{b} - (\mathbf{a} \cdot \mathbf{b})\mathbf{c} \tag{8.35}$$

we have:

$$(\mathbf{B} \times \mathbf{v}) \times \mathbf{B} = -\mathbf{B} \times (\mathbf{B} \times \mathbf{v}) = -(\mathbf{B} \cdot \mathbf{v})\mathbf{B} + B^2\mathbf{v} = -B^2\mathbf{v}_{\parallel} + B^2\mathbf{v} = B^2\mathbf{v}_{\perp} \tag{8.36}$$

where $\mathbf{v}_{\perp}$ and $\mathbf{v}_{\parallel}$ are the parallel and normal plasma velocity components to $\mathbf{B}$, respectively.

Let's set $\mathbf{w} = \mathbf{v}_{\perp}$; $\mathbf{w}$ can be interpreted, in the case $\sigma \to \infty$, as the speed of movement of the lines of force, frozen in plasma. We have:

$$\mathbf{w} = \frac{(\mathbf{B} \times \mathbf{v}) \times \mathbf{B}}{B^2} . \tag{8.37}$$

The freezing condition ($\sigma \to \infty$) applied to Ohm's law gives $\mathbf{E} = \mathbf{B} \times \mathbf{v}$ waves:

$$\mathbf{w} = \frac{\mathbf{E} \times \mathbf{B}}{B^2} . \tag{8.38}$$

For the moment, let us drop the hypothesis that $\sigma \to \infty$ and see, in all generality, the motion of the plasma with respect to the lines of force. The equation of motion is:

$$\rho \frac{d\mathbf{v}}{dt} = \mathbf{F} + \mathbf{j} \times \mathbf{B} \tag{8.39}$$

where $\mathbf{F}$ is the resultant of all non-electromagnetic forces. Eliminating $\mathbf{j}$ with Ohm's law we have:

$$\rho \frac{d\mathbf{v}}{dt} = \mathbf{F} + \sigma(\mathbf{E} + \mathbf{v} \times \mathbf{B}) \times \mathbf{B} . \tag{8.40}$$

It is:

$$\sigma(\mathbf{E} + \mathbf{v} \times \mathbf{B}) \times \mathbf{B} = \sigma \mathbf{E} \times \mathbf{B} - \sigma B^2 \mathbf{v}_{\perp} . \tag{8.41}$$

With this result and remembering the definition of $\mathbf{w}$ the equation of motion becomes:

$$\rho \frac{d\mathbf{v}}{dt} = \mathbf{F} - \sigma B^2 (\mathbf{v}_{\perp} - \mathbf{w}) . \tag{8.42}$$

From this relationship we deduce that:

– the motion parallel to $\mathbf{B}$ is determined only by the non-electromagnetic force $\mathbf{F}$;
– the velocity of the plasma motion in a direction perpendicular to $\mathbf{B}$ decays from an arbitrary initial value in a time of the order of $\tau = \rho/\sigma B^2$ to the value $\mathbf{v}_{\perp} = \mathbf{w} + \mathbf{F}_{\perp}/\sigma B^2$, where $\mathbf{F}_{\perp}$ is the $\mathbf{F}$ component normal to $\mathbf{B}$;
– in the limit of infinite conductivity we have $\mathbf{v}_{\perp} \to \mathbf{w}$, i.e. the plasma motion is integral to the lines of force, and in this case $\tau \to 0$;

- the term proportional to B^2 in the equation of motion corresponds to a frictional force that tends to prevent the motion of the plasma perpendicular to the lines of force

$$\mathbf{F}_{mv} = \sigma B^2 (\mathbf{v}_\perp - \mathbf{w}) \tag{8.43}$$

whereby the quantity σB^2 is called magnetic viscosity.

Another example of motion of a plasma that is easily treatable is Hartmann motion (see Appendix C).

8.4   Virial Theorem

In Chap. 2, Sect. 2.5 we derived the virial theorem, in the absence of a magnetic field, starting from Euler's equation. We will now study the effect of the magnetic field starting from the momentum conservation equation (8.22) in which we will also introduce the autogravitational force and neglect the viscous term, i.e.

$$\rho \frac{d\mathbf{v}}{\partial t} = - \operatorname{grad} p - \operatorname{grad} \left(\frac{B^2}{2\mu_0} \right) + \frac{1}{\mu_0} (\mathbf{B} \cdot \operatorname{grad}) \mathbf{B} - \rho \operatorname{grad} \Phi . \tag{8.44}$$

Multiplying scalar by $\mathbf{r}$ and integrating on the volume V we have:

$$\int_V \rho \mathbf{r} \cdot \frac{d\mathbf{v}}{dt} dV = - \int_V \mathbf{r} \cdot \operatorname{grad} \left(p + \frac{B^2}{2\mu_0} \right) dV$$
$$+ \frac{1}{\mu_0} \int_V \mathbf{r} \cdot (\mathbf{B} \cdot \operatorname{grad}) \mathbf{B} \, dV - \int_V \rho \mathbf{r} \cdot \operatorname{grad} \Phi \, dV . \tag{8.45}$$

Without repeating the calculation of the integrals already considered in the Appendix to Chap. 2, we evaluate the integrals that contain the magnetic terms. consider

$$\int_V \mathbf{r} \cdot \operatorname{grad} \left(\frac{B^2}{2\mu_0} \right) dV . \tag{8.46}$$

The method of calculating this integral is the same as that used to evaluate the integral

$$\int_V \mathbf{r} \cdot \operatorname{grad} p \, dV \tag{8.47}$$

for which we obtain:

$$\int_V \mathbf{r} \cdot \operatorname{grad} \left(\frac{B^2}{2\mu_0} \right) dV = \int_S \frac{B^2}{2\mu_0} \mathbf{r} \cdot d\boldsymbol{\sigma} - 3 \int_V \frac{B^2}{2\mu_0} dV \tag{8.48}$$

where S is the outline of V. Let's consider now:

$$\frac{1}{\mu_0} \int_V \mathbf{r} \cdot (\mathbf{B} \cdot \text{grad}) \mathbf{B} \, d V \tag{8.49}$$

where is it

$$
\begin{aligned}
\mathbf{r} \cdot (\mathbf{B} \cdot \text{grad}) \mathbf{B} &= \mathbf{r} \cdot (\mathbf{B} \cdot \text{grad}) \sum_j B_j \mathbf{c}_j \\
&= \sum_j \mathbf{r} \cdot \mathbf{c}_j (\mathbf{B} \cdot \text{grad}) B_j \\
&= \sum_j x_j \, \text{div} \, (B_j \mathbf{B}) \\
&= \sum_j [\, \text{div} \, (x_j B_j \mathbf{B}) - B_j^2] \\
&= \text{div} \, [(\mathbf{r} \cdot \mathbf{B}) \mathbf{B}] - B^2
\end{aligned}
\tag{8.50}
$$

and therefore:

$$\frac{1}{\mu_0} \int_V \mathbf{r} \cdot (\mathbf{B} \cdot \text{grad}) \mathbf{B} \, d V = \frac{1}{\mu_0} \int_S (\mathbf{r} \cdot \mathbf{B}) \mathbf{B} \cdot d\boldsymbol{\sigma} - \frac{1}{\mu_0} \int_V B^2 \, d V . \tag{8.51}$$

In stationary conditions the virial theorem acquires the form:

$$2K + 2U + M + W - \int_S p\mathbf{r} \cdot d\boldsymbol{\sigma} + \frac{1}{\mu_0} \int_S (\mathbf{r} \cdot \mathbf{B}) \mathbf{B} \cdot d\boldsymbol{\sigma} - \int_S \frac{B^2}{2\mu_0} \mathbf{r} \cdot d\boldsymbol{\sigma} = 0 \tag{8.52}$$

where M and W are the magnetic and gravitational energy respectively.

8.5   Application of the Virial Theorem

To apply (8.52) it is necessary to evaluate whether the surface integrals are zero or not. In the case of a uniform magnetic field, the magnetic terms make no overall contribution to the virial theorem; this becomes clear remembering that the virial theorem is deduced starting from the equation of motion, in which, in this case, the magnetic force vanishes. Let us consider the case of a uniform magnetic field in which a spherical gaseous cloud is immersed. We verify that the sum of the contributions of the magnetic terms is zero. Let us then consider the case of a uniform field $\mathbf{B}_0$, in which a spherical cloud of radius R_0 is immersed. We show that:

$$M + \frac{1}{\mu_0} \int_S (\mathbf{r} \cdot \mathbf{B}_0) \mathbf{B}_0 \cdot d\boldsymbol{\sigma} - \int_S \frac{B^2}{2\mu_0} \mathbf{r} \cdot d\boldsymbol{\sigma} = 0 . \tag{8.53}$$

We assume the z axis in the direction of $\mathbf{B}_0$, so $\mathbf{B}_0 = B_0\mathbf{k}$. The first term then turns out to be:

$$M = \int_V \frac{B_0^2}{2\mu_0}\,\mathrm{d}V = \frac{2\pi B_0^2 R_0^4}{3\mu_0 R_0}\,. \tag{8.54}$$

Being

$$\mathbf{r} \cdot \mathbf{B} = r_0\hat{\mathbf{n}} \cdot \mathbf{B}_0 = R_0 B_0 \cos\theta \tag{8.55}$$

$$\mathbf{B}_0 \cdot \mathrm{d}\boldsymbol{\sigma} = B_0 \cos\theta\,\mathrm{d}\sigma \tag{8.56}$$

where $\hat{\mathbf{n}}$ is the vector of R_0 and θ the angle that $\mathbf{R}_0$ forms with the direction of $\mathbf{B}$, the second term is:

$$\frac{1}{\mu_0}\int_S R_0(\hat{\mathbf{n}} \cdot \mathbf{B}_0)^2\,\mathrm{d}\sigma = \frac{B_0^2 R_0}{\mu_0}\int_S \cos^2\theta\,\mathrm{d}\sigma\,. \tag{8.57}$$

We can evaluate $\mathrm{d}\sigma$ as the lateral surface of a cylinder whose base has radius $R_0 \sin\theta$ and height $R_0\,\mathrm{d}\theta$, so:

$$\frac{1}{\mu_0}\int_S (\mathbf{r} \cdot \mathbf{B}_0)\mathbf{B}_0 \cdot \mathrm{d}\boldsymbol{\sigma} = \frac{2B_0^2 R_0}{\mu_0}\int_0^{\pi/2} 2\pi\cos^2\theta R_0^2 \sin\theta\,\mathrm{d}\theta = \frac{4\pi B_0^2 R_0^4}{3\mu_0 R_0}\,. \tag{8.58}$$

In the same way we obtain that the third term is equivalent to:

$$\int_S \frac{B_0^2}{2\mu_0}\mathbf{r} \cdot \mathrm{d}\boldsymbol{\sigma} = \frac{2\pi B_0^2 R_0^4}{\mu_0 R_0}\,. \tag{8.59}$$

You will then ultimately have:

$$M + \frac{1}{\mu_0}\int_S (\mathbf{r} \cdot \mathbf{B}_0)\mathbf{B}_0 \cdot \mathrm{d}\boldsymbol{\sigma} - \int_S \frac{B_0^2}{2\mu_0}\mathbf{r} \cdot \mathrm{d}\boldsymbol{\sigma} = \frac{2\pi B_0^2 R_0^4}{\mu_0 R_0}\left(\frac{1}{3} + \frac{2}{3} - 1\right) = 0\,. \tag{8.60}$$

In a cloud, however, the magnetic field is not uniform resulting from a distortion of the general galactic field generated during the cloud formation process. Let us consider the process of formation of a spherical cloud from the ambient medium of density ρ_0 and pervaded by a uniform magnetic field $\mathbf{B}_0$. The $\mathbf{B}_0$ field is uniform and exerts no force. Under the action of an instability the gas can condense and, due to the freezing of the magnetic lines of force, these are dragged and the field is distorted in a direction perpendicular to $\mathbf{B}_0$: thus a magnetic force $(1/\mu_0)(\mathrm{rot}\,\mathbf{B} \times \mathbf{B})$ is generated which prevents further lateral contractions.

Within the sphere of radius R the density is described approximately by:

$$\rho = \rho_0\left(\frac{R}{r_0}\right)^{-3} \tag{8.61}$$

where R_0 is the starting radius. Outside the sphere the contraction is non-uniform and gives rise to a monotonically decreasing density field; at great distances we have $\rho = \rho_0$.

This density field describes a quasi-uniform spherical cloud superimposed on a uniform background. Admitting freezing of the lines of force, the magnetic field is defined by the conditions:

(a) for $r \leq R$ the lines of the field are almost straight with

$$\mathbf{B} = \mathbf{B}_0 \left(\frac{R_0}{R} \right)^2 \tag{8.62}$$

(b) for $R \leq r \leq R_0$, the lines of force are radial with

$$B = B_0 \left(\frac{R_0}{r} \right)^2 \tag{8.63}$$

(c) for $r \geq R_0$, the field is uniform in intensity

$$\mathbf{B} = \mathbf{B}_0 . \tag{8.64}$$

We apply the virial theorem to the sphere of radius R_0, neglecting the contribution of kinetic energy and thermal energy. The magnetic energy is the sum of the energies of the gas within R and that contained in the spherical shell between the radii R and R_0, i.e. :

$$M = M_1 + M_2 \tag{8.65}$$

with

$$M_1 = \int_V \frac{B^2}{2\mu_0} \, dV = \int_0^R \frac{B_0^2}{2\mu_0} \left(\frac{R_0}{R} \right)^4 4\pi r^2 \, dr = \frac{2\pi B_0^2 R_0^4}{3\mu_0 R} \tag{8.66}$$

is

$$M_2 = \int_{V'} \frac{B^2}{2\mu_0} \, dV = \frac{2\pi B_0^2 R_0^4}{\mu_0} \left(\frac{1}{R} - \frac{1}{R_0} \right) = \frac{2\pi B_0^2 R_0^4}{\mu_0 R} \tag{8.67}$$

where the term $1/R_0$ has been neglected compared to $1/R$. Therefore:

$$M = \frac{8\pi B_0^2 R_0^4}{3\mu_0 R} . \tag{8.68}$$

The surface terms are the same as those calculated in the uniform case:

$$\frac{4\pi B_0^2 R_0^4}{3\mu_0 R_0} - \frac{2\pi B_0^2 R_0^4}{\mu_0 R_0} = \frac{2\pi B_0^2 R_0^4}{3\mu_0 R_0} \tag{8.69}$$

and they are negligible compared to M. We then write the expression of the virial theorem:

$$M + W = -\frac{1}{k}\frac{GM^2}{R} + \frac{8\pi B_0^2 R_0^4}{3\mu_0 R} \tag{8.70}$$

where k is a dimensionless factor, of the order of unity, which depends on the geometry of the cloud and on the details of the distribution of matter: for the simplest model, a uniformly dense spheroid with a uniform internal field, k is 1, while for a very flattened body k amounts to about $1/4$.

The application of the virial theorem allows us to introduce a critical mass M_{cr} defined by the magnetic flux $\Phi_m = \pi R^2 B$ trapped in the cloud:

$$GM_{cr}^2 = k\frac{8\pi B_0^2 R_0^4}{3\mu_0}. \tag{8.71}$$

This critical mass is independent of the radius R (this is because, in the hypothesis of freezing of the magnetic field, both the gravitational and the magnetic energy have the same dependence on R).

For the conservation of the magnetic flux Φ_{cr} we have:

$$GM_{cr}^2 = \frac{8k\Phi_{cr}^2}{3\pi\mu_0}. \tag{8.72}$$

Note that, since the magnetic force is perpendicular to the field, the equilibrium to the gravitational forces by the magnetic field can only be exercised in the directions perpendicular to the field, while in the direction of the field itself the magnetic forces cannot balance the gravitational forces.

Since on the other hand the clouds are in equilibrium in all three dimensions, we must think that in the direction of the field the equilibrium to the gravitational forces is achieved by means of mass motions (described by the generic term of turbulence). This turbulence can be interpreted by means of an Alfvén wave spectrum (see Chap. 9).

In the spirit of the virial theorem, if a cloud has a mass greater than the critical mass M_{cr} the gravitational forces prevail over the magnetic ones and there is a contraction, even if the freezing of the magnetic field continues to apply. This is called the supercritical regime and the cloud is unstable. If, on the other hand, the mass of the cloud is less than M_{cr}, the contraction is prevented by the magnetic force, if the conservation of the magnetic flux is assumed. The eventual evolution of the clouds in such conditions is determined by the ambipolar diffusion and is called subcritical regime.

8.6 Ambipolar Diffusion

In a fully ionised plasma the magnetic field interacts directly with all the particles that are charged and then, under the conditions that we have seen, the lines of force are tied (frozen) to the plasma. If the gas is partially ionised, the magnetic field interacts directly only with the charged component and indirectly with the neutral particles through collisions of these with the charged particles. These collisions allow an exchange of momentum and, if the number of collisions per unit of volume and per unit of time between the particles of the two types is sufficiently high, the field can still freeze with the gas as a whole. This possibility disappears when, due to the low number of collisions between charged and neutral particles, the magnetic field becomes no longer coupled with the fluid. This happens if the ionisation fraction is too low. In this case, while the lines of force remain frozen to the gas of charged particles, there is a drift of the neutral particles with respect to the field and to the charged particles. This phenomenon is called *ambipolar diffusion* and its name derives from a similar phenomenon, observed in laboratory, in which electrons and positive ions, kept close by electrostatic forces, *drift* with respect to the neutral gas. In the interstellar medium the magnetic forces, instead of electrostatic ones, keep ions and electrons together, given the negligible role played by the electric field.

Let's study this process quantitatively. As neutral particles we take atoms or molecules of H with mass m_n; as charged particles we take the nuclei of C^{12} of mass m_i. Collisions between neutral particles and electrons can be neglected as regards the exchange of momentum. We can assume $m_i \gg m_n$. In this case we cannot limit ourselves to treating the fluid as a single continuous system, precisely because we want to highlight the separation of the fluid made up of charged particles from that made up of neutral particles. We will therefore consider two different fluids interacting with each other through collisions and for each of them we will write a different equation of motion. On the other hand, the separation of electrons from ions does not need to be taken into account; in fact, making use of:

$$\text{rot } \mathbf{B} = \mu_0 \mathbf{j} \tag{8.73}$$

The current is produced by the relative motion of the ions with respect to the electrons:

$$\mathbf{j} = n_e e(\mathbf{u}_i - \mathbf{u}_e) \tag{8.74}$$

and therefore the relative velocity required to maintain the magnetic field is:

$$||\mathbf{u}_i - \mathbf{u}_e|| \approx \frac{B}{\mu_0 n_e e L} \tag{8.75}$$

where L is the characteristic dimension. With $B = 10^{-10}$ T $= 10^{-6}$ G e $n_e = 10^{-3}$ cm^{-3} we find $||\mathbf{u}_i - \mathbf{u}_e|| = 10^{-5}$ cm/s, which is small enough to ignore the separation between two types of charged particles. Therefore we will consider a single charged fluid and since the electrons give a negligible contribution to the plasma density we can indicate with m_i the mass of the charged particles and with $\mathbf{u}_i$ their speed.

The number of collisions per second per cm^3 between ions and neutral particles is given by:

$$r = n_n n_i \langle \sigma v \rangle \tag{8.76}$$

where v is the relative velocity, σ is the collision cross section and $\langle \sigma v \rangle$ is the average of σv weighted on the velocity distribution.

For each collision the momentum exchanged is:

$$\frac{m_i m_n}{m_i + m_n}(\mathbf{u}_i - \mathbf{u}_n) \tag{8.77}$$

therefore the momentum exchanged per second and per cm^3 is:

$$\mathbf{f}_c = n_i n_n \langle \sigma v \rangle \frac{m_i m_n}{m_i + m_n}(\mathbf{u}_i - \mathbf{u}_n) \tag{8.78}$$

and it is equivalent to a force per unit of volume (dynamical friction). The amount τ_{in} defined by:

$$\frac{1}{\tau_{in}} = n_i \langle \sigma v \rangle \tag{8.79}$$

is the collision time of a neutral particle in the sea of ions. From the assumption $m_i \gg m_n$ we have:

$$\frac{m_i m_n}{m_i + m_n} \approx \frac{m_n}{1 + \frac{m_n}{m_i}} = m_n \tag{8.80}$$

and then

$$\mathbf{f}_c = \frac{n_n}{\tau_{in}} m_n (\mathbf{u}_i - \mathbf{u}_n) = \frac{\rho_n}{\tau_{in}}(\mathbf{u}_i - \mathbf{u}_n) . \tag{8.81}$$

The equation of motion for the charged fluid is then:

$$\rho_i \left[\frac{\partial \mathbf{u}_i}{\partial t} + (\mathbf{u}_i \cdot \text{grad})\mathbf{u}_i \right] = -\text{grad } p_i - \rho_i \text{ grad } \Phi + \frac{1}{\mu_0}(\text{rot } \mathbf{B} \times \mathbf{B}) + \mathbf{f}_c \tag{8.82}$$

where Φ is the gravitational potential; the equation for the neutral particle fluid is:

$$\rho_n \left[\frac{\partial \mathbf{u}_n}{\partial t} + (\mathbf{u}_n \cdot \text{grad})\mathbf{u}_n \right] = -\text{grad } p_n - \rho_n \text{ grad } \Phi - \mathbf{f}_c . \tag{8.83}$$

In equilibrium conditions the accelerations cancel each other out, moreover in the fluid equation of charged particles, being in the conditions in which the ambipolar diffusion is effective, $\rho_i \ll \rho_n$, the pressure can be neglected and also the gravitational force can be left out. From (8.82) we then obtain the equilibrium condition:

$$\frac{1}{\mu_0}(\operatorname{rot} \mathbf{B}) \times \mathbf{B} = -\mathbf{f}_c \tag{8.84}$$

i.e.:

$$\frac{\rho_n}{\tau_{in}}(\mathbf{u}_i - \mathbf{u}_n) = -\frac{1}{\mu_0}(\operatorname{rot} \mathbf{B}) \times \mathbf{B} \tag{8.85}$$

and from this relation we deduce the ambipolar diffusion rate:

$$\mathbf{v}_{AD} = \mathbf{u}_i - \mathbf{u}_n = -\frac{\tau_{in}}{\rho_n \mu_0}(\operatorname{rot} \mathbf{B}) \times \mathbf{B} . \tag{8.86}$$

The evolution of the magnetic field can be derived by exploiting the hypothesis that it is frozen to ions, hence:

$$\frac{\partial \mathbf{B}}{\partial t} + \operatorname{rot}(\mathbf{B} \times \mathbf{u}_i) = 0 . \tag{8.87}$$

Obtaining $\mathbf{u}_i$ from (8.86) we have:

$$\frac{\partial \mathbf{B}}{\partial t} + \operatorname{rot}(\mathbf{B} \times \mathbf{u}_n) = \operatorname{rot}\left[\frac{\tau_{in}}{\rho_n \mu_0} \mathbf{B} \times (\mathbf{B} \times \operatorname{rot} \mathbf{B})\right] \tag{8.88}$$

and operating on the second member of (8.88) making use of vector identities, we arrive, after long calculations omitted here, to an equation of the type:

$$\frac{\partial \mathbf{B}}{\partial t} + \operatorname{rot}(\mathbf{B} \times \mathbf{u}_n) = \frac{\tau_{in} B^2}{\rho_n \mu_0} \nabla^2 \mathbf{B} + \frac{\tau_{in} B^2}{\rho_n \mu_0} \mathcal{B} \tag{8.89}$$

where is it

$$\mathcal{B} = \mathbf{B}(\operatorname{grad} B)^2 + (\mathbf{B} \cdot \operatorname{grad})\left(\frac{B^2}{2}\right) - (\mathbf{B} \cdot \operatorname{grad})\operatorname{grad}\left(\frac{B^2}{2}\right) - \operatorname{rot}[\mathbf{B} \times [(\mathbf{B} \cdot \operatorname{grad})\mathbf{B}]] \tag{8.90}$$

The (8.89) is a nonlinear diffusion equation. Let's compare it with the equation (7.64) which describes the evolution of the magnetic field. If $\eta_m \to 0$ ($\sigma \to \infty$) we have the freezing condition. Similarly, if the second member of (8.89) is negligible, there is coupling of the neutral particles to the magnetic field. Therefore, for ambipolar diffusion to be negligible it must be:

$$\frac{\tau_{in}}{\rho_n \mu_0} \ll 1 \tag{8.91}$$

i.e. the collision time must be very small and this happens if the density of both types of colliding particles is large enough.

If this relationship is not verified there is a drift of one type of particles with respect to the other type and the field. By comparing (8.89) with the equation (7.64) the ambipolar diffusion coefficient D is given by:

$$D = \frac{\tau_{in} B^2}{\rho_n \mu_0} .$$
(8.92)

The amount

$$v_{Alf} = \sqrt{\frac{B^2}{\rho_n \mu_0}}$$
(8.93)

has the dimensions of a velocity and is called Alfven velocity (see Chap. 9, Sects. 9.1 and 9.2); then D can be expressed by:

$$D = \tau_{in} v_{Alf}^2 .$$
(8.94)

From (8.89) we get the ambipolar diffusion time τ_{AD}:

$$\tau_{AD} = \frac{L^2}{D}$$
(8.95)

where L is a characteristic length. From this, introducing the expression of D we have:

$$\tau_{AD} = \frac{1}{\tau_{in}} \left(\frac{L}{v_{Alf}} \right)^2 .$$
(8.96)

The Alfven speed is the propagation speed of a particular type of magnetohydrodynamic waves (see Chap. 9) and the quantity L/v_{Alf} is the time τ_{Alf} of propagation of such waves in a region of dimension L waves:

$$\tau_{AD} = \frac{\tau_{Alf}^2}{\tau_{in}} .$$
(8.97)

From (8.86), by setting $\| \text{rot } \mathbf{B} \| \approx B/L$ an evaluation of the ambipolar diffusion velocity in km / s is obtained

$$v_{AD} = 0.4 \left[\frac{v_{Alf}}{2 \text{ km/s}} \right]^2 \left[\frac{\tau_{in}}{10^4 \text{ yr}} \right] \left[\frac{0.1 \text{ pc}}{L} \right]$$
(8.98)

and similarly for the diffusion time we have from (8.96):

$$\tau_{AD} = 2 \cdot 10^5 \left[\frac{2 \text{ km/s}}{v_{Alf}} \right]^2 \left[\frac{10^4 \text{ yr}}{\tau_{in}} \right] \left[\frac{L}{0.1 \text{ pc}} \right]^2$$
(8.99)

where τ_{AD} is expressed in years.

Ambipolar diffusion plays an important role in star formation. We saw in the previous paragraph that, in the hypothesis of freezing of the magnetic field, in the subcritical regime there is no gravitational instability. In molecular clouds, and especially in the densest parts, neutral particles are much more numerous than ions; the ionisation fraction n_i/n_n is between 10^{-9} and 10^{-3}. Under such conditions there is not sufficient dynamic coupling between the neutral particles on the one hand and the ions and the magnetic field on the other. In this case, the ions and magnetic field do not follow the motion of the neutral particles.

If ambipolar diffusion is effective we can imagine this scenario: suppose that a perturbation alters the equilibrium and the action of the gravitational field produces a condensation in the neutral material, which, being insensitive to the magnetic field and little affected by collisions with ions, is free to contract. On the other hand, ions are blocked by the magnetic field and do not follow neutral particles.

Therefore in the new condensation the magnetic field is reduced because the lines of force have drifted with respect to the neutral material. Increasing the density of the material further decreases the ionization fraction and the ambipolar diffusion will be further accentuated. In this condensation, equilibrium is first guaranteed by the gaseous pressure and the region occupied by the neutral gas behaves approximately like an isothermal sphere in equilibrium; subsequently to the growth of its mass it becomes unstable and begins the collapse. This appears to be the mechanism for the formation of low-mass single stars, while the supercritical regime would give rise to clusters or associations. In fact, for a density of $n = 10^3 - 10^4$ cm^{-3} and a magnetic field of 30μG the (8.71) gives a critical mass of about $2 \cdot 10^3$ M$_\odot$.

One of the problems that ambipolar diffusion allows to solve is the reduction of the magnetic fields that would occur with the conservation of the magnetic flux. As will be seen below, a dark molecular cloud is not homogeneous and has clumps or cores with dimensions of the order of 0.1 pc and mass of 1 M$_\odot$, for whose magnetic field is of the order of 30μG. If from these densities stars were formed with a frozen magnetic field, there would be enormous magnetic fields for the stars. In fact from:

$$B_0 R_0^2 = B R^2 \tag{8.100}$$

for a star with a radius of 10^{11} cm we would have $B \approx 3 \cdot 10^8$ G. The values of the stellar magnetic fields are 10^4 G or less. Therefore during the star formation process the field must decouple from matter: the mechanism may be ambipolar diffusion.

8.7   Synthetic Summary

This chapter described some properties of the motion of plasma particles in presence of a magnetic field, using the magneto-hydro-dynamical approximation. First, starting from the magnetic force tensor, we have clarify that magnetic force has two component, a pressure (magnetic pressure) perpendicular to the direction of the field, and a tension along the field lines of force. A new set of fluid dynamics equation is derived, which include magnetic terms, together with a new formulation of energy conservation. A new expression is also derived for the virial theorem, and an update instability criterium is obtained, which include the B field. A new critical mass, M_{cr} is introduced which supersedes Jeans mass and which will used later on, when discussing the physics of molecular clouds and star formation. Finally, we saw that ambipolar diffusion can turn relevant in the higher density and low er ionisation ISM components.

Appendices

Appendix A: Energy Conservation

We compute the partial derivative of the energy density with respect to the time of (8.27):

$$\frac{\partial U}{\partial t} = \frac{1}{2}v^2\frac{\partial \rho}{\partial t} + \rho \mathbf{v} \cdot \frac{\partial \mathbf{v}}{\partial t} + \epsilon\frac{\partial \rho}{\partial t} + \rho\frac{\partial \epsilon}{\partial t} + \frac{\partial}{\partial t}\left(\frac{B^2}{2\mu_0}\right). \tag{8.101}$$

so using the continuity equation we have:

$$\frac{\partial U}{\partial t} = -\left(\frac{1}{2}v^2 + \epsilon\right)\text{div}\,(\rho\mathbf{v}) + \rho\mathbf{v} \cdot \frac{\partial \mathbf{v}}{\partial t} + \rho\frac{\partial \epsilon}{\partial t} + \frac{\partial}{\partial t}\left(\frac{B^2}{2\mu_0}\right). \tag{8.102}$$

From the first law of thermodynamics we derive:

$$\rho\frac{\partial \epsilon}{\partial t} = \rho T\frac{\partial s}{\partial t} - \frac{p}{\rho}\text{div}\,(\rho\mathbf{v}) \tag{8.103}$$

and being $\epsilon + (p/\rho) = w$ we have:

$$\frac{\partial U}{\partial t} = -\left(\frac{1}{2}v^2 + w\right)\text{div}\,(\rho\mathbf{v}) + \rho\mathbf{v} \cdot \frac{\partial \mathbf{v}}{\partial t} + \rho T\frac{\partial s}{\partial t} + \frac{\partial}{\partial t}\left(\frac{B^2}{2\mu_0}\right). \tag{8.104}$$

With the equation of motion (8.22), neglecting viscosity, we have:

$$\frac{\partial U}{\partial t} = -\left(\frac{1}{2}v^2 + w\right)\text{div}\,(\rho\mathbf{v}) + \rho\mathbf{v} \cdot \text{grad}\left(\frac{v^2}{2} + w\right)$$
$$- \mathbf{v} \cdot (\mathbf{j} \times \mathbf{B}) + \rho T\frac{\partial s}{\partial t} + \frac{\partial}{\partial t}\left(\frac{B^2}{2\mu_0}\right). \tag{8.105}$$

From $T \, \mathrm{d}s = \mathrm{d}\epsilon + p \, \mathrm{d}V$ and from:

$$\mathrm{d}\, w = \mathrm{d}\epsilon + p \, \mathrm{d}\left(\frac{1}{\rho}\right) + \frac{1}{\rho} \, \mathrm{d}\, p \tag{8.106}$$

we have:

$$\mathrm{grad}\, p = \rho \,\mathrm{grad}\, w - \rho T \,\mathrm{grad}\, s \tag{8.107}$$

which, replaced in the previous one, gives:

$$\frac{\partial U}{\partial t} = -\,\mathrm{div}\left[\rho\mathbf{v}\left(\frac{1}{2}v^2 + w\right)\right] + \rho\mathrm{T}\left(\frac{\partial s}{\partial t} + \mathbf{v}\cdot\mathrm{grad}\, s\right) + \mathbf{v}\cdot(\mathbf{j}\times\mathbf{B}) + \frac{\partial}{\partial t}\left(\frac{\mathrm{B}^2}{2\mu_0}\right).$$
$$\tag{8.108}$$

Using (8.23) to delete $\mathrm{d}s/\mathrm{d}t$, and (7.100) to transform the last term in previous equation, we finally have:

$$\frac{\partial U}{\partial t} = -\,\mathrm{div}\left[\rho\mathbf{v}\left(\frac{1}{2}v^2 + w\right) - \mathrm{K}\,\mathrm{grad}\,\mathrm{T} + \frac{1}{\mu_0}\mathbf{B}\times(\mathbf{v}\times\mathbf{B}) - \frac{1}{\mu_0^2\sigma}\mathbf{B}\times\mathrm{rot}\,\mathbf{B}\right]$$
$$+ \mathbf{v}\cdot(\mathbf{j}\times\mathbf{B}) + \frac{1}{\mu_0}\mathbf{v}\cdot(\mathbf{B}\times\mathrm{rot}\,\mathbf{B}).$$
$$\tag{8.109}$$

     The last Maxwell equation in the MHD approximation shows that the last two terms are equal and opposite. It is thus obtained

$$\frac{\partial U}{\partial t} = -\,\mathrm{div}\,\mathbf{g} \tag{8.110}$$

being:

$$\mathbf{g} = \rho\mathbf{v}\left(\frac{1}{2}v^2 + w\right) - \mathrm{K}\,\mathrm{grad}\,\mathrm{T} + \frac{1}{\mu_0}\left[\mathbf{B}\times(\mathbf{v}\times\mathbf{B}) - \frac{1}{\mu_0\sigma}\mathbf{B}\times\mathrm{rot}\,\mathbf{B}\right].$$
$$\tag{8.111}$$

Appendix B: Parallel Stationary Motion

For a conductive, incompressible and non-viscous fluid the equation of motion is (8.22). Suppose that the electrical conductivity has a finite and variable value from point to point, therefore the equation of the magnetic field to be used will be the (7.84). Based on identity:

$$\mathrm{rot}\,(\mathbf{a}\times\mathbf{b}) = \mathbf{a}\,\mathrm{div}\,\mathbf{b} - \mathbf{b}\,\mathrm{div}\,\mathbf{a} + (\mathbf{b}\cdot\mathrm{grad}\,)\mathbf{a} - (\mathbf{a}\cdot\mathrm{grad}\,)\mathbf{b} \tag{8.112}$$

yes you have

$$\text{rot}\,(\mathbf{v} \times \mathbf{B}) = \mathbf{v}\,\text{div}\,\mathbf{B} - \mathbf{B}\,\text{div}\,\mathbf{v} + (\mathbf{B} \cdot \text{grad}\,)\mathbf{v} - (\mathbf{v} \cdot \text{grad}\,)\mathbf{B}\,. \tag{8.113}$$

But since the fluid is incompressible, $\text{div}\,\mathbf{B} = 0$ and $\text{div}\,\mathbf{v} = 0$, therefore the equation of the magnetic field becomes:

$$\frac{\partial \mathbf{B}}{\partial t} = (\mathbf{B} \cdot \text{grad}\,)\mathbf{v} - (\mathbf{v} \cdot \text{grad}\,)\mathbf{B} - \text{rot}\,(\eta_m\,\text{rot}\,\mathbf{B})\,. \tag{8.114}$$

Let's now consider a motion that has these requirements:

(a) is stationary, i.e. $\partial/\partial t = 0$;
(b) $\mathbf{v}$ and $\mathbf{B}$ are parallel, so $\mathbf{B} = \lambda\mathbf{v}$.

We will show that in such conditions the equations can formally be reduced to those of hydrodynamics in the absence of the magnetic field. In the hypotheses made, the following applies:

$$\text{div}\,\mathbf{B} = \text{div}\,(\lambda\mathbf{v}) = \lambda\,\text{div}\,\mathbf{v} + \mathbf{v} \cdot \text{grad}\,\lambda = 0 \tag{8.115}$$

from which it follows:

$$\mathbf{v} \cdot \text{grad}\,\lambda = 0\,. \tag{8.116}$$

This relationship implies that, along a flow line, λ does not change value, although from one flow line to another its value may be different.

We introduce in the equation of the magnetic field the hypothesis of stationarity of motion and the expression of $\mathbf{B}$ in terms of λ, i.e.

$$\lambda(\mathbf{v} \cdot \text{grad}\,)(\mathbf{v}) - (\mathbf{v} \cdot \text{grad}\,)(\lambda\mathbf{v}) - \text{rot}\,[\eta_m\,\text{rot}\,(\lambda\mathbf{v})] = 0 \tag{8.117}$$

so if we consider

$$(\mathbf{v} \cdot \text{grad}\,)(\lambda\mathbf{v}) = \mathbf{v}(\mathbf{v} \cdot \text{grad}\,\lambda) + \lambda(\mathbf{v} \cdot \text{grad}\,)\mathbf{v} \tag{8.118}$$

the equation is reduced to:

$$\text{rot}\,[\eta_m\,\text{rot}\,(\lambda\mathbf{v})] = 0\,. \tag{8.119}$$

With the same positions (stationarity and $\mathbf{B}$ parallel to $\mathbf{v}$) the equation of motion becomes:

$$\rho(\mathbf{v} \cdot \text{grad}\,)\mathbf{v} = -\,\text{grad}\,\left(p + \frac{\lambda^2 v^2}{2\mu_0}\right) + \frac{\lambda}{\mu_0}(\mathbf{v} \cdot \text{grad}\,)(\lambda\mathbf{v}) \tag{8.120}$$

but for what has already been found we have:

$$\left(\rho\frac{\lambda^2}{\mu_0}\right)(\mathbf{v}\cdot\text{grad})\mathbf{v} = -\text{grad}\left(p+\frac{\lambda^2 v^2}{2\mu_0}\right).\qquad(8.121)$$

We introduce a fictitious density:

$$\rho^* = \rho - \frac{\lambda^2}{\mu_0}\qquad(8.122)$$

and a pressure which is the sum of the gaseous and magnetic pressure:

$$p^* = p + \frac{\lambda^2 v^2}{2\mu_0}.\qquad(8.123)$$

The equations become:

$$\text{div }\mathbf{v} = 0\qquad(8.124)$$

$$\rho^*(\mathbf{v}\cdot\text{grad})\mathbf{v} = -\text{grad }p^*\qquad(8.125)$$

which are the stationary form of the fluid dynamics equations for an incompressible fluid. It also turns out:

$$\mathbf{v}\cdot\text{grad }\rho^* = \mathbf{v}\cdot\text{grad }\rho\,\mathbf{v}\cdot\text{grad}\left(\frac{\lambda^2}{\mu_0}\right) = -\frac{2\lambda}{\mu_0}\mathbf{v}\cdot\text{grad }\lambda = 0.\qquad(8.126)$$

In the stationary case the term $\mathbf{v}\cdot\text{grad }\rho$ is null according to the continuity equation. From this we conclude that ρ^* is constant along a flow line. Once the value of λ has been specified on each flow line and appropriate boundary conditions are imposed, the equations of motion are solved.

For the analogy to be complete it is necessary in particular that:

$$\text{rot}\,(\eta_\text{m}\,\text{rot }\mathbf{B}) = 0\qquad(8.127)$$

is identically satisfied. This condition occurs in two cases:

(a) the fluid is a perfect conductor, that is $\eta_\text{m} = 0$, for which we are brought back to the solution of the equation of motion for an incompressible fluid:

$$\rho^*(\mathbf{v}\cdot\text{grad})\mathbf{v} = -\text{grad }p^*\qquad(8.128)$$

(b) η_m and λ are constants and we have

$$\mathbf{v} = \text{grad }\Phi\qquad(8.129)$$

where Φ is a harmonic function (the motion is called potential in this case as the speed can be derived from the gradient of a scalar function), so that the equation of the magnetic field is identically satisfied being:

$$\text{rot}\,(\lambda \mathbf{v}) = \lambda\,\text{rot}\,\mathbf{v} + \mathbf{v}\,\text{grad}\,\lambda \qquad (8.130)$$

$$\mathbf{v} = \lambda\,\text{rot}\,\mathbf{v} = \lambda\,\text{rot}\,\text{grad}\,\phi = 0 \qquad (8.131)$$

therefore any potential motion of an incompressible fluid in the absence of a magnetic field represents a solution of the equations of a parallel stationary magnetohydrodynamic motion for a fluid with uniform electrical conductivity.

Appendix C: Hartmann Motion

This example serves to illustrate the competition between freezing of force lines, diffusion, and the behavior imposed by boundary conditions.

Let us consider an incompressible, viscous and conductive fluid with electrical conductivity σ immersed in a uniform magnetic field $\mathbf{B}_0$ oriented like the z axis. The fluid moves in the direction of the x axis and is bounded by two infinite non-conductive surfaces a $z = 0$ and $z = a$. The surfaces are moving in the direction of the x axis, with velocity v_1 and v_2 respectively.

We will look for a stationary solution that represents a motion in the direction of the x axis, in which the different quantities depend only on z. In the case where $\sigma \to \infty$, $\mathbf{E}$ will have a component along the y axis: let's assume it is the only one. It must be constant, so $\mathbf{E} = E_0\mathbf{j}$. Since the moving fluid tends to drag lines of force with it, we predict that there is a $B_x(z)$ component in the direction of the x axis, so:

$$\mathbf{B} = B_x(z)\hat{\mathbf{i}} + B_0\hat{\mathbf{k}} \qquad (8.132)$$

and therefore the continuity equation is reduced to:

$$\text{div}\,\mathbf{v} = 0 . \qquad (8.133)$$

This is satisfied identically by a direct velocity along the x axis which depends only on z, as it is in the present case. Given the geometry of the problem it results:

$$(\mathbf{v} \cdot \text{grad}\,)\mathbf{v} = \left[v_x(z)\hat{\imath} \cdot \left(\hat{\imath}\frac{\partial}{\partial x} + \hat{\jmath}\frac{\partial}{\partial y} + \hat{k}\frac{\partial}{\partial z} \right) \right] v_x(z)\hat{\imath} = v_x(z)\hat{\imath}\frac{\partial v_x}{\partial x} = 0 \quad (8.134)$$

and the equation of motion is:

$$\text{grad}\,p = \mathbf{j} \times \mathbf{B} + \eta\nabla^2\mathbf{v} . \qquad (8.135)$$

The only component of $\mathbf{j}$ other than 0 is the one along the y axis. From Ohm's law $j_y(z)$ holds:

$$j_y(z) = \sigma[E_0 - B_0 v(z)] \tag{8.136}$$

where $v(z)$ indicates the only non-zero component of $\mathbf{v}$, that is, the one along the x axis. Projecting the Eq. (8.135) onto the axes, we obtain the system:

$$\frac{\partial p}{\partial x} = \sigma B_0 (E_0 - B_0 v) + \eta \frac{\partial^2 v}{\partial z^2} \tag{8.137}$$

$$\frac{\partial p}{\partial y} = 0 \tag{8.138}$$

$$\frac{\partial p}{\partial z} = \sigma B_x (E_0 - B_0 v) \tag{8.139}$$

The magnetic force along the z axis is balanced by the pressure gradient. Assuming that there is no pressure gradient along the x axis from the (8.137) we have:

$$\frac{\partial^2 v}{\partial z^2} - \left(\frac{M}{a}\right)^2 v = -\left(\frac{M}{a}\right)^2 \frac{E_0}{B_0} \tag{8.140}$$

where is it

$$M = \left(\frac{\sigma B_0^2 a^2}{\eta}\right)^{\frac{1}{2}} \tag{8.141}$$

it is called Hartmann's number. The square M^2 measures the ratio between the order of magnitude of the magnetic friction force and that of the dynamic friction force:

$$\frac{F_{vm}}{F_{vd}} = \frac{\sigma B^2 v}{\eta \nabla^2 v} = \frac{\sigma B^2 a^2}{\eta} = M^2 . \tag{8.142}$$

The solution of (8.141) subject to the boundary conditions $v(0) = v_1$ and $v(a) = v_2$ is:

$$v(z) = \frac{v_1}{\sinh M} \sinh\left(M\frac{az}{a}\right) + \frac{v_2}{\sinh M} \sinh\left(\frac{Mz}{a}\right)$$
$$+ \frac{E_0}{B_0}\left[1 - \frac{\sinh(M\frac{az}{a}) + \sinh\frac{Mz}{a}}{\sinh M}\right] . \tag{8.143}$$

In the limiting case $B_0 \to 0$, $M \to 0$ we get:

$$v(z) = v_1 + \frac{z}{A}(v_2 - v_1) \tag{8.144}$$

(see Chap. 3, Sect. 3.8).

Chapter 9
Magnetohydrodynamic Waves

9.1 Classification of MHD Waves

In fluid dynamics, the only wave motions of small amplitude are plane longitudinal waves (sound waves) which propagate with speed:

$$c = \left(\frac{dp}{d\rho}\right)_{ad} = \left(\frac{\gamma p}{\rho}\right)^{\frac{1}{2}} . \tag{9.1}$$

In magnetohydrodynamics, different types of waves are possible. One of these is associated with the transverse motion of the magnetic induction lines; the magnetic tension tends to bring them back into rectilinear form, however causing transverse oscillations. This type of magnetohydrodynamic wave, called Alfven wave, is the only one possible in an incompressible fluid. If we assume infinite electrical conductivity, the magnetic lines of force are frozen to the particles of the fluid. Therefore we can think that the force tubes are associated with a mass having a linear density equal to the density of the fluid. Furthermore, it has been seen that the magnetic force can be decomposed into a hydrostatic pressure $B^2/2\mu_0$ and a tension B^2/μ_0 along the lines of force. The pressure is balanced by the pressure of the surrounding regions and what remains is the tension along the induction lines which acts to bring them back to the equilibrium position through a series of oscillations.

Therefore, if the fluid is incompressible, the force tubes behave like vibrating material cords with linear density and subject to a tension B^2/μ_0 and by this analogy we can expect that when the fluid is perturbed the force tubes experience transverse oscillations, the phase velocity being given by

$$v = \sqrt{\frac{B^2}{\mu_0 \rho}} . \tag{9.2}$$

In a compressible fluid, in addition to the Alfven waves, just mentioned, other waves can be excited. If the speed of the particles and the direction of propagation are

© The Author(s), under exclusive license to Springer Nature Switzerland AG 2021   179
G. Carraro, *Astrophysics of the Interstellar Medium*,
UNITEXT for Physics, https://doi.org/10.1007/978-3-030-75293-4_9

both parallel to **B** ordinary sound waves can be excited because the motion parallel to the lines of force is not affected by the magnetic field. The propagation speed is the speed of sound c.

If, on the other hand, the speed of the particles is parallel to the direction of propagation of the waves (longitudinal waves) but both are perpendicular to **B**, another type of compression wave can be excited in which the hydrostatic part of the magnetic pressure is added to the fluid ordinary pressure. Assuming also in this case an adiabatic relationship between pressure and density, the speed of the waves will be:

$$v^2 = c^2 + \frac{B^2}{2\mu_0\rho} \, . \tag{9.3}$$

All of these waves are called magnetohydrodynamic waves (MHD). In general, their behavior is very complicated and they can be analyzed in transverse and longitudinal waves only in particular cases.

A classification of waves can be based on recall forces; this is not an exhaustive classification but is limited to the cases that interest us. The main recall forces are:

- gas pressure,
- gravity force,
- magnetic force.

If the acting recall force is only one we have *pure modes*.

(A) Acoustic waves. The recall force is the gaseous pressure and, being this isotropic, the oscillation is longitudinal. The dispersion relationship is:

$$k^2 = \frac{\omega^2}{v_s^2} \, . \tag{9.4}$$

(B) Gravity waves. The recall force is Archimedes' buoyant force. Waves of gravity and convection are to be considered as the stable manifestation and the unstable manifestation of the same modes. Their propagation is isotropic.

(C) Alfven waves. The recall force is the magnetic tension (this can be seen by deriving the wave equation with the condition of constant density as the magnetic pressure is eliminated). There is a perfect analogy with the excited waves in a spring under tension. The waves are transverse and propagate along the magnetic field with speed:

$$v_{\text{Alf}} = \frac{B_0}{\mu_0\rho_0} \, . \tag{9.5}$$

The density perturbation is zero and the dispersion relation is:

$$\omega^2 = \frac{(kB_0)^2}{\mu_0\rho_0} \, . \tag{9.6}$$

If there are two or three recalling forces we have *mixed modes*; in this case the oscillations have a complicated dependence on the period and the wavelength and, in general, a classification in longitudinal and transverse waves is not possible. In the case of two recall forces we have:

(A) Heavy-acoustic waves: combined action of pressure and gravity. The acoustic modes and the modes of gravity are distinct. In the acoustic modes we have the waves influenced by gravity that appear in the derivation of the Jeans criterion: the phase velocity is a function of the wavelength and at high frequencies they tend to the purely acoustic limit.

(B) Magneto-hydrodynamic waves: combined action of pressure and magnetic force. There are three flavours :

   – pure Alfven mode,
   – MHD slow mode,
   – MHD fast mode.

   and in the weak field limit ($\beta = c/v_A \gg 1$) or strong field ($\beta \ll 1$) one mode between the slow and fast modes is dominated by pressure as a recall force and the other by force magnetic (including magnetic pressure).

(C) Heavy-magnetic waves: combined action of the force of gravity and magnetic force.

9.2 Alfven Waves

Let us consider the case of the Alfven waves. We set up an incompressible, non-viscous fluid, with infinite electrical conductivity, uniform and initially at rest. It is pervaded by a uniform magnetic field $\mathbf{B}_0$. The equations of continuity, motion and field are automatically satisfied as each of the terms appearing there is null. Suppose the system undergoes a small perturbation that sets it in motion with $\mathbf{v}$ velocity. Let also be:

$$p' = p_0 + p \qquad \mathbf{B}' = \mathbf{B}_0 + \mathbf{b} \tag{9.7}$$

while it will be $\rho' = \rho_0$ for the incompressibility hypothesis; the perturbations $p, \mathbf{v}, \mathbf{b}$ are small so that terms higher than the first can be neglected. The perturbed system satisfies the equations:

$$\rho_0 \left[\frac{\partial \mathbf{v}}{\partial t} + (\mathbf{v} \cdot \text{grad})\mathbf{v} \right] = -\text{grad} \left[p' + \frac{B'^2}{2\mu_0} \right] + \frac{1}{\mu_0}(\mathbf{B}' \cdot \text{grad})\mathbf{B}' \tag{9.8}$$

$$\frac{\partial \mathbf{B}'}{\partial t} = \text{rot}(\mathbf{v} \times \mathbf{B}') \tag{9.9}$$

$$\text{div } \mathbf{v} = 0 \tag{9.10}$$

$$\text{div } \mathbf{B}' = 0 .\tag{9.11}$$

Taking into account the conditions of the unperturbed system and linearizing the equations we obtain:

$$\rho_0 \frac{\partial \mathbf{v}}{\partial t} = -\text{grad} \left(p + \frac{\mathbf{B}_0 \cdot \mathbf{b}}{\mu_0} \right) + \frac{1}{\mu_0} (\mathbf{B}_0 \cdot \text{grad}) \mathbf{b}\tag{9.12}$$

$$\frac{\partial \mathbf{b}}{\partial t} = \text{rot}(\mathbf{v} \times \mathbf{B}_0)\tag{9.13}$$

$$\text{div } \mathbf{b} = 0 .\tag{9.14}$$

We take the z axis parallel to $\mathbf{B}_0$, so:

$$\mathbf{B}_0 = B_0 \mathbf{k}\tag{9.15}$$

and therefore we have:

$$\frac{1}{\mu_0} (\mathbf{B}_0 \cdot \text{grad}) \mathbf{b} = \frac{B_0}{\mu_0} \frac{\partial \mathbf{b}}{\partial z} .\tag{9.16}$$

Equation (9.15) can then be written as:

$$\rho_0 \frac{\partial \mathbf{v}}{\partial t} = -\text{grad} \left(p + \frac{\mathbf{B}_0 \cdot \mathbf{b}}{\mu_0} \right) + \frac{B_0}{\mu_0} \frac{\partial \mathbf{b}}{\partial z} .\tag{9.17}$$

Taking the divergence of this equation we obtain:

$$\rho_0 \frac{\partial}{\partial t} (\text{div } \mathbf{v}) = -\text{div grad} \left(p + \frac{\mathbf{B}_0 \cdot \mathbf{b}}{\mu_0} \right) + \frac{B_0}{\mu_0} \frac{\partial}{\partial z} (\text{div } \mathbf{b}) .\tag{9.18}$$

but, being div $\mathbf{v} = 0$ and div $\mathbf{b} = 0$, we have:

$$\nabla^2 \left(p + \frac{\mathbf{B}_0 \cdot \mathbf{b}}{\mu_0} \right) = 0 .\tag{9.19}$$

The solution of this equation, finite everywhere and without singularity, is:

$$p + \frac{\mathbf{B}_0 \cdot \mathbf{b}}{\mu_0} = \text{cost.}\tag{9.20}$$

Introducing this result in (9.17) yields:

$$\rho_0 \frac{\partial \mathbf{v}}{\partial t} = \frac{B_0}{\mu_0} \frac{\partial \mathbf{b}}{\partial z} .\tag{9.21}$$

Let us now consider the equation of the magnetic field (9.13). It is:

$$\mathbf{v} \times \mathbf{B}_0 = v_y B_0 \hat{\mathbf{i}} - v_x B_0 \hat{\mathbf{j}} \tag{9.22}$$

In all its full splendor, the rot looks like this:

$$
\begin{aligned}
\mathrm{rot}(\mathbf{v} \times \mathbf{B}_0) &= \hat{\mathbf{i}} \frac{\partial}{\partial z}(v_x B_0)) + \hat{\mathbf{j}} \frac{\partial}{\partial z}(v_y B_0) + \hat{\mathbf{k}} \left[-\frac{\partial}{\partial x}(v_x B_0) - \frac{\partial}{\partial y}(v_y B_0) \right] \\
&= B_0 \left[\frac{\partial}{\partial z}(v_x \hat{\mathbf{i}}) + \frac{\partial}{\partial z}(v_y \hat{\mathbf{j}}) + \frac{\partial}{\partial z}(v_z \hat{\mathbf{k}}) \right] - B_0 \left[\frac{\partial v_x}{\partial x} + \frac{\partial v_y}{\partial y} + \frac{\partial v_z}{\partial z} \right] \\
&= B_0 \frac{\partial \mathbf{v}}{\partial z} - \mathbf{B}_0 \mathrm{div}\, \mathbf{v}
\end{aligned}
\tag{9.23}
$$

and therefore we have:

$$\frac{\partial \mathbf{b}}{\partial t} = B_0 \frac{\partial \mathbf{v}}{\partial z} . \tag{9.24}$$

Equations (9.21) and (9.24) can then be combined together to finally obtain the equations which $\mathbf{v}$ and $\mathbf{b}$ has to satisfy:

$$\frac{\partial^2 \mathbf{b}}{\partial z^2} - \frac{1}{v_{\mathrm{Alf}}^2} \frac{\partial^2 \mathbf{b}}{\partial t^2} = 0 \tag{9.25}$$

$$\frac{\partial^2 \mathbf{v}}{\partial z^2} - \frac{1}{v_{\mathrm{Alf}}^2} \frac{\partial^2 \mathbf{v}}{\partial t^2} = 0 \tag{9.26}$$

with

$$v_{\mathrm{Alf}}^2 = \frac{B_0^2}{\mu_0 \rho} . \tag{9.27}$$

9.3 MHD Waves: General Discussion

Suppose that the conditions set out in the previous paragraph are verified, except for the incompressibility; therefore now we have:

$$\mathrm{div}\, \mathbf{v} \neq 0 . \tag{9.28}$$

Initially both $\mathbf{B}_0 = \mathrm{cost.}$, $P_0 = \mathrm{cost.}$, $\rho_0 = \mathrm{cost}$ and $\mathbf{b} = 0$. The perturbed and linearized equations are:

$$\frac{\partial \rho}{\partial t} + \rho_0 \mathrm{div}\, \mathbf{v} = 0 \tag{9.29}$$

$$\frac{\partial \mathbf{v}}{\partial t} = -\frac{c^2}{\rho_0}\text{grad}\rho - \frac{1}{\mu_0\rho_0}(\mathbf{B}_0 \times \text{rotb}) \qquad (9.30)$$

$$\frac{\partial \mathbf{b}}{\partial t} = \text{rot}(\mathbf{v} \times \mathbf{B}_0) \qquad (9.31)$$

where we made use of the adiabatic relation $p = c^2\rho$ to eliminate p: this is justified by the fact that any cause of dissipation is absent and therefore the motion occurs with $s = \text{cost.}$. Of the system of these three equations we look for solutions such as:

$$\mathbf{v} = \mathbf{V}\exp[i(\mathbf{k} \cdot \mathbf{r} - \omega t)] \qquad (9.32)$$

$$\rho = R\exp[i(\mathbf{k} \cdot \mathbf{r} - \omega t)] \qquad (9.33)$$

$$\mathbf{b} = \mathbf{B}\exp[i(\mathbf{k} \cdot \mathbf{r} - \omega t)] \ . \qquad (9.34)$$

We introduce these expressions in (9.29); being:

$$\frac{\partial \rho}{\partial t} = -i\omega\rho \ , \qquad (9.35)$$

and therefore div $\mathbf{v}$ becomes:

$$
\begin{aligned}
\text{div } \mathbf{v} &= \mathbf{V}\text{grad}[\exp[i(\mathbf{k} \cdot \mathbf{r} - \omega t)]] + \exp[i(\mathbf{k} \cdot \mathbf{r} - \omega t)]\text{div } \mathbf{V} \\
&= \mathbf{V}\exp[i(\mathbf{k} \cdot \mathbf{r} - \omega t)]i\,\text{grad}(\mathbf{k} \cdot \mathbf{r}) \\
&= \mathbf{v} \cdot \mathbf{k} \ .
\end{aligned}
\qquad (9.36)
$$

This way equation (9.29) becomes:

$$\omega\rho = \rho_0\mathbf{k} \cdot \mathbf{v} \ . \qquad (9.37)$$

By operating analogously on (9.30) we obtain:

$$\frac{\partial \mathbf{v}}{\partial t} - i\omega\mathbf{v} \qquad (9.38)$$

$$\text{grad}\rho = i\rho\mathbf{k} \ . \qquad (9.39)$$

Remembering the identity

$$\text{rot}(\psi\mathbf{a}) = \text{grad}\psi \times \mathbf{a} + \psi\text{rot}a \qquad (9.40)$$

with ψ scalar function, we obtain:

$$\text{rotb} = \text{grad}[i\mathbf{k} \cdot \mathbf{r}] \times \mathbf{b} = i\mathbf{k} \cdot \mathbf{b} \qquad (9.41)$$

$$\mathbf{B}_0 \times \mathrm{rot}\mathbf{b} = i\mathbf{B}_0 \times (\mathbf{k} \times \mathbf{b}) \tag{9.42}$$

and ultimately:

$$\omega\mathbf{v} = -\frac{c^2}{\rho_0}\rho\mathbf{k} - \frac{1}{\mu_0\rho_0}\mathbf{B}_0 \times (\mathbf{k} \times \mathbf{b}) \ . \tag{9.43}$$

From the equation of the magnetic field Eq. 9.31:

$$\frac{\partial\mathbf{b}}{\partial t} = -i\omega\mathbf{b} \tag{9.44}$$

we also have:

$$\mathrm{rot}(\mathbf{v} \times \mathbf{B}_0) = \mathrm{grad}\,\exp[i\mathbf{k} \cdot \mathbf{r}] \times (\mathbf{V} \times \mathbf{B}_0) = i\mathbf{k} \times (\mathbf{v} \times \mathbf{B}_0) \ . \tag{9.45}$$

In this way the equation of the magnetic field Eq. 9.31 becomes:

$$-\omega\mathbf{b} = \mathbf{k} \times (\mathbf{v} \times \mathbf{B}_0) \ . \tag{9.46}$$

Taking ρ from Eq. 9.37 and replacing in Eq. 9.43 gives:

$$-\omega\mathbf{v} + \frac{c^2}{\omega}(\mathbf{k} \cdot \mathbf{v})\mathbf{k} = -\frac{1}{\mu_0\rho_0}\mathbf{B}_0 \times (\mathbf{k} \times \mathbf{b}) \ . \tag{9.47}$$

The Eqs. 9.47 and 9.46 constitute the system of equations that generally describe MHD waves.

We adopt a Cartesian reference system with the axis x in the direction of the propagation direction identified by the wave vector $\mathbf{k}$, and with the plane xy coinciding with the plane of the vectors $\mathbf{k}$ and $\mathbf{B}_0$. Therefore the components of $\mathbf{B}_0$ and of $\mathbf{k}$ are respectively $(\mathbf{B}_{0x}, \mathbf{B}_{0y}, 0)$ and $(\mathbf{k}, 0, 0)$. It is:

$$\mathbf{v} \times \mathbf{B}_0 = -v_z B_{0y}\mathbf{c}_1 + v_z B_{0x}\mathbf{c}_2 + (v_x B_{0y} - v_y B_{0x})\mathbf{c}_3 \tag{9.48}$$

$$\mathbf{k} \times (\mathbf{v} \times \mathbf{B}_0) = -k(v_x B_{0y} - v_y B_{0x})\mathbf{c}_2 + kv_z B_{0x}\mathbf{c}_3 \ . \tag{9.49}$$

We project Eq. 9.46 onto the axes: we have

$$-\omega b_x = \mathbf{k} \times (\mathbf{v} \times \mathbf{B}_0) \cdot \mathbf{c}_1 = 0 \tag{9.50}$$

i.e. $b_x = 0$,

$$ub_y = v_x B_{0y} - v_y B_{0x} \tag{9.51}$$

where $u = \omega/k$,

$$ub_z = -v_z B_{0x} \ . \tag{9.52}$$

As for Eq. 9.47 we have:

$$(\mathbf{k} \cdot \mathbf{v})\mathbf{k} = k^2 v_x \mathbf{c}_1 \tag{9.53}$$

$$\mathbf{k} \times \mathbf{b} = -kb_z \mathbf{c}_2 + kb_y \mathbf{c}_3 \tag{9.54}$$

$$\mathbf{B}_0 \times (\mathbf{k} \times \mathbf{b}) = kb_y B_{0y} \mathbf{c}_1 - kb_y B_{0x} \mathbf{c}_2 - kb_z B_{0x} \mathbf{c}_3 . \tag{9.55}$$

The projections of this equation onto the axes are:

$$v_x \left(u - \frac{c^2}{u} \right) = \frac{B_{0y}}{\mu_0 \rho_0} b_y \tag{9.56}$$

$$uv_y = -\frac{B_{0x}}{\mu_0 \rho_0} b_y \tag{9.57}$$

$$uv_z = -\frac{B_{0x}}{\mu_0 \rho_0} b_z . \tag{9.58}$$

In summary, six scalar equations (9.50), (9.51), (9.52), (9.56), (9.57) and (9.58) have been obtained for six unknowns: b_x , b_y, b_z, v_x, v_y, and v_z.

9.4   MHD Waves with Generic Propagation Direction

According to Eq. 9.50, the perturbed magnetic field $\mathbf{b}$ has a zero component along the x axis and is therefore perpendicular to the direction of propagation. The other equations can be grouped into two groups, the first consisting of

$$ub_z = -B_{0x} v_z \tag{9.59}$$

$$uv_z = -\frac{B_{0x}}{\mu_0 \rho_0} b_z \tag{9.60}$$

and the second consisting of

$$ub_y = B_{0y} v_x - B_{0x} v_y \tag{9.61}$$

$$uv_y = -\frac{B_{0x}}{\mu_0 \rho_0} b_y \tag{9.62}$$

$$v_x \left(u - \frac{c^2}{u} \right) = \frac{B_{0y}}{\mu_0 \rho_0} b_y . \tag{9.63}$$

Since the variables are separate, the two systems are independent and the perturbations described by the first group are independent from those described by the
second group. Recalling that we have chosen the reference system so that $\mathbf{k} = k\mathbf{c}_1$
and keeping in mind the Eq. 9.37 we have:

$$\rho = \frac{\rho_0}{\omega}(\mathbf{k} \cdot \mathbf{v}) = \frac{\rho_0}{u}v_x \tag{9.64}$$

that is, the density perturbation is associated with the wave described by the second
group of equations, while the one described by the first group does not correspond
to any variation in density: the fluid in this case behaves as if it were incompressible.

Let us consider the equations of the first group. The compatibility condition,
expressed by the cancellation of the determinant of the coefficients, allows us to
determine the phase velocity that results:

$$u_1 = \frac{B_{0x}}{\sqrt{\mu_0 \rho_0}} \, . \tag{9.65}$$

In this type of wave, the b_z component oscillates simultaneously with v_z:

$$v_z = -\frac{u_1}{B_{0x}} b_z \, . \tag{9.66}$$

From Eq. 9.65 we get:

$$\omega = \frac{1}{\sqrt{\mu_0 \rho_0}}(\mathbf{B}_0 \cdot \mathbf{k}) \tag{9.67}$$

said dispersion relation. This relationship depends on the direction of the wave vector
with respect to $\mathbf{B}_0$. The physical velocity of propagation is the group velocity:

$$\frac{\partial \omega}{\partial \mathbf{k}} = \frac{\mathbf{B}_0}{\sqrt{\mu_0 \rho_0}} \tag{9.68}$$

which does not depend on the direction of $\mathbf{k}$. This is the propagation speed of the
wave. The direction of the wave, conceived as the direction of the group velocity,
coincides with the direction of $\mathbf{B}_0$. This wave therefore has the characteristics of an
Alfvén wave: it is transverse, it propagates with velocity $\mathbf{v}_A$ and does not involve
variations in the density of the fluid.

Let us now consider the second group of equations. The compatibility of the
system involves the cancellation of the determinant of the system, which leads to the
following equation:

$$u^4 - c^2 u^2 - \frac{B_0^2}{\mu_0 \rho_0} u^2 + \frac{B_{0x}^2}{\mu_0 \rho_0} c^2 = 0 \, . \tag{9.69}$$

Since $B_{0x} = B_0 \cos\theta = B_0 n$ this can be written in the form:

$$u^4 - (v_A^2 + c^2)u^2 + v_A c^2 n^2 = 0 \tag{9.70}$$

with $v_A^2 = B_0^2/(\mu_0 \rho_0)$. The solutions of this equation are:

$$u^2 = \frac{1}{2}\left[(v_A^2 + c^2) \pm \sqrt{(v_A^2 + c^2)^2 - 4v_A^2 c^2 n^2}\right]. \tag{9.71}$$

The solution corresponding to the negative sign describes a wave which is called slow wave, the one corresponding to the positive sign is called fast wave. In both, the oscillating quantities are b_y, v_x and v_y as well as ρ; the vectors $\mathbf{v}$ and $\mathbf{b}$, being $b_z = v_z = 0$, are contained in the plane of the vectors $\mathbf{B_0}$ and $\mathbf{k}$.

If the direction of propagation is not parallel or perpendicular to $\mathbf{B_0}$ ($\theta \neq 0$ and $\theta \neq \pi/2$) the waves do not separate into longitudinal and transverse waves (obviously this concerns the waves corresponding to the second group, because the wave with $u = u_1$ is always transversal).

The limiting cases of the weak magnetic field and the strong magnetic field are dealt with in the appendix.

9.5   Waves with Particular Direction of Propagation

A) The direction of propagation of the wave, identified by the wave vector $\mathbf{k}$ is parallel to $\mathbf{B_0}$ ($\theta = 0$). It will then be $\mathbf{k} = (k, 0, 0)$ and $\mathbf{B_0} = (B_0, 0, 0)$. The equations of the first group remain unchanged, those of the second group become:

$$u b_y = -v_y B_0 \tag{9.72}$$

$$u v_y = -\frac{B_0}{\mu_0 \rho_0} b_y \tag{9.73}$$

$$v_x \left(u - \frac{c^2}{u}\right) = 0. \tag{9.74}$$

The cancellation of the determinant of the system leads to the equation:

$$\left(u - \frac{c^2}{u}\right)\left(\frac{B_0^2}{\mu_0 \rho_0} - u^2\right) = 0 \tag{9.75}$$

which has the solutions $u = c$ and $u = v_A$. In correspondence to the eigenvalue v_A the system that defines the eigenvectors becomes

$$v_A = b_y - B_0 v_y \tag{9.76}$$

$$v_A b_y = -\frac{B_0}{\mu_0 \rho_0} b_y \tag{9.77}$$

$$v_x \left(v_A - \frac{c^2}{v_A} \right) = 0 . \tag{9.78}$$

It describes a transverse wave ($v_x = 0$) which does not produce density perturbations (since $\rho = \rho_0/\omega \mathbf{k} \cdot \mathbf{v} = 0$), and which propagates with v_A speed: it is therefore an Alfvén wave.

Corresponding to the eigenvalue c the system becomes:

$$cb_y = -v_y B_0 \tag{9.79}$$

$$cv_y = -\frac{B_0}{\mu_0 \rho_0} b_y \tag{9.80}$$

$$v_x \left(c - \frac{c^2}{c} \right) = 0 . \tag{9.81}$$

This system has solution $v_x \neq 0$, $b_y = v_y = 0$ and describes a longitudinal wave that propagates, with speed equal to that of sound, in the direction of the field $\mathbf{B}_0$ and to it is not associated with any perturbation of the magnetic field, while according to Eq. 9.37 there is a perturbation of the density.

B) The propagation direction is perpendicular to the field $\mathbf{B}_0$ ($\theta = \pi/2$). In this case it will be $\mathbf{k} = (k, 0, 0)$ and $\mathbf{B}_0 = (0, B_0, 0)$. While the first system presents only the identically zero solution, the equations of the second group become:

$$ub_y = v_x B_0 \tag{9.82}$$

$$uv_y = 0 \tag{9.83}$$

$$v_x \left(u - \frac{c^2}{u} \right) = \frac{B_0}{\mu_0 \rho_0} b_y . \tag{9.84}$$

By canceling the determinant of the system we obtain the equation:

$$u \left(-u^2 + x^2 + \frac{B_0^2}{\mu_0 \rho_0} \right) = 0 \tag{9.85}$$

which has the solutions:

$$u = 0 \qquad u = \sqrt{c^2 + V_A^2} . \tag{9.86}$$

The modes corresponding to the second system are now reduced to only one, whose propagation speed is $u = \sqrt{c^2 + V_A^2}$.

Fig. 9.1 Example of a
magneto-sonic wave

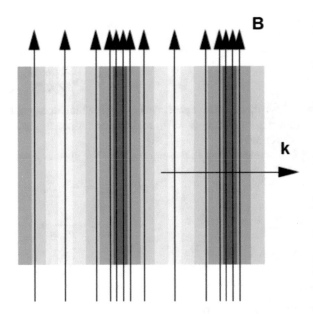

Since $v_y = 0$ (and $v_x \neq 0$, $b_y \neq 0$) the magnetosonic wave is longitudinal with a
direction of propagation perpendicular to $\mathbf{B}_0$ and the perturbed magnetic field obeys:

$$\mathbf{b} = \frac{v}{u}\mathbf{B}_0 = \frac{kv}{\omega}\mathbf{B}_0 . \tag{9.87}$$

The trend of the magnetic induction lines is represented in Fig. 9.1: these waves produce compressions and rarefactions of the field lines without changing their direction
(the only non-zero component of $\mathbf{b}$ is the parallel one a $\mathbf{B}_0$). For comparison, Fig. 9.2
shows the situation of Alfven waves, in which the induction lines oscillate laterally.

9.6   Alfven Waves Attenuation

The considerations made up to this point ignore the dissipative effects that are present
if the electrical conductivity of the medium is not infinite, or if the viscosity is not
negligible. Let us limit ourselves for simplicity to the case of Alfven waves for an
incompressible completely ionised plasma. Let's compare the relative weight of the
two dissipative terms:

$$\frac{\nu\nabla^2\mathbf{v}}{\eta_m\nabla^2\mathbf{B}} = \frac{v}{B}\mu_0\sigma\nu . \tag{9.88}$$

In the case of conductive liquids such as mercury and in the case of stellar gas it
results $\mu_0\sigma\nu \ll 1$, and ohmic dissipation is then prevalent, while in the case of the
interstellar medium we have $\mu_0\sigma\nu \gg 1$ and in this case the effect of viscosity is

Fig. 9.2 Example of an Alfven magnetic wave

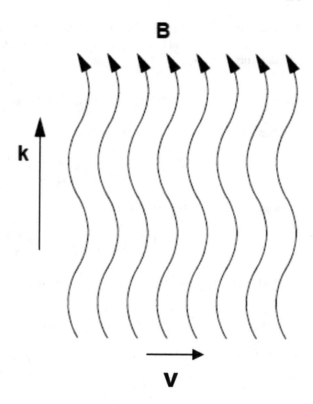

more important. Therefore in developing the method of small perturbations in the equation of conservation of momentum the term of viscosity should be included:

$$\rho \left[\frac{\partial \mathbf{v}}{\partial t} + (\mathbf{v} \cdot \text{grad})\mathbf{v} \right] = -\text{grad}\left(p + \frac{B^2}{2\mu_0} \right) + \frac{1}{\mu_0}(\mathbf{B} \cdot \text{grad})\mathbf{B} + \eta \nabla^2 \mathbf{v} \quad (9.89)$$

where η is the dynamic viscosity coefficient. In the equation of the magnetic field, however, the diffusion term is ignored:

$$\frac{\partial \mathbf{B}}{\partial t} = \text{rot}(\mathbf{v} \times \mathbf{B}) \quad (9.90)$$

where η_m is the magnetic diffusivity. Repeating the derivation of section Eq. 9.3 we arrive at the equations:

$$\left(\frac{\partial}{\partial t} - \nu \nabla^2 \right) \mathbf{v} = \frac{B_0}{\mu_0 \rho_0} \frac{\partial \mathbf{b}}{\partial z} \quad (9.91)$$

$$\frac{\partial \mathbf{b}}{\partial t} = B_0 \frac{\partial \mathbf{v}}{\partial z} . \quad (9.92)$$

The unperturbed magnetic field $\mathbf{B}_0$ is assumed to be parallel to the z axis.

To obtain an equation in a single unknown, we derive Eq. 9.91 with respect to z and multiply it by B_0

$$B_0 \frac{\partial^2 \mathbf{v}}{\partial t \partial z} = \nu B_0 \frac{\partial}{\partial z} \nabla^2 \mathbf{v} + v_A^2 \frac{\partial^2 \mathbf{b}}{\partial z^2} \tag{9.93}$$

we derive Eq. 9.92 with respect to t and obtain

$$\frac{\partial^2 \mathbf{b}}{\partial t^2} = B_0 \frac{\partial^2 \mathbf{b}}{\partial t \partial z} \tag{9.94}$$

we replace the Eq. 9.94 in the Eq. 9.93

$$\frac{\partial^2 \mathbf{b}}{\partial t^2} = \nu B_0 \frac{\partial}{\partial z} \nabla^2 \mathbf{v} + v_A^2 \frac{\partial^2 \mathbf{b}}{\partial z^2} \tag{9.95}$$

we apply the ∇^2 operator to Eq. 9.92 and multiply it by ν

$$\nu \nabla^2 \frac{\partial \mathbf{b}}{\partial t} = \nu B_0 \frac{\partial}{\partial z} \nabla^2 \mathbf{b} \tag{9.96}$$

and eliminating $\nabla^2 \mathbf{v}$ with this in Eq. 9.95 we finally obtain

$$\frac{\partial^2 \mathbf{b}}{\partial t^2} - \nu \frac{\partial}{\partial t} \nabla^2 \mathbf{b} = v_A^2 \frac{\partial^2 \mathbf{b}}{\partial z^2} \,. \tag{9.97}$$

Similarly, a similar equation is obtained for $\mathbf{v}$.

We are looking for a solution of this equation that represents a plane wave propagating along the z axis

$$\mathbf{b} = \mathbf{B} \exp[i(kz - \omega t)] \tag{9.98}$$

and replacing in Eq. 9.97 gives the condition

$$\omega^2 + i\nu k^2 \omega - k^2 v_A^2 = 0 \,. \tag{9.99}$$

The term $\nu k^2 \omega$ is small compared to $k^2 v_A^2$ and therefore we can set $k^2 v_A^2 = \omega_0^2$ which would be the value of of ω^2 in the absence of dissipation. The solution of the equation is:

$$\omega = \frac{1}{2}(-\nu k^2 i \pm \sqrt{-\nu^2 k^4 + 4\omega_0^2}) \,. \tag{9.100}$$

The first term within the square root can be neglected and therefore we obtain:

$$\omega = -\frac{1}{2}\nu k^2 i \pm \omega_0 \,. \tag{9.101}$$

The time dependence of the solution is then:

$$\exp[-i\omega t] = \exp[\pm i\omega_0 t]\exp[-\omega t] \qquad (9.102)$$

where the attenuation coefficient ω is given by:

$$\omega = \frac{\nu k^2}{2} = \frac{\nu\omega_0^2}{2v_A^2} . \qquad (9.103)$$

The attenuation therefore increases rapidly with frequency while decreasing as the intensity of the magnetic field increases. The attenuation time defined as the time required for the intensity to decrease by a factor of $1/and$ of the initial value is:

$$\tau = \frac{1}{\theta} = \frac{2v_A^2}{\nu\omega_0^2} . \qquad (9.104)$$

In the case of a *gas that is only partially ionized* the attenuation mechanisms due to viscosity and finite electrical conduction are negligible compared to other causes. Among these we have:

(i) *Friction between ions and neutral particles.* If in the period of a wave the collisions between ions and neutral particles are rare, the wave sets in motion only the ionized gas and the occasional collisions of an ion with a neutral atom act as a source of friction and damping of the wave. Waves that do not undergo attenuation are therefore those for which the period of oscillation is less than the characteristic time for the exchange of momentum between the two types of particles.

(ii) *Transformation of MHD waves into shock waves due to profile steepening.* This phenomenon occurs because the speed of advancement of the wave varies, which changes the profile until discontinuities are created. As the wave profile becomes steep, dissipation increases and the wave rapidly decays.

(iii) *Coupling of waves to acoustic modes.* Decay of an Alfven wave into another Alfven wave of larger wavelength and a magneto-acoustic wave, the latter dissipating energy. This situation results in a cascade of wave energy down to the smallest wave numbers.

Calculations show that the damping times of magneto-acoustic waves are much shorter than those of Alfven waves, which are of the order of 10^6 years and therefore the expected wave field in molecular clouds is essentially a wave field of Alfven.

9.7 Excitation of MHD Waves

In general, any motion of the gas except that which takes place parallel to the magnetic field generates MHD waves. A source produces a mixture of modes and a range of directions of propagation. The main mechanisms of production of MHD waves are:

(i) Collapse of a rotating clump connected to the rest of the molecular cloud by the magnetic field. The rotation produces torsional magnetic motions as the collapsing fragment loses angular momentum. The torsion of the magnetic field generates waves.

(ii) Expansion of HII regions (or similarly solar wind bubbles). Let us consider the expansion phase in which the HII region expands with the velocity a $v \approx v_s$ (speed of sound) in a uniform magnetic field. At this stage the field can be compressed until the magnetic energy density is large enough to balance the gas pressure in the HII region. Due to the inertia, the expansion continues until the magnetic pressure cancels the inertia. But the excess magnetic pressure produces a contraction. There is therefore an oscillation mechanism that generates magneto-acoustic waves and the energy comes from the ionizing flux.

(iii) Collisions between clouds or clumps in which the associated magnetic fields intervene. This mechanism allows the transfer of the orbital motion of clouds into chaotic internal motions. Due to the interaction and entanglement of the lines of force, large perturbations of the magnetic field are induced at the periphery of the cloud, which excite a system of MHD waves trapped within the cloud. The calculations of Falgarone and Puget (1986) find that the internal velocity dispersion is comparable with that observed.

(iv) Large-scale non-radial oscillations of clouds, resulting from the star formation process.

9.8 Magnetohydrodynamic Shock Waves

The magnetic field affects the motion of an ionised gas significantly, and therefore the properties of shock waves will be altered as well when considering the magnetic field. In this case, in formulating the conditions at the discontinuity surface, with respect to the fluid dynamic case, the pressure and magnetic energy must be taken into account.

We assume that:

(a) shock waves are flat, which is reasonable in most astrophysical situations because any curvatures are very small and the plane wave approximation is sufficient,
(b) shock waves are stationary,
(c) the magnetic field is perpendicular to the fluid velocity $\mathbf{v}$ and therefore is parallel to the discontinuity,
(d) the electrical conductivity is infinite, and, finally,
(e) radiation processes and heat conduction can be neglected.

Let us take a reference system in which the impact front is at rest and the x axis is normal to the discontinuity plane. In the case of a normal wave the components v_y and v_z are zero and for purely hydrodynamic waves the boundary (jump) conditions are:

$$[\rho v] = 0 \qquad (9.105)$$

$$[p + \rho v^2] = 0 \tag{9.106}$$

$$\left[\rho v \left(\frac{1}{2} v^2 + w \right) \right] = 0 \tag{9.107}$$

where the notation $[z]$ means the difference $z_2 - z_1$.

Let's see how these conditions are modified by the presence of a magnetic field parallel to the discontinuity surface.

(a) mass flow conservation: remains unchanged and can be written in the usual form:

$$\rho_1 v_1 = \rho_2 v_2 . \tag{9.108}$$

(b) conservation of the flow of the momentum: in it appears the hydrostatic pressure, in the MHD case we will also have to introduce the magnetic pressure $p_m = B^2/2\mu_0$, hence the condition (9.107) should be replaced with the

$$p_1 + \rho_1 v_1^2 + \frac{B_1^2}{2\mu_0} = p_2 + \rho_2 v_2^2 + \frac{B_2^2}{2\mu_0} . \tag{9.109}$$

(c) conservation of the energy flow: the condition (9.107) is obtained from the general conservation equation (1.56) which is of the form:

$$\frac{\partial U}{\partial t} + \operatorname{div} \mathbf{g} = 0 . \tag{9.110}$$

In our case we will use (8.26) which is formally equal to (1.56) where U and $\mathbf{g}$ are given by (8.26) and (8.27). For the assumptions made it results:

$$\mathbf{g} = \rho \mathbf{v} \left(\frac{1}{2} v^2 + w \right) + \frac{1}{\mu_0} \mathbf{B} \times (\mathbf{v} \times \mathbf{B}) \tag{9.111}$$

therefore the condition Eq. 9.107 must be replaced with

$$\left[\rho \mathbf{v} \left(\frac{1}{2} v^2 + w \right) + \frac{1}{\mu_0} \mathbf{B} \times (\mathbf{v} \times \mathbf{B}) \right] \cdot \hat{\mathbf{i}} = 0 \tag{9.112}$$

where $\hat{\mathbf{i}}$ is the vector of the normal to the plane of the discontinuity. We have seen that $[\rho v] = 0$ so the previous relation can be written in the form:

$$\left[\frac{1}{2} v^2 + w + \frac{1}{\rho v} \frac{1}{\mu_0} \mathbf{B} \times (\mathbf{v} \times \mathbf{B}) \cdot \hat{\mathbf{i}} \right] = 0 . \tag{9.113}$$

Keeping vector identity in mind

$$\mathbf{a} \times (\mathbf{b} \times \mathbf{c}) = \mathbf{b} \cdot (\mathbf{a} \cdot \mathbf{c}) - \mathbf{c} \cdot (\mathbf{a} \cdot \mathbf{b}) \tag{9.114}$$

we have:

$$\mathbf{B} \times (\mathbf{v} \times \mathbf{B}) \cdot \hat{\mathbf{i}} = [\mathbf{v} \cdot (\mathbf{B} \cdot \mathbf{B}) - \mathbf{B} \cdot (\mathbf{v} \cdot \mathbf{B})] \cdot \hat{\mathbf{i}} = \mathbf{v} \cdot B^2 \hat{\mathbf{i}} = vB^2 \tag{9.115}$$

being $\mathbf{v} \cdot \mathbf{B} = 0$ by hypothesis. The Eq. 9.113 therefore becomes:

$$\left[\frac{1}{2} v^2 + w + \frac{B^2}{\mu_0 \rho} \right] = 0 \tag{9.116}$$

and remembering the (5.26) we have:

$$\frac{1}{2} v_1^2 + \frac{\gamma}{\gamma - 1} \frac{p_1}{\rho_1} + \frac{B_1^2}{\mu_0 \rho_1} = \frac{1}{2} v_2^2 + \frac{\gamma}{\gamma - 1} \frac{p_2}{\rho_2} + \frac{B_2^2}{\mu_0 \rho_2} \,. \tag{9.117}$$

These conditions are not sufficient to determine the quantities ρ, $\mathbf{v}$, p and $\mathbf{B}$ behind the shock as a function of the same quantities before the shock. A further condition can be deduced from the equation of the magnetic field which in the hypotheses made (stationary motion, infinite conductivity) is reduced to:

$$\mathrm{rot}(\mathbf{v} \times \mathbf{B}) = 0 \,. \tag{9.118}$$

Since $\mathbf{B}$ is parallel to the impact plane we can impose a further condition on the reference system by choosing the y axis parallel to $\mathbf{B}$. It will then be $\mathbf{B} = (0, B, 0)$ and $\mathbf{v} = (v, 0, 0)$, therefore:

$$\mathbf{v} \times \mathbf{B} = vB\mathbf{k} \tag{9.119}$$

is

$$\mathrm{rot}(\mathbf{v} \times \mathbf{B}) = \mathrm{rot}(vB\mathbf{k}) = \mathrm{grad}(vB) \times \mathbf{k} \,. \tag{9.120}$$

Using the vectorial identity:

$$\mathrm{rot}(f\mathbf{u}) = (\mathrm{grad} f) \times \mathbf{u} + f \mathrm{rot}\mathbf{u} \,, \tag{9.121}$$

we obtain

$$\mathrm{rot}(\mathbf{v} \times \mathbf{B}) = [\mathrm{grad}(vB)] \times \mathbf{k} = \frac{\partial(vB)}{\partial y} \hat{\mathbf{i}} - \frac{\partial(vB)}{\partial x} \hat{\mathbf{j}} = 0 \,. \tag{9.122}$$

From

$$\frac{\partial}{\partial x}(vB) = 0 \tag{9.123}$$

integrating between a point 1 and a point 2 on the opposite sides to the plane of the discontinuity we have:

$$v_1 B_1 = v_2 B_2 \tag{9.124}$$

which represents the further condition required to determine the state of the gas on one side of the shock as a function of the state on the other side. From Eqs. 9.108 and 9.124 we get:

$$\frac{B_1}{\rho_1} = \frac{B_2}{\rho_2}. \tag{9.125}$$

This condition, which has already been obtained in Sect. 7.6, expresses the freezing of the lines of force of the magnetic field, as it is obvious since this result was obtained from the equation of the magnetic field assuming negligible ohmic dissipation, as was also supposed in the present case.

Using a rather long proof given in Appendix B it can be shown that:

(1) The condition for the occurrence of an MHD shock wave is given by:

$$v_1^2 > v_{A1}^2 + c_1^2 \tag{9.126}$$

where v_{A1} is the Alfvén velocity in front of the shock front. The critical speed is not in this case the speed of sound, and therefore it is convenient to introduce the magnetic Mach number defined by:

$$\text{Ma}_1^* = \frac{v_1}{\sqrt{v_{A1}^2 + c_1^2}}. \tag{9.127}$$

(2) The presence of the transverse magnetic field reduces the compression. The ratio ρ_2/ρ_1 in the presence of the magnetic field is smaller than that in its absence. The increase in pressure produced by the passage of the shock wave is also smaller in the presence of a transverse magnetic field.

(3) Let us consider now the case of strong shock waves: we have that

  (a) the density is the same as it would be in the absence of the magnetic field; it turns out in fact

$$\frac{\rho_2}{\rho_1} \to \frac{\gamma+1}{\gamma-1} \tag{9.128}$$

  (b) for the pressure we have

$$\frac{p_2}{p_1} \to \infty \tag{9.129}$$

  (c) for temperature, from the equation of state we have

$$\frac{T_2}{T_1} = \frac{V_2}{V_1}\frac{p_2}{p_1} = \frac{\rho_1}{\rho_2}\frac{p_2}{p_1} \to \infty \tag{9.130}$$

  (d) thermal energy grows without limits, being $E_{th} \sim T$;
  (e) for magnetic energy results

$$\left(\frac{B_2^2}{2\mu_0}\right) \Big/ \left(\frac{B_1^2}{2\mu_0}\right) \left(\frac{B_2}{B_1}\right)^2 = X^2 = \left(\frac{\gamma+1}{\gamma-1}\right)^2 \tag{9.131}$$

therefore in strong shocks the magnetic pressure can be neglected and the connection conditions are reduced to normal fluid dynamics.

We evaluate how much kinetic energy is dissipated. From Eq. 9.117 we obtain in the case $\gamma = 5/3$

$$\Delta E_c = \frac{1}{2}v_1^2 - \frac{1}{2}v_2^2 = \frac{5}{2}\left(\frac{p_2}{\rho_2} - \frac{p_1}{\rho_1}\right) + \left(\frac{B_2^2}{\mu_0\rho_2} - \frac{B_1^2}{\mu_0\rho_1}\right) \tag{9.132}$$

so in the case of strong waves, i.e. when

$$p_2 \gg p_1 \qquad V_1 = 4V_2 \tag{9.133}$$

we have that:

$$\Delta E_c = \frac{5}{2\rho_1}\left(\frac{p_2}{4} - p_1\right) + \frac{3B_1^2}{\mu_0\rho_1} \tag{9.134}$$

that is, the presence of the magnetic field slightly modifies the dissipation of the kinetic energy.

The case of a weak shock wave is different, where the initial kinetic energy is comparable with the magnetic energy. In this case the dissipation of kinetic energy is considerably reduced by the presence of the magnetic field and much of the kinetic energy is instead transformed into magnetic energy.

If the cooling time is small compared to the shock advance time, cooling processes become important. The conditions Eqs. 9.108, 9.109 and 9.125 remain valid, while the condition on energy is replaced by the condition

$$G - L = 0 \tag{9.135}$$

which determines the temperature (if there are variations in the chemical composition, two of these equations will be needed to determine the temperature T_1 and the temperature T_2).

The case of normal plane shock waves with a magnetic field parallel to the gas velocity is straightforward to solve because in this case the magnetic forces have no influence on the motion of the fluid perpendicular to the discontinuity.

9.9   Synthetic Summary

Small perturbations in a charged fluid can generate waves. In general, very different and complex waves can arise, depending on the orientation between the direction of motion of the fluid particles, the direction of the perturbation, and the direction of the magnetic field. We have treated with particular emphasis some particular directions of the propagation wave. Magneto-sonic waves are those which propagated in a direction perpendicular to the magnetic field, and in that case the propagation speed is a combination of both the sound speed and the Alfven speed. When the propagation is parallel to the magnetic field B, Alfven waves arise, namely waves that propagate with the Alfven velocity. We finally consider the effect of the magnetic field when shock waves are generated, namely when the fluid moves with supersonic velocity. The field B can interact with the shock altering its properties only when it is perpendicular to the shock direction, and parallel to the front shock.

Appendices

Appendix A: Waves in Weak and Strong Magnetic Fields

Case with weak magnetic field. Assuming that $v_A \ll c$, the characteristic equation (9.70) has solutions:

$$u^2 = \frac{1}{2} \left[(v_A^2 + c^2) \pm \sqrt{(v_A^2 + c^2)^2 - 4V_A^2 c^2 n^2} \right] \tag{9.136}$$

$$= \frac{1}{2} \left[(v_A^2 + c^2) \pm c^2 \sqrt{\left(\frac{v_A}{c}\right)^4 + 1 + 2\left(\frac{v_A}{c}\right)^2 (1 - 2n^2)} \right] . \tag{9.137}$$

Let's write:

$$\left(\frac{v_A}{c}\right)^2 = x \ll 1. \tag{9.138}$$

Then, from the series development of $f(x) = \sqrt{x^2 + 1 + 2x(1 - 2n^2)}$, i.e.

$$f(x) \approx 1 + (1 - 2n^2)x \tag{9.139}$$

one gets:

$$u^2 = \frac{1}{2} \{(v_A^2 + c^2) \pm c^2 [1 + 1(1 - 2n^2)x]\} . \tag{9.140}$$

Still exploiting $v_A^2 \ll c^2$ we have two solutions, the first

$$u_1^2 = \frac{1}{2} [v_A^2 + c^2 + c^2 + (1 - 2n^2)v_A^2] \sim c^2 \tag{9.141}$$

that gives

$$u_1 \approx c \tag{9.142}$$

(fast wave), and the second

$$u_2^2 = \frac{1}{2}[v_A^2 + c^2 - c^2 - (1 - 2n^2)v_A^2] \sim n^2 v_A^2 \tag{9.143}$$

that gives

$$u_2 \approx v_A n \tag{9.144}$$

(slow wave). Let's examine the characteristics of each of the two waves.

(a) *Fast wave*. From the Eqs. 9.62 and 9.63 we obtain:

$$\left|\frac{b_y}{v_x}\right| = \frac{\mu_0 \rho_0}{|B_{0y}|}\left|u - \frac{c^2}{u}\right| \tag{9.145}$$

$$\left|\frac{v_y}{v_x}\right| \approx \left|\frac{B_{0x}}{B_{0y}}\right|\left|1 - \frac{c^2}{u^2}\right| \tag{9.146}$$

remembering the condition $v_A^2 \gg c^2$ we obtain:

$$\left|\frac{b_y}{v_x}\right| = \frac{\mu_0 \rho_0}{|B_{0y}|}\left|c - \frac{c^2}{c}\right| \approx 0 \tag{9.147}$$

$$\left|\frac{v_y}{v_x}\right| \approx \left|\frac{B_{0x}}{B_{0y}}\right|\left|1 - \frac{c^2}{c}\right| \approx 0 . \tag{9.148}$$

The fast wave behaves approximately like a longitudinal sound wave ($v_x \ll v_y$): the perturbation of the magnetic field is small compared to that of the velocity ($b_y \ll v_x$) and the propagation speed tends to c.

(b) *Slow wave*. Based on Eq. 9.146 we have:

$$\left|\frac{v_y}{v_x}\right| \approx \frac{B_{0x}}{B_{0y}}\left|1 - \frac{c^2}{v_A^2 n^2}\right| \approx \frac{B_{0x}}{B_{0y}}\frac{c^2}{v_A^2 n^2} \gg 1 \tag{9.149}$$

i.e. $v_y \gg v_x$ and $b_y \gg v_x$. In a weak magnetic field the slow mode is very similar to an Alfven wave: in fact it is an almost transverse wave with negligible density perturbation and with a propagation speed equal to the Alfven speed.

Case with strong magnetic field. Supposed $v_A \gg c$, the solutions of the characteristic equation of the system of the second group are given by:

$$u^2 = \frac{1}{2}\left[(v_A^2 + c^2) \pm v_A^2\sqrt{\left(\frac{c}{v_A}\right)^4 + 1 + 2\left(\frac{c}{v_A}\right)^2(1 - 2n^2)}\right] . \tag{9.150}$$

Proceeding as in the previous case, we obtain:

$$u^2 = \frac{1}{2}\{(v_A^2 + c^2) \pm [v_A^2 + (1 - 2n^2)c^2]\} \tag{9.151}$$

from which we have:

$$u_1^2 = \frac{1}{2}[v_A^2 + c^2 + v_A^2 + (1 - 2n^2)c^2] \sim v_A^2 \tag{9.152}$$

i.e.:

$$u = u_1 = v_A \tag{9.153}$$

(fast wave) e

$$u_2^2 = \frac{1}{2}[v_A^2 + c^2 - v_A^2 - (1 - 2n^2)c^2] \sim n^2 c^2 \tag{9.154}$$

i.e.:

$$u = u_2 = nc \tag{9.155}$$

(slow wave). In both ways (fast and slow) no similarities can be identified with particular cases.

Appendix B: MHD Shockwaves

From the Eqs. 9.108, 9.109 and 9.117, by setting:

$$X = \frac{\rho_2}{\rho_1} = \frac{v_1}{v_2} = \frac{B_2}{B_1} \tag{9.156}$$

$$Y = \frac{p_2}{p_1} \tag{9.157}$$

$$N^2 = \gamma \mathrm{Ma}_1^2 = \gamma \frac{v_1^2}{c_1^2} \tag{9.158}$$

$$Q = \frac{B_1^2}{2\mu_0} p_1 \tag{9.159}$$

the equations are obtained:

$$N^2 \left(1 - \frac{1}{X}\right) = (Y - 1) + Q(X^2 - 1) \tag{9.160}$$

$$N^2 \left(1 - \frac{1}{X^2}\right) = \frac{2\gamma}{\gamma - 1}\left(\frac{Y}{X} - 1\right) + 4Q(X - 1) \tag{9.161}$$

Eliminating Y and taking into account that a solution of the cubic equation that is obtained is $X = 1$, which does not correspond to a shock wave (being $\rho_1 = \rho_2$), the following second degree equation remains :

$$Q(2 - \gamma)X^2 + \left[\gamma(Q + 1) + \frac{1}{2}(\gamma - 1)N^2\right]X - \frac{1}{2}(\gamma + 1)N^2 = 0 \,. \qquad (9.162)$$

Since $\gamma < 2$, the coefficient of X^2 is positive like that of the term in X while the known term is negative: therefore one solution is positive and the other is negative (it can be verified that the solutions are real); the negative one has no physical meaning and must be discarded. Furthermore, to have a shock (with a corresponding increase in entropy within the shock front due to energy dissipation) it must be $X > 1$ and therefore you must have:

$$C > A + B \qquad (9.163)$$

where A, B and C are the coefficients of X^2, X and the known term, respectively. Therefore:

$$\frac{1}{2}(\gamma + 1)N^2 > Q(2 - \gamma) + \gamma(Q + 1) + \frac{1}{2}(\gamma - 1)N^2 \qquad (9.164)$$

from which

$$N^2 > 2Q + \gamma \qquad (9.165)$$

which with the definitions of Q and N becomes:

$$\gamma\left(\frac{v_1^2}{c_1^2}\right) > \frac{2B_1^2}{2\mu_0 \rho_1} + \gamma \qquad (9.166)$$

and from which we finally deduce:

$$v_1^2 > v_{A1}^2 + c_1^2 \qquad (9.167)$$

which is a condition for the occurrence of an MHD shock wave. We note that v_{A1} is the Alfven velocity in front of the shock front: the critical velocity in this case is not the velocity of sound.

We now verify that the introduction of the transverse magnetic field reduces the intensity of the shock. Let X_n be the solution of the equation (9.162) in the absence of a magnetic field ($Q = 0$)

$$\left[\gamma + \frac{1}{2}(\gamma + 1)\right]X_n - \frac{1}{2}(\gamma + 1)N^2 = 0 \,. \qquad (9.168)$$

Subtracting this equation from (9.162) we have:

$$\left[\frac{1}{2}(\gamma - 1)N^2 + \gamma\right](X - X_n) = -QX[(2 - \gamma)X + \gamma] \tag{9.169}$$

what places:

$$B = QX[(2 - \gamma)X + \gamma] \tag{9.170}$$

$$A = \frac{1}{2}(\gamma - 1)N^2 + \gamma \tag{9.171}$$

becomes:

$$X = X_n - \frac{B}{A}. \tag{9.172}$$

Since $\gamma < 2$ and $X > 1$ it will be:

$$B > 0 \qquad A > 0 \tag{9.173}$$

and therefore:

$$X < X_n. \tag{9.174}$$

The presence of the transverse magnetic field therefore reduces the compression. Denoting the positive root of (9.162) by X, we have:

$$Y = 1 + N^2\left(1 - \frac{1}{X}\right) - Q(X^2 - 1). \tag{9.175}$$

We denote by Y_n the value of Y in the absence of a magnetic field, so Y_n satisfies the:

$$Y_n = 1 + N^2\left(1 - \frac{1}{X_n}\right). \tag{9.176}$$

Subtracting the latter from the penultimate one has:

$$(Y - Y_n) = -N^2\left(\frac{1}{X} - \frac{1}{X_n}\right) - Q(X^2 - 1) \tag{9.177}$$

and being $X < X_n$ it results:

$$Y < Y_n \tag{9.178}$$

that is, the pressure increase produced by the passage of the shock wave is less in the presence of a transverse magnetic field. Since it has been seen that the condition for the shock to develop is that the speed of the gas exceeds the speed of the magnetosonic waves, it is convenient to introduce the magnetic Mach number defined by:

$$\mathrm{Ma}_1^* = \frac{v_1}{\sqrt{v_{A1}^2 + c_1^2}} \tag{9.179}$$

results:

$$\text{Ma}_1^{*2} = \frac{N^2}{2Q + \gamma} \tag{9.180}$$

being

$$\frac{\gamma v_{A1}^2}{c_1^2} = 2Q . \tag{9.181}$$

The Eq. 9.162 becomes:

$$Q(2 - \gamma)X^2 + \left[\gamma(Q + 1) + \frac{1}{2}(\gamma - 1)(2Q + \gamma)\text{Ma}_1^{*2}\right]X - \frac{1}{2}(\gamma - 1)(2Q + \gamma)\text{Ma}_1^{*2} = 0 \tag{9.182}$$

The positive root depends on Q that is the ratio of magnetic pressure to gas pressure and on Ma_1^*.

Let's consider the case of strong shocks ($\text{Ma}_1^* > 1$):

(a) *density*: the Eq. 9.162 can be approximated with

$$\frac{1}{2}(\gamma - 1)(2Q + \gamma)\text{Ma}_1^{*2}X - \frac{1}{2}(\gamma + 1)(2Q + \gamma)\text{Ma}_1^{*2} = 0 \tag{9.183}$$

from which:

$$X = \frac{\rho_2}{\rho_1} \rightarrow \frac{\gamma + 1}{\gamma - 1} . \tag{9.184}$$

therefore in a strong shock the density is the same as in the absence of the magnetic field;

(b) *pressure*: from the equation

$$Y = 1 + N^2 \left(1 - \frac{1}{X}\right) - Q(X^2 - 1) \tag{9.185}$$

substituting N as a function of Ma_1^*:

$$Y = 1 + (2Q + \gamma)\text{Ma}_1^{*2} \left(1 - \frac{1}{X}\right) - Q(X^2 - 1) \tag{9.186}$$

in the hypothesis $\text{Ma}_1^* \gg 1$ can be approximated with:

$$Y = (2Q + \gamma)\text{Ma}_1^{*2} \left(1 - \frac{1}{X}\right) \tag{9.187}$$

that is

$$\frac{p_2}{p_1} \rightarrow \infty \tag{9.188}$$

(c) *temperature*: from the equation of state

$$\frac{T_2}{T_1} = \frac{V_2}{V_1}\frac{p_2}{p_1} = \frac{\rho_1}{\rho_2}\frac{p_2}{p_1} \to \infty \tag{9.189}$$

(d) *thermal energy*: grows without limits being $E_{th} \sim T$;
(e) *magnetic energy*: results

$$\left(\frac{B_2^2}{2\mu_0}\right)\Big/\left(\frac{B_1^2}{2\mu_0}\right) = \left(\frac{B_2}{B_1}\right)^2 = X^2 = \left(\frac{\gamma+1}{\gamma-1}\right)^2 \tag{9.190}$$

therefore in strong shocks the magnetic pressure can be neglected and the connection conditions are reduced to normal fluid dynamics.

We evaluate how much kinetic energy is dissipated. From Eq. 9.117 we get in the case $\gamma = 5/3$:

$$\Delta E_c = \frac{1}{2}v_1^2 - \frac{1}{2}v_2^2 = \frac{5}{2}\left(\frac{p_2}{\rho_2} - \frac{p_1}{\rho_1}\right) + \left(\frac{B_2^2}{\mu_0\rho_2} - \frac{B_1^2}{\mu_0\rho_1}\right) \tag{9.191}$$

that in the case of strong waves, i.e.

$$p_2 \gg p_1 \qquad V_1 = 4V_2 \tag{9.192}$$

from

$$\Delta E_c = \frac{5}{2\rho_1}\left(\frac{p_2}{4} - p_1\right) + \frac{3B_1^2}{\mu_0\rho_1} \tag{9.193}$$

therefore the presence of the magnetic field slightly modifies the dissipation of the kinetic energy.

Chapter 10
Dust from the Interstellar Medium

10.1 Introduction

The presence of dust in the interstellar medium was highlighted by Trumpler who, in an investigation of the diameters of Galactic clusters, found a poorly convincing result: their diameters were an increasing function of distance. The origin of this result was linked to the determination of the distances made with photometry, a distance that was overestimated precisely for the absorbing action of solid dust particles. The study of the spatial distribution of these led to the conclusion that dust was associated with the interstellar gas.

This solid component constitutes approximately 1% of the total mass of the interstellar medium and its main manifestation is the selective absorption and scattering of stellar radiation, as a result of which the starlight reaches us attenuated and reddened. Furthermore, the surface of the grains is the site of formation of the H_2 and of other molecules in molecular clouds and the grains intervene in the cooling and heating processes of the clouds as well.

The dust grains have dimensions of the order of 10^{-4} cm and smaller and in the modelling they are reproduced by a power law distribution :

$$n(a) \propto a^{-p} \qquad a_l \leq a \leq a_u \tag{10.1}$$

with $p \approx 3.5$, $a_l = 0.001\mu$ and $a_u = 1\mu$. From the chemical point of view they are made up of silicates, amorphous carbon, graphite and many times have a complicated structure with mantles of volatile compounds (ice or organic material) that surround the most refractory part. The existence of two populations of grains (silicates and graphite) is necessary in modelling to justify the extinction in the optical region and in the UV, as we are going to see later.

© The Author(s), under exclusive license to Springer Nature Switzerland AG 2021
G. Carraro, *Astrophysics of the Interstellar Medium*,
UNITEXT for Physics, https://doi.org/10.1007/978-3-030-75293-4_10

10.2   The Extinction of Radiation

Electromagnetic radiation with $\lambda \leq 912$ Å is mainly absorbed by neutral hydrogen. For longer wavelengths, the main responsible for the attenuation of the radiation is dust through two phenomena: absorption with a coefficient $k_{a,\lambda}$ and scattering with a coefficient $k_{s,\lambda}$. The overall effect is called extinction and is characterized by a coefficient:

$$k_{e,\lambda} = k_{a,\lambda} + k_{s,\lambda} . \tag{10.2}$$

The extinction coefficient can also be written in the form:

$$k_{e,\lambda} = n_g \sigma_a(\lambda) + n_g \sigma_s(\lambda) = n_g \sigma_e(\lambda) \tag{10.3}$$

where n_g is the number of grains per cm^3.

The extinction of radiation is studied using the equation of radiative transport:

$$\frac{\mathrm{d}I_\lambda}{\mathrm{d}s} = -k_{e,\lambda} I_\lambda + j_\lambda^* + j_\lambda^s \tag{10.4}$$

where j_λ^* is the coefficient of the emission of stellar sources, while j_λ^s is the emission coefficient due to the diffusion of the radiation in the direction of the line of sight, i.e.

$$j_\lambda^s = k_{s,\lambda} \int I_\lambda(\mathbf{k}') F(\mathbf{k}, \mathbf{k}') \mathrm{d}\omega' \tag{10.5}$$

where the function $F(\mathbf{k}, \mathbf{k}')$ is the phase function and gives the distribution of scattered photons as a function of the scattering angle.

In the case of a point source external to the absorbing material, as when a single star is taken into consideration, both j_λ^* and j_λ^s are neglected, the latter because if the absorbing material is far away from the source the diffuse radiation conveyed in the direction of the line of sight is negligible. Instead, $I_\lambda(0) \neq 0$ should be set. This way Eq. 10.4 is reduced to:

$$\frac{\mathrm{d}I_\lambda}{\mathrm{d}s} = -k_{e,\lambda} I_\lambda \tag{10.6}$$

which by integration gives:

$$I_\lambda(s_0) = I_\lambda(0) \exp\left[-\int_0^{s_0} k_{e,\lambda} \mathrm{d}s\right] \tag{10.7}$$

where s_0 is the geometric thickness of the absorbent zone. If we indicated with $\sigma_g = \pi a^2$ the geometric cross section, we can introduce the efficiency factors using the following definitions:

$$Q_{a,\lambda} = \frac{\sigma_{a,\lambda}}{\sigma_g} \qquad Q_{s,\lambda} = \frac{\sigma_{s,\lambda}}{\sigma_g} \qquad Q_{e,\lambda} = \frac{\sigma_{e,\lambda}}{\sigma_g} \tag{10.8}$$

with

$$\sigma_{e,\lambda} = \sigma_{a,\lambda} + \sigma_{s,\lambda} \; . \tag{10.9}$$

We will therefore have:

$$k_{e,\lambda} = n_g \sigma_e(\lambda) = n_g Q_e(\lambda) \sigma_g \tag{10.10}$$

and the solution Eq. 10.7 becomes:

$$I_\lambda(s_0) = I_\lambda(0) \exp\left[-\int_0^{s_0} n_g Q_e(\lambda) \sigma_g ds \right] \; . \tag{10.11}$$

If the interposed absorbing material is homogeneous, the integral is simplified:

$$\int_0^{s_0} n_g Q_e(\lambda) \sigma_g ds = Q_e(\lambda) \sigma_g N_g \tag{10.12}$$

and the intensity is expressed by:

$$I_\lambda(s_0) = I_\lambda(0) \exp[-Q_e(\lambda) \sigma_g N_g] \tag{10.13}$$

where N_g is the column density of the grains, that is the number of grains contained in a column of unit base and of height s_0.

We introduce the concept of optical depth defined as:

$$d\tau = -k_e ds \; . \tag{10.14}$$

The integration of this equation yields $\tau = -ks + a$, where a is an integration constant This is determined by imposing that $\tau = 0$ for $s = s_0$ and then $a = ks_0$. Therefore, $\tau = ks_0 = \tau_0$ for $s = 0$. Eventually, Eq. 10.13 becomes:

$$I_\lambda(s_0) = I_\lambda(0) \exp[-\tau_{0,\lambda}] \; . \tag{10.15}$$

Extinction is defined as the quantity:

$$A(\lambda) = -2.5 \log \frac{I_\lambda(s_0)}{I_\lambda(0)} = 1.086 \tau_{0,\lambda} \tag{10.16}$$

and based on this definition:

$$A(\lambda) = -2.5 \log I_\lambda(s_0) - [-2.5 \log I_\lambda(0)] = m(\lambda) - m_0(\lambda) \tag{10.17}$$

being:

$$m(\lambda) = -2.5 \log I_\lambda(s_0) + \text{cost.} \tag{10.18}$$

and an analogous relation for $m_0(\lambda)$.

The extinction at a given wavelength is therefore the difference between the magnitude of a star and the magnitude that the same star would have in the absence of dust.

10.3   Observational Determination of Stellar Extinction

The study of extinction has a twofold interest: on the one hand it is necessary to correct the observational data of stars and galaxies from the effects of extinction and redness, on the other hand its study allows to clarify the properties of dust in the different environments of the interstellar medium and clarify the processes that cause their evolution. The measurement of the extinction of a star is performed by the method of the pairs of stars, based on the Eq. 10.17: each pair is formed by two stars of the same spectral type, one un-reddened (this is deduced from the fact that its observed color index $(B - V)$ coincides with that inferred from the spectral type), and the other, which is the star under examination, reddened by extinction.

At the wavelength λ for the star in question it results:

$$m(\lambda) = m_0(\lambda) + A(\lambda) \tag{10.19}$$

where $m(\lambda)$ is the apparent magnitude of the star under examination and $m_0(\lambda)$ would be its apparent magnitude in the absence of dust. But $m_0(\lambda)$ is also the apparent magnitude of the comparison star, except for a distance-dependent constant that can be measured. An equal relation can also be written for the V band:

$$m(V) = m_0(V) + A(V) \, . \tag{10.20}$$

Making the difference between the two relations we obtain:

$$m(\lambda) - m(V) = m_0(\lambda) - m_0(V) + A(\lambda) - A(V) \tag{10.21}$$

hence we can introduce the excess of color $E(\lambda - V)$ of the color index $(\lambda - V)$:

$$E(\lambda - V) = [m(\lambda) - m(V)] - [m_0(\lambda) - m_0(V)] = A(\lambda) - A(V) \, . \tag{10.22}$$

The excess of color, i.e. the difference between the color index of the red star and the color index of the non-red star, is equal to the difference between the extinctions at the wavelengths considered.

As λ varies, we obtain the $E(\lambda - V)$ function which defines the extinction curve of the line of sight in the direction of the star. The extinction curve is typically normalised dividing $E(\lambda - V)$ by $E(B - V)$:

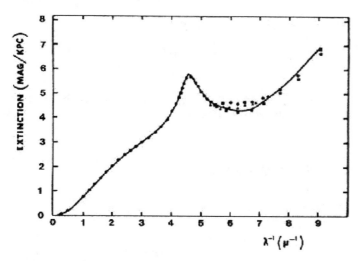

Fig. 10.1 Adaptation of Whittet (1988) Milky Way average extinction curve

$$\epsilon(\lambda - V) = \frac{E(\lambda - V)}{E(B - V)}. \tag{10.23}$$

The stars whose extinction measurements are made must be bright. Normally, the derivation of the extinction curve is made for stars of type O, B and WR.

10.4 The Extinction Curve of the Diffuse Medium in the Galaxy

An average extinction curve in our galaxy was derived from the analysis of the extinction curve relating to numerous lines of sight that mainly cross regions of diffuse gas (HI) (see Fig. 10.1). The general characteristics are the following: the absorption is very small in the IR, it increases as the wavelength decreases and has a maximum (bump) in the UV at a wavelength of 2175 Å, it has a minimum for $\lambda \sim$ 1800 Å and then goes back to the extreme UV. The ratio $R_V = A(V)/E(B - V)$ is (3.1). The average curve is reproduced by means of a model that requires a mixture of grains of silicates and graphite: graphite is responsible for the bump at $\lambda = 2170$ Å.

The figure shows Whittet's (1988) determination: the quantity ϵ is expressed as a function of $x = 1/\lambda$; the extinction curve can also be expressed by the quantity $A(\lambda)/A(V)$. We pass to the quantity $\epsilon(\lambda - V)$ previously introduced by:

$$\epsilon(\lambda - V) = R_V \left[\frac{A(\lambda)}{A(V)} - 1 \right] \tag{10.24}$$

where the R_V parameter is defined according to:

$$R_V = \frac{A(V)}{E(B - V)} \, .$$

(10.25)

For the popular medium, the ratio R_V is (3.1). Extinction curves, however can have different shapes from direction to direction (see Fig. 10.2).

10.5   Methods for Determining R_V

According to the definition, $\epsilon(\lambda - V)$ is:

$$\epsilon(\lambda - V) = \frac{A(\lambda) - A(V)}{E(B - V)}$$

(10.26)

When the extinction vanishes at large wavelengths, we have:

$$R_V = \lim_{\lambda \to \infty} \frac{A(\lambda) - A(V)}{E(B - V)} = -\frac{A(V)}{E(B - V)} \, .$$

(10.27)

The R_V ratio can therefore be deduced by extrapolating the extinction observed at small wavelengths. In particular it can be deduced by means of the approximate relation:

$$R_V \approx 1.1 \frac{E(V - K)}{E(B - V)}$$

(10.28)

which presupposes the knowledge of extinction, as well as in the B and V bands, that in the infrared band K ($\lambda_{\text{eff}} = 2.19\mu$).

Similar relations are:

$$R_V \approx 1.39 \frac{E(V - J)}{E(B - V)}$$

(10.29)

for the J band ($\lambda_{\text{eff}} = 1.22\mu$)

$$R_V \approx 1.19 \frac{E(V - H)}{E(B - V)}$$

(10.30)

for the H band ($\lambda_{\text{eff}} = 1.63\mu$)

$$R_V \approx 1.07 \frac{E(V - L)}{E(B - V)}$$

(10.31)

for the L band ($\lambda_{\text{eff}} = 3.45\mu$)

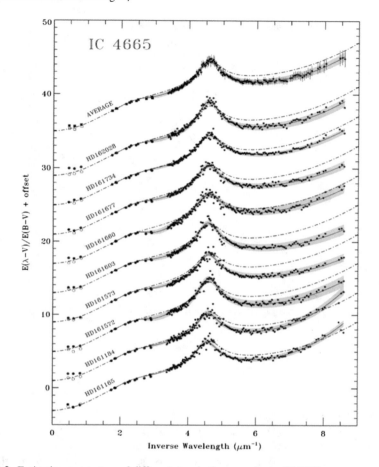

Fig. 10.2 Extinction curves toward different stars in the open cluster IC 4665

Within our galaxy the extinction curves towards most directions show a behaviour similar to that of the average curve described above. There are, however, extinction curves that have different characteristics:

(1) first the value of R_V is different from the value of (3.1) and covers a wide range from about 2 to over 6,
(2) the rise in the UV extreme can be lower or much higher than that of the diffuse medium curve,
(3) the bump can be more or less intense and more or less large.

Curves with these characteristics refer to lines of sight that pass through dense media such as molecular clouds or regions hit by very intense radiation fields, such as HII regions, or that have been hit by shock waves caused by stellar winds or supernova explosions .

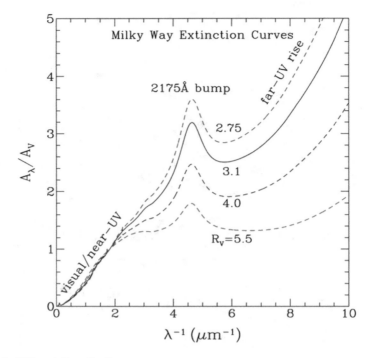

Fig. 10.3 Different R_V extinction curves

Curves which deviate from the mean curve generally have values of R_V different from (3.1) (see Fig. 10.3). In their seminal paper, Cardelli, Clayton and Mathis (1988) found that extinction curves from the near infrared to ultraviolet region can be presented as a uniparametric family with parameter R_V. In fact they find that the $A(\lambda)/A(V)$ extinction can be expressed with the following relations:

$$\frac{A(\lambda)}{A(V)} = a(x) + \frac{b(x)}{R_V} \tag{10.32}$$

with $x = 1/\lambda$ e

$$a(x) = 1.802 - 0.316x - 0.104/[(x - 4.67)^2 + 0.341] + F_a(x) \tag{10.33}$$

$$b(x) = -3.090 + 1.825x - 1.206/[(x - 4.62)^2 + 0, 263]F_b + (x) \tag{10.34}$$

$$F_a(x) = -0.04473(x - 5.9)^2 - 0.009779(x - 5.6)^3 \quad \text{if } x \geq 5.9 \tag{10.35}$$

$$F_b(x) = 0.2160(x - 5.6)^2 + 0.1207(x - 5.9)^3 \quad \text{if } x \geq 5.9 \tag{10.36}$$

is

$$F_a = F_b = 0 \quad \text{if } x < 5.9 \,. \tag{10.37}$$

The CCM relationship, varying by R_V, is applicable to a wide range of environments including lines of sight that cross the diffuse medium and lines through dark clouds or star-forming regions. The existence of this relationship suggests that the environmental processes that modify the grains are efficient and act on all grains.

Since the value of R_V is an approximate measure of the average grain size, the CCM relation provides the physical basis for understanding the difference in the extinction curves.

10.6 The Equilibrium Temperature of the Grains

The emission of infrared radiation from dust heated up by young hot stars is a powerful diagnostic tool for studying star formation in starburst galaxies. The infrared luminosity of a galaxy will depend on the amount of dust, their physical properties, and their temperature which will be determined by the stellar radiation field. Here we outline the principles for determining the temperature of dust grains immersed in a radiation field $I(\lambda)$.

The temperature of the grains is determined by the balance between the radiation absorption process and their thermal emission. The energy gained per cm^3 and per second following the absorption of radiation, whose specific energy density is $u(\lambda) = I(\lambda)/c$, is expressed by:

$$G = \int_0^\infty c u(\lambda) n_g Q_a(\lambda) \sigma_g \mathrm{d}\lambda \tag{10.38}$$

where σ_g is the geometric cross section of the grains and Q_a is the absorption factor previously introduced. The density $u(\lambda)$ is a function of the environment and is evaluated taking into account the emission of the star, the dilution factors and possibly the dust absorptions (here the latter factor is ignored); the method for its treatment is indicated in the next chapter.

The grains, having temperature T_g, radiate electromagnetic energy and their emission coefficient can be obtained from the Kirchhoff principle:

$$j(\lambda) = k_a(\lambda) B_\lambda(T_g) \tag{10.39}$$

where B_λ is the Planck function.

To obtain the lost energy per cm^3 and per second it is necessary to integrate the expression of $j(\lambda)$ over the total solid angle and all wavelengths:

$$L = 4\pi \int_0^\infty n_g Q_a(\lambda) \sigma_g B_\lambda(T_g) \mathrm{d}\lambda \,. \tag{10.40}$$

The energy balance is obtained by comparing Eqs. 10.38 and 10.40:

$$\int_0^\infty u(\lambda)Q_a(\lambda)d\lambda = \frac{4\pi}{c}\int_0^\infty Q_a(\lambda)B_\lambda(T_g)d\lambda . \qquad (10.41)$$

This equation defines the temperature T_g of the grains. The absorption factor of the grains $Q_a(\lambda)$ is, in general, a decreasing function of λ and the two integrals in (10.41) give the main contributions in different spectral regions. As a first approximation we can put:

$$Q_A(\lambda) \propto \frac{A}{\lambda} \qquad (10.42)$$

with $A = $ constant; this approximation is correct in the range of the visual and IR wavelengths.

Let us assume that the stellar radiation field is produced by a star with $T_S = 3 \cdot 10^4$ K, whose radius R_S is equal to 9 $R_\odot$. Let's also suppose to consider a grain placed at the distance r from the star. To evaluate the radiation field at this distance, it is necessary to introduce the dilution factor W. The average intensity is, by definition:

$$J(\nu) = \frac{1}{4\pi}\int I(\nu)d\Omega \qquad (10.43)$$

and being $I(\nu) = B(\nu, T_S)$:

$$J(\nu) = \frac{1}{4\pi}B(\nu, T_S)\frac{\pi R_S^2}{r^2} = B(\nu, T_S)W . \qquad (10.44)$$

The dilution factor is therefore the ratio between the solid angle Ω subtended by the star and the total solid angle:

$$W = \frac{\Omega}{4\pi} = \frac{R_S^2}{4r^2} . \qquad (10.45)$$

The energy density of the radiation field will be that of a black body with temperature $T = T_S$ and dilution factor W:

$$u(\lambda) = \frac{8\pi hc}{\lambda^5}W\left(\exp\left[\frac{hc}{k\lambda T_S}\right] - 1\right)^{-1} . \qquad (10.46)$$

and therefore for the second member of the balance equation we have:

$$B_\lambda(T_g) = \frac{c}{4\pi}u(\lambda) = \frac{2hc^2}{\lambda^5}\left(\exp\left[\frac{hc}{k\lambda T_g}\right] - 1\right)^{-1} \qquad (10.47)$$

and Eq. 10.41 becomes:

Table 10.1 Temperature of grains around a star with $T_S = 3 \cdot 10^4$ K

r [pc]	W	T_g
0.001	$1.0 \cdot 10^{-8}$	754.6
0.005	$4.4 \cdot 10^{-10}$	403.5
0.010	$1.0 \cdot 10^{-10}$	300.0
0.050	$4.4 \cdot 10^{-12}$	160.6
0.100	$1.0 \cdot 10^{-12}$	119.4
0.500	$4.4 \cdot 10^{-14}$	64.0
1.000	$1.0 \cdot 10^{-14}$	47.6

$$\int_0^\infty \frac{A}{\lambda} \frac{8\pi hc}{\lambda^5} W \left(\exp\left[\frac{hc}{kT_S\lambda} \right] - 1 \right)^{-1} d\lambda = \frac{4\pi}{c} \int_0^\infty \frac{A}{\lambda} \frac{2hc^2}{\lambda^5} \left(\exp\left[\frac{hc}{kT_g\lambda} \right] - 1 \right)^{-1} d\lambda. \qquad (10.48)$$

Both integrals are proportional to T^5 and hence :

$$T_g = T_S W^{1/5}. \qquad (10.49)$$

Table 10.1 reports the temperatures of the grains at different distances from the star. In reality, it is necessary to take into account a more realistic description of the grains, remembering that there are at least two populations of grains of different chemical composition and with defined size distributions: the properties of the grains are a function of all these parameters.

Let us consider the temperature of the grains inside a molecular cloud; in general the radiation will come both from the outside and from sources immersed in the cloud. In the absence of internal sources, the temperature of the grains inside the cloud decreases with the optical depth because the external UV radiation is attenuated by the grains themselves and transformed into far IR radiation which is poorly absorbed by the grains. This effect is the basis of a very simple calculation which shows that there is a reduction of a few degrees down to a central value of 8 K for a spherical cloud with an extinction of 10 mag in the visual.

If there is a source of radiation in the center of the cloud, the problem of radiation transport must be addressed, but now the problem is more complicated because the grains close to the source are sufficiently hot so that the radiation they emit can be absorbed by the grains located more externally. This is the problem of the infrared spectrum of protostellar sources, which will be addressed in one of the following chapters.

10.7 The Temperature of the Small Grains

The considerations developed in the previous section assume that two conditions are met:

(a) the thermal energy of the grain must be much greater than the energy of a single photon, so that the absorption does not cause significant changes in temperature;
(b) the energy of an absorbed photon must be rapidly distributed throughout the entire volume of the grain, and the time for this to happen is the energy diffusion time $\tau = a^2/D$ where a is the grain size and D is the diffusion coefficient which is typically $D < 10^{-9}$ cm^2/s.

The diffusion time is less than 1/1000 of a second and therefore it makes sense to speak of the temperature of the grain after the absorption of the photon. To verify when the first condition is satisfied we must evaluate the thermal energy of a grain. For this we consider Debye's theory. A solid is made up of a significant number of atoms constrained in the equilibrium position by intense forces of an electrical nature. The only individual motions of the atoms are small vibrations around the equilibrium configuration. The coupling between the atoms is so strong that the oscillation of one of them also disturbs the others, giving rise to a collective vibrational excitation. These collective vibrations excite standing waves, the frequencies of which depend on the size of the body. If N is the number of atoms, the degrees of freedom are $3N$; there is a relationship between the degrees of freedom and the maximum frequency ν_0.

An important parameter is the Debye temperature:

$$\Theta_D = \frac{h\nu_0}{k} \tag{10.50}$$

where ν_0 is the maximum oscillation frequency. In the case of the grains of our interest $\Theta_D \approx 500$ K. The heat capacity is given by the relation:

$$C_V = 9R \left(\frac{T}{\Theta_D}\right)^3 \int_0^{\Theta_D/T} \frac{x^4 e^x}{(e^x - 1)^2} dx . \tag{10.51}$$

When $T \gg \Theta_D$, C_V tends to the asymptotic value $3R$ (Dulong-Petit law) and in this case the thermal energy is:

$$E_{Th} = 3kNT . \tag{10.52}$$

The absorption of a photon of energy and will determine an increase in internal energy:

$$\Delta E_{Th} = \int_{T_1}^{T_2} 3kNdT = \epsilon \tag{10.53}$$

and the temperature variation is:

$$\Delta T = T_2 - T_1 = \frac{\epsilon}{3kN} = 4 \cdot 10^3 \frac{\epsilon[\text{eV}]}{N} . \tag{10.54}$$

If $a = 0.001\mu$, $N = 40$ and $\epsilon = 10$ eV, then $\Delta T = 1000$ K (but remember that it must be $T_1 > \Theta_D$ in order to apply these relations).

Table 10.2 Temperature increase for the absorption of a 10 eV photon, H_i is the initial thermal energy

$a\ [\mu]$	T_f	H_i	ΔT
0.001	156.0	$1.7 \cdot 10^{-4}$	146.0
0.005	46.7	0.02	36.7
0.010	27.9	0.17	17.9
0.050	11.0	21.1	1.0
0.100	10.0	169.0	0.0

When $T < \Theta_D$, on the other hand, the approximation can be adopted

$$C_V = 67.6a^3 T^3 . \tag{10.55}$$

The expression of E_{Th} is obtained by integrating 7.55 and it results:

$$E_{Th} = 16.9a^3 T^4 \tag{10.56}$$

where a is the grain radius expressed in microns. For the absorption of a photon of energy ε eV we have:

$$\Delta E_{Th} = \int_{T_i}^{T_f} 67.6a^3 T^3 dT = \varepsilon \tag{10.57}$$

is

$$T_f = \left(T_i^4 + \frac{\varepsilon}{16.9a^3} \right)^{1/4} . \tag{10.58}$$

Suppose that a grain with temperature $T_i = 10$ K and radius a absorbs a photon of energy 10 eV (it is the energy of a photon with a wavelength of 1240 Å). Since $T_i \ll \Theta_D$, we use Eq. 10.58 to derive the temperature increase. The result for different radius values is presented in Table 10.2.

From these estimates it results that if the grain size exceeds several hundredths of μ, the thermal energy of the grain is greater than the energy of a UV photon and its absorption does not cause significant increases in temperature, but if the grain is smaller the variations are relevant. The time needed for the grain to relax at the original temperature can be evaluated with the

$$t_c = \frac{E_{Th}}{R} \tag{10.59}$$

where E_{Th} is the internal energy and R the cooling rate:

$$R = 4\pi a^2 \epsilon \sigma T^4 \tag{10.60}$$

where ϵ is the emissivity which is on the order of $3 \cdot 10^{-3}$. A relaxation time of 3 seconds is achieved.

Using the results of Habing (1968) for the evaluation of the average density of the radiation field we obtain a flux of energetic photons intercepted by a grain of radius $a = 5 \cdot 10^{-6}$ cm which is worth $3 \cdot 10^{-2} \text{s}^{-1}$, therefore the interval between two successive captures is such that large temperature fluctuations can be produced.

Intermediate-sized grains (on the order of $3 - 30$ nm), which have specific cooling times and heats of intermediate values, cool sufficiently slowly so that they can absorb several photons between complete cooling periods. Based on these considerations, Desert et al. (1986) calculated the temperature distribution of grains that absorb many photons (without however reaching an equilibrium situation). These grains emit at shorter wavelengths than large (cold) grains.

10.8   IR Emission of Galaxies

The infrared spectrum (or spectral energy distribution, SED) due to the emission of the dust has two maxima (see Fig. 10.4): one at about 100μ and one between 4 and 20μ. It can be explained by the contribution of different dust components. The main contribution to the heating of the dust comes from the general stellar radiation field (ISRF) and the more massive stars of the disk population. These components of the dust are:

- cold dust ($T \approx 15 - 25K$, emission peak at about 110μ, far-IR) associated with atomic gas or with regions of molecular gas, whose grains are heated by the general radiation field and whose overall brightness is $5.7 \cdot 10^9 L_\odot$;
- warm dust ($T \approx 30 - 40K$, emission peak at 60μ, mid-IR) associated with ionized gas in HII regions of low density and heated by O and B stars (mainly O), whose brightness overall is $7.6 \cdot 10^9 L_\odot$;
- hot dust ($T \approx 250 - 500K$, emission peak at 5 - 10μ, near-IR) due to small grains (VSG) heated by the general field, or by normal grains heated by M stars, whose overall brightness is $2 \cdot 10^9 L_\odot$.

The correlation between the luminosity in the FIR (far-IR) and the luminosity $L(\alpha)$ of the Hα line for a sample of galaxies from the Virgo cluster has been studied and a good non-linear correlation has been found between the two, while the correlation between $L(\alpha)$ and the IR brightness of the hot component only (i.e. in the mid-IR) was found to be very good and also linear, indicating that the brightness in the mid-IR and that of line Hα are caused by the same: the action of massive stars.

In conclusion, only a portion of the luminosity in the IR can be indicative of the extent of star formation at the present (this means over the last million years). If we consider the sequence of spiral galaxies, we find that the ratio between the brightness due to cold dust and that due to hot dust is decreasing from Sa towards Sm. This is understandable if we consider that late-type spirals currently have a higher star

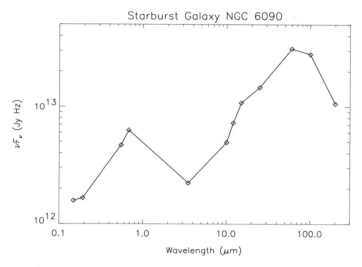

Fig. 10.4 Typical SED of a starburst galaxy

formation rate than in the past. In starburst galaxies the role of hot dust is even more relevant.

10.9 Synthetic Summary

About 1% of the ISM is in solid state. It is composed of small particles of different chemical composition which absorbs and scatter radiation. What we commonly call extinction is produced by dust grains. Extinction is very important to derive properties like distance to astronomical objects and to correctly interpret the spectral energy distribution of young distant galaxies.

Chapter 11
HII Regions

11.1 Introduction

On a large scale the interstellar medium receives energy from the density waves of the spiral arms, while on a smaller scale the energy supply comes from the stars and in particular from the most massive stars ($M > 10M_\odot$). There are three mechanisms and in each of them the mechanism of energy transfer is a shock wave; specifically the energy emitted by the stars can be:

(1) in the form of ultraviolet electromagnetic radiation which ionizes the surrounding environment and generates the HII regions,
(2) in the form of kinetic energy of matter ejected at high speed, i.e. the stellar wind,
(3) in the form of the energy produced by a supernova explosion.

To get an idea of the amount of energy involved in these phenomena we consider a B0 star with a mass of about $18M_\odot$. Over the course of its Main Sequence life, this star radiates $5 \cdot 10^{51}$ erg of ultraviolet electromagnetic energy, and photons are capable of ionising hydrogen whose ionization potential is $E_H = 13.6\,\text{eV}$; the energy produced by its stellar wind amounts to $4 \cdot 10^{49}$ erg; the final act, the supernova explosion, releases an energy equal to 10^{51} erg. The conversion efficiency of these forms of energy into kinetic energy of the interstellar medium varies according to the process considered.

In this and the next two chapters we will analyse these different mechanisms and calculate their efficiency. The injection of this energy into a thermally unstable medium such as interstellar gas is responsible for the highly inhomogeneous nature of this. In turn, the inhomogeneity has a noticeable effect on how energy flows in the interstellar medium. After influencing the gas and its structure, this energy is eventually radiated into intergalactic space.

© The Author(s), under exclusive license to Springer Nature Switzerland AG 2021   223
G. Carraro, *Astrophysics of the Interstellar Medium*,
UNITEXT for Physics, https://doi.org/10.1007/978-3-030-75293-4_11

Table 11.1 S_{UV}^* is the number of UV photons ($\lambda < 912$ Å) emitted per second; R_s is the radius of the region that the star is able to ionise if the density is 10^2 cm^{-3}

$M/M_\odot$	Spectr.	T_{eff} [K]	S_{UV}^* [s^{-1}]	R_s [pc]
100	–	52500	$2.0 \cdot 10^{50}$	9.6
80	–	51300	$1.4 \cdot 10^{50}$	8.5
60	–	47900	$7.5 \cdot 10^{49}$	6.9
50	O5 V	41200	$5.6 \cdot 10^{49}$	6.3
23	O8 V	36300	$5.0 \cdot 10^{48}$	2.8
17.5	B0 V	33500	$1.6 \cdot 10^{48}$	1.9
8.3	B3 V	25000	$1.0 \cdot 10^{46}$	0.3

11.2   Ionisation Balance

HII regions are areas where hydrogen is ionised; the cause of the gas ionisation is obvious since they are always associated with stars of the first spectral types (O and B). The effective temperature of these stars is high (between 20,000 and 50,000 K) and, according to Wien's displacement law, a substantial fraction of their radiation is emitted in the form of UV photons capable of ionising H and He. Stars of more advanced spectral types do not have sufficient UV flux to ionise a significant amount of gas.

The energy used to ionise the gas is also the region's heating source. Photons with energy $h\nu \geq E_H = 13.6$ eV determine the ionisation of the atoms of H, while the difference $h\nu - E_H$ is distributed in the form of kinetic energy of the released electron. This, through repeated impacts, leads to equipartition conditions. In addition to elastic collisions, free electrons are affected by inelastic collisions with protons which determine the recombination of the electron. The energy of the photons produced in the recombination comes from the sum of two contributions: the kinetic energy of the electron and the binding energy corresponding to the level in which the electron is located. The captured electron can cascade down to the fundamental level: thus the lines of the different series are produced and in particular of Balmer and Lyman ones.

There is ionisation equilibrium, namely the number of photo-ionizations balances the number of recombinations and this equilibrium determines the degree of ionisation.

In the following we will study in detail the evolution of an HII region; for reasons of simplicity we will analize ideal situations that will allow us to highlight the most important aspects. The main simplifications are:

(1) the gas consists of hydrogen only, in which case the density of the protons is equal to the density of the electrons, and the ionisation fraction is defined as

$$x = \frac{n_p}{n_p + n_{HI}} = \frac{n_p}{n} \tag{11.1}$$

where n_{HI} is the number of neutral atoms per cm^3 and n is the number of nuclei corresponding to neutral atoms or ions, and furthermore the region where the gas is completely ionised is defined without ambiguity, as we will show shortly;
(2) only the fundamental level of H is populated;
(3) the environment in which the HII regions originate is made up of the densest parts of the molecular clouds; we will assume a constant density n, a typical value being $n = 10^2 \mathrm{cm}^{-3}$;
(4) we first ignore the absorption of UV radiation by dust, and we evaluate the equilibrium condition of the ionization by calculating the photo-ionization and recombination rates;

If, on the other hand, the gas is composed of some He as well, and if the effective temperature of the star is less than 50,000 K, the region where He is ionized is significantly smaller than the region where the H is ionized. Besides, we expect He to be ionized only once, because the ionisation potential of HeII is particularly high and the photons are not sufficiently energetic.

Ionization rate. If we denote by $J(\nu)d\nu$ the flux of photons with frequency in the interval $(\nu, \nu + d\nu)$ the number of ionizations per cm^3 and per second is:

$$r_i = n_H \int_{\nu_0}^{\infty} \alpha(\nu) J(\nu) d\nu \tag{11.2}$$

where $\nu_0 = E_H / h$ and $\alpha(\nu)$ is the cross section for the transition of the electron from the ground state to a free state; this cross section depends on the photon frequency and decreases as the frequency increases:

$$\alpha(\nu) \propto \left(\frac{\nu_0}{\nu}\right)^3 . \tag{11.3}$$

A good approximation is to consider the value corresponding to the threshold frequency ν_0:

$$\alpha(\nu) \approx \alpha(\nu_0) = \alpha_0 = 6.8 \cdot 10^{-18} \mathrm{cm}^2 . \tag{11.4}$$

In this approximation we have:

$$r_0 = n_H \alpha_0 \int_{\nu_0}^{\infty} J(\nu) d\nu = n(1 - x) J \alpha_0 \tag{11.5}$$

where J is the total flux of ionizing photons.

Recombination rate. The recombination can take place at a level characterised by any value of the principal quantum number n. The number of recombinations per unit of time and per unit of volume will be proportional to $n_p n_e$ which in our case (the gas consists only of H) is equal to $x^2 n^2$. Furthermore, the probability of recombination on the jth level (β_j) depends on the temperature T, since it depends on the distribution of the relative velocities of the electron with respect to the proton.

The recombination rate is given by:

$$r_r = \sum_j n_p n_e \beta_j(T) \tag{11.6}$$

where the summation is extended to all levels of the hydrogen atom. The energy of the emitted photon will be:

$$h\nu = E_c + \frac{E_H}{n^2} = E_H \left(\frac{E_c}{E_H} + \frac{1}{n^2} \right) \tag{11.7}$$

so setting $T = 10^4$ K it results:

$$\frac{E_c}{E_H} = \frac{3}{2} \frac{kT}{E_H} = 0.1 \tag{11.8}$$

and therefore we have:

$$h\nu = E_H \left(0.1 + \frac{1}{n^2} \right) . \tag{11.9}$$

For $n = 1$ we have $h\nu = 1.1 E_H > E_H$, and for $n > 2$ we have $h\nu < E_H$. Therefore the photons emitted in the recombination at the fundamental level, and only those, are able to re-ionize the H. Given the high cross section of UV photon capture by neutral H, it can be assumed that recombination at the fundamental level is locally balanced by immediate ionization (on-the-spot hypothesis). So when evaluating r_r using (11.6), $\beta(j)$ will contain all levels except the first:

$$r_r = \sum_{j \neq 1} n_p n_e \beta_j(T) = x^2 n^2 \beta^{(2)} \tag{11.10}$$

with

$$\sum_{j \neq 1} \beta_j(T) = \beta^{(2)} \tag{11.11}$$

where $\beta^{(2)}$ is a function of T, as we said for each of the β_j. An approximate expression of this function is given by:

$$\beta^{(2)}(T) = 2 \cdot 10^{-10} T^{-3/4} \text{cm}^3 \text{s}^{-1} \tag{11.12}$$

The UV radiation field has two components: one is emitted by the exciting star and the other, diffuse component, originates in the ionised gas. The latter, since the thermal emission in the UV region is negligible, is due to recombination at the fundamental level and is reabsorbed by an immediate photoionisation.

Ignoring dust absorption, the variation in the flux J of UV radiation is due to the photoionisation process:

$$\frac{d}{dr}(4\pi r^2 J) = -4\pi r^2 n(1-x)\alpha_0 J \tag{11.13}$$

and being:

$$4\pi r^2 J = S_{UV}(r) \tag{11.14}$$

we have:

$$\frac{dS_{UV}}{dr} = -n(1-x)\alpha_0 S_{UV} . \tag{11.15}$$

The amount

$$d\tau_H = n(1-x)\alpha_0 dr \tag{11.16}$$

is the optical depth relative to neutral hydrogen over the distance dr. By integrating one gets:

$$S_{UV}(r) = S_{UV}^* \exp(-\tau_H) \tag{11.17}$$

with

$$\tau_H = \int_0^r n(1-x)\alpha_0 dr . \tag{11.18}$$

Clearly, S_{UV}^* is the flux at the star surface. The relation (11.17), using (11.14), can then be written as:

$$J(r) = \frac{S_{UV}^*}{4\pi r^2} \exp(-\tau_H) . \tag{11.19}$$

The variation of J is therefore due to two causes: (a) the geometric dilution, (b) the absorption of photons by the atoms of H. At small distances from the star the dilution effect prevails, while at large distances this factor becomes negligible with respect to absorption and a geometry with parallel planes can be adopted.

The ionization equilibrium is expressed by the condition $r_i = r_r$ which can be written in the form:

$$\frac{x^2}{1-x} = \frac{J\alpha_0}{n\beta^{(2)}} . \tag{11.20}$$

We can show with an approximate numerical example that in an HII region H is completely ionized: in fact the right hand side of equation (11.20) is a number much larger than 1. To estimate J we suppose that it varies as the inverse of the square of the distance (dilution), i.e. we neglect the reduction due to absorption. Therefore we have:

$$J = \frac{S_{UV}^*}{4\pi r^2} . \tag{11.21}$$

If we take $n = 10^2$ cm^{-3}, $r = 1$ pc, $S^*_{UV} = 10^{49}$ s^{-1}, we have $J(r) = 8.84 \cdot 10^{10}$ cm^{-2} s^{-1}. From the ionization balance equation, using $T = 10^4$ K ($\beta^{(2)} = 2 \cdot 10^{-13}$) we obtain:

$$\frac{x^2}{1 - x} = 3 \cdot 10^4 \gg 1 . \tag{11.22}$$

We can re-arrange the left-hand side as:

$$\frac{x^2}{1 - x} = \frac{x^2 + 1 - 1}{1 - x} = \frac{1}{1 - x} - (1 + x) \tag{11.23}$$

and, since $x \leq 1$ certainly the term $(1 + x)$ is small when compared with $1/(1 - x)$. Therefore, neglecting it, we have:

$$1 - x = 3 \cdot 10^{-5} \ll 1 . \tag{11.24}$$

This implies $x \approx 1$, i.e. the gas is completely ionized. Neglecting the absorption of the atoms of H in the evaluation of J was a good approximation. In fact, assuming the value of $(1 - x)$ just obtained, we get $\tau_H = 0.06$ and $\exp(-\tau_H) = 0.94$.

11.3   Energy Balance and Temperature of an HII Region

We evaluate the temperature of the ionized gas by determining the balance between the heating and cooling processes, in the hypothesis previously adopted that the gas consists only of H. As we will see shortly, the temperature can be determined by the condition $G - L = 0$ instead of the more general relation (4.11).

Given that the energy $Q(\nu)$ transferred to the gas by an ionization produced by a photon of frequency ν is:

$$Q(\nu) = h(\nu - \nu_0) , \tag{11.25}$$

the average value $\langle Q \rangle$ will depend on the temperature of the exciting star and is obtained by averaging the distribution of the UV photons emitted by the star:

$$\langle Q \rangle = \frac{\int_{\nu_0}^{\infty} h(\nu - \nu_0) S^*(\nu) d\nu}{\int_{\nu_0}^{\infty} S^*(\nu) d\nu} . \tag{11.26}$$

We can approximate the emission of the star with that of a black body having a temperature T^* equal to the effective temperature of the star:

$$S^*(\nu) = 4\pi R_*^2 \frac{e_T}{h\nu} \tag{11.27}$$

where e_T is the specific emissive power that is given by Plank's law, so

$$S^*(\nu) = 4\pi R_*^2 \frac{8\pi\nu^2}{c^3} \left(\exp\left[\frac{h\nu}{kT^*}\right] - 1 \right)^{-1}. \tag{11.28}$$

On the other hand, for each frequency of the integration interval it results:

$$\frac{h\nu}{kT^*} \geq \frac{h\nu_0}{kT^*} = \frac{E_H}{kT^*} > 1 \tag{11.29}$$

and therefore we can approximate:

$$S^*(\nu) = 4\pi R_*^2 \frac{8\pi\nu^2}{c^3} \exp\left[-\frac{h\nu}{kT^*}\right]. \tag{11.30}$$

Substituting this expression in (11.26) and introducing the variable change $w = \nu - \nu_0$, we are reduced to calculating only integrals of the type:

$$\int_0^\infty w^p \exp[-\alpha w]dw \tag{11.31}$$

which for $p \in \mathbb{N}$ have a solution

$$\frac{p!}{a^{p+1}}. \tag{11.32}$$

We therefore have:

$$\langle Q \rangle = kT^* \tag{11.33}$$

and therefore the energy gained by the gas per unit of time and per unit of volume is:

$$G = \langle Q \rangle r_i. \tag{11.34}$$

The energy loss is represented by the kinetic energy $E_c = 3kT/2$ of the recombining electron, so the cooling rate per unit of volume will be:

$$L = \frac{3}{2}kTr_r. \tag{11.35}$$

By equating losses and gains and remembering the condition of the ionization balance ($r_i = r_r$) we obtain:

$$T = \frac{2}{3}\frac{\langle Q \rangle}{k} = \frac{2}{3}T^*. \tag{11.36}$$

To evaluate the temperature of the ionized gas we used the condition $G - L = 0$ instead of the more general equation of the balance (4.11). The procedure is correct, and the temperature is determined only by the radiative processes, if the cooling time is less than the dynamical time. It is:

Table 11.2 Temperature of the ionized gas as a function of the effective temperature of the star

T^* [K]	T [K]
25000	7.250
30000	7.600
35000	7.900
40000	8.200
45000	8.500
50000	8.700

$$\tau_{\text{cool}} = \frac{3kTn}{2L} = \frac{1}{n\beta^{(2)}} \approx 1600[\text{yr}] \qquad (11.37)$$

which is smaller, as we will see later, than the dynamic time.

For T^* in the range $(25000 - 53000)$, T falls approximately in the range $(15000 - 35000)$, but observations show $T < 10^4$ K, that is, it is shifted towards lower temperatures. For a correct treatment it is necessary to take into account the emission due to the collisional excitation of elements, which we have neglected, such as OII, OIII and NII: although they are much less abundant they are very effective in cooling, and the amount of radiation emitted in the such elements exceeds that emitted in the lines of the hydrogen series.

Under the physical conditions of the HII regions, elements such as C, O and N are partially ionized; for example the ionization potential of OI is very close to that of hydrogen, the ionization potential of OII is 35.1 eV. Therefore we expect to find O once ionized, and a twice ionized fraction in the regions where the hottest stars are present. The transitions undergoing these elements are prohibited (magnetic dipole or electric quadrupole transitions). In interstellar conditions, the excited levels are populated by collisions of the ion in the ground state with electrons at the expense of the electron's kinetic energy.

By balancing the heating process already considered and the cooling process due to these mechanisms, we find a gas temperature in the HII region of 8000 K, when the exciting star has an effective temperature of 35000 K. As the temperature changes T^* the values indicated in Table 11.1 are obtained. The comment previously made about the method of evaluating the temperature by radiative processes alone remains valid even taking into account the new cooling mechanism, since τ_{cool} is reduced.

In addition to the temperature of the exciting star, the temperature of the ionized gas is highly dependent on the chemical composition, which is obvious since the main cooling agents are C, O and N. Detailed calculations of the heat balance for different compositions show that for a change of a factor of 10 in metallicity there is a temperature change of about 7000 K in the direction of a higher temperature for the poorer metal compositions. These results are also confirmed by observations of the temperatures of the galactic HII regions, measured by radio observations, as a function of the galacto-centric distance (see Fig. 11.1 for the case of NGC 300). The

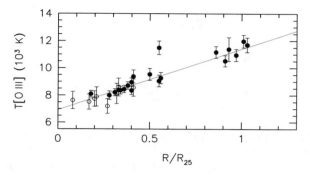

Fig. 11.1 Radial temperature gradient in NGC 300

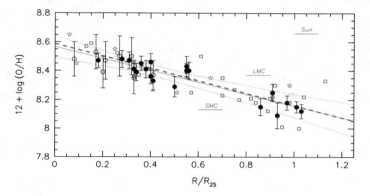

Fig. 11.2 Radial abundance gradient in NGC 300

dependence on metallicity manifests itself in the correlation between metallicity and distance as revealed by Fig. 11.2, again for NGC 300, from [1].

11.4   The Strömgren Sphere

Since the electrons recombine and therefore the hydrogen atoms must be continuously ionized, the UV flux of a star keeps only a finite region ionized. The volume of gas that the star is able to ionize is that in which the total number of recombinations per second equals the flux of photons emitted by the star and, for a given density, is identified by the relationship:

$$S_{UV}^* = \int_0^{R_s} 4\pi r^2 x^2 n^2 \beta^{(2)} \, dr \tag{11.38}$$

and since $x = 1$ and for $n = \text{cost.}$:

$$S_{UV}^* = \frac{4\pi}{3} R_s^3 n^2 \beta^{(2)} \tag{11.39}$$

from this we get the radius R_s:

$$R_s = \left(\frac{3 S_{UV}^*}{4\pi n^2 \beta^{(2)}} \right)^{1/3} \tag{11.40}$$

where R_s is called the radius of the Strömgren sphere. As we will see shortly, it gives the size of the HII region as a function of the density of the ionized gas, which varies with time. If $S_{UV}^* = 10^{49}$ s $^{-1}, n = 10^2$ cm^{-3} e $T = 10^4$ K, it turns out $R_s = 3.5$ pc. Table 11.2 shows the radii of the spheres of Strömgren assumed $n = 10^2$ cm; to have the radii corresponding to different values of the gas density n just multiply the data of the table by $(n/10^2)^{-2/3}$.

In deriving R_s it has been implicitly assumed that J decreases as the distance from the exciting star increases until it is canceled out due to the absorption of photons. This would imply that x varies continuously from 1 to 0, while $x = 1$ is assumed everywhere. In reality, the transition from fully ionized gas to neutral gas occurs over a distance much shorter han R_s. The thickness of the transition zone is determined by the absorption of photons and corresponds to the mean free path:

$$l \approx \frac{1}{(1 - x)n\alpha_0} . \tag{11.41}$$

Putting $x = 0.5$, an intermediate value between the minimum and maximum, we have:

$$l \approx \frac{0.1}{n}[\text{pc}]. \tag{11.42}$$

For $n = 10^2$cm^{-3} we have $l = 10^{-3}$ pc, which is much smaller than the size of the ionized volume.

11.5   Effect of Dust on the HII Region

The radius of the Strömgren sphere is influenced by the presence of dust grains since, due to the absorption of UV photons by the dust, fewer atoms can be ionized and consequently the Strömgren sphere is expected to squeeze.

Let's analyze this effect. Ignoring the absorption of both powders and hydrogen, the photo-ionization rate is given by (11.5), in which the J flux can be expressed as S_{UV}^*:

$$r_i = n(1 - x)\alpha_0 \frac{S_{UV}^*}{4\pi r^2} . \tag{11.43}$$

Considering the absorption of dust and hydrogen the previous relationship will be written:

$$r_i = n(1-x)\alpha_0 \frac{S_{UV}^* \exp[-\tau]}{4\pi r^2} \tag{11.44}$$

where τ is the optical depth due to hydrogen and grains:

$$\tau = \tau_H + \tau_g . \tag{11.45}$$

By imposing the equilibrium condition $r_i = r_r$, we have:

$$\frac{1-x}{x^2} = \frac{4\pi n \beta^{(2)}}{\alpha_0 S_{UV}^*} r^2 \exp[\tau] \tag{11.46}$$

which can be put in the form:

$$\frac{1-x}{x^2} = 3\left(\frac{r}{R_s}\right)^2 \frac{\exp[\tau]}{n\alpha_0 R_s} . \tag{11.47}$$

The quantity $\tau_{H,s}^{(n)} = n\alpha_0 R_s$, is the optical depth over the distance R_s due to hydrogen if this were all neutral. Let's set $y = r/R_s$. With this position (11.47) becomes:

$$\frac{1-x}{x^2} = \frac{3y^2 \exp[\tau]}{\tau_{H,s}^{(n)}} . \tag{11.48}$$

We also have:

$$\frac{d}{dy}\tau_H = R_s \frac{d}{dr}[(1-x)n\alpha_0 R_s] = (1-x)n\alpha_0 R_s \tag{11.49}$$

or:

$$\frac{d\tau_H}{dy} = (1-x)\tau_{H,s}^{(n)} . \tag{11.50}$$

Substituting the expression of $(1-x)$ from (11.48) into (11.50) we obtain:

$$\frac{d\tau_H}{dy} = 3x^2 y^2 \exp[\tau_H + \tau_g] . \tag{11.51}$$

Let's set $x = 1$, multiply by $\exp[-\tau_H]$, and integrate between 0 and $y_g = R_{s,g}/R_s$, where $R_{s,g}$ is the radius of the Strömgren sphere in presence of dust. We have:

$$\int_0^{\tau_{H,i}} \exp[-\tau_H] d\tau_H = 3 \int_0^{y_g} y^2 \exp[\tau_g] dy \tag{11.52}$$

where $\tau_{H,i}$ is the optical depth of H over the distance $R_{s,g}$.

The integral in the left hand side of (11.52) amounts to about 1 being $\tau_{H,i}$ large, since all photons are absorbed within the ionized region: at distance $R_{s,g}$ all photons are absorbed by gas and dust. Here $\tau_{H,i}$ refers only to gas and not to dust, but as $\tau_H > \tau_g$ (see below) the conclusion is also valid for $\tau_{H,i}$.

Furthermore, supposing that the dust is uniformly distributed, we have $\tau_g = y\tau_{s,g}$ with $\tau_{s,g} = n_g Q_e \sigma_g R_{s,g}$ and therefore $\tau_{s,g}$ is the optical depth of the grains over a distance equal to the radius of Strömgren, calculated in the absence of dust. It will then be:

$$\int_0^{y_g} y^2 \exp[y\tau_{s,g}]\mathrm{d}y = \frac{1}{3}. \tag{11.53}$$

For a given value of $\tau_{s,g}$ you can choose y_g so that the integral matches (11.53); this way Table 11.3 is constructed. It lists the ratio $R_{s,g}/R_s$ between the radius of the Strömgren sphere in the presence of dust and its radius in the absence of dust as a function of the optical thickness of the dust at UV wavelengths .

To evaluate the effect of dust grains we still have to calculate the optical depth due to the powders. It is:

$$\frac{\tau_{s,g}}{\tau_{H,s}^{(n)}} = \frac{\tau_{s,g}}{n\alpha_0 R_s} = \frac{\tau_{s,g}}{\alpha_0 N_H(R_s)} \tag{11.54}$$

with $N_H(R_s) = nR_s$. From the definition of extinction:

$$A_\lambda = -2.5 \log \frac{I_\lambda}{I_\lambda(0)} = 1.086\tau_g(\lambda) \tag{11.55}$$

hence for $\lambda = 912$ Å we have:

$$\tau_{s,g} = \frac{1}{1.086} \frac{A_{912}}{E(B-V)} E(B-V) \tag{11.56}$$

making use of the mean extinction curve relative to our galaxy we obtain $A_{912}/E(B-V) = 13$ for which:

$$\tau_{s,g} = \frac{13}{1.086} E(B-V) \tag{11.57}$$

therefore:

$$\frac{\tau_{s,g}}{\tau_{H,s}^{(n)}} = 12 \frac{E(B-V)}{\alpha_0 N_H(R_s)}. \tag{11.58}$$

The ratio $E(B-V)/N_H(R_s)$ is also known from observations and for the Galaxy is $1.7 \cdot 10^{-22}$. We then have:

$$\frac{\tau_{s,g}}{\tau_{H,s}^{(n)}} = 3 \cdot 10^{-4}. \tag{11.59}$$

Table 11.3 Numerical solution of the integral in Eq. (11.53)

$\tau_{s,g}$	y_g	$\tau_{s,g}$	y_g
0.1	0.98	3.0	0.62
0.2	0.96	4.0	0.56
0.4	0.91	6.0	0.47
0.6	0.87	8.0	0.42
0.8	0.84	10.0	0.37
1.0	0.81	15.0	0.30
1.5	0.75	20.0	0.25
2.0	0.70	40.0	0.15

Obviously, for objects belonging to other galaxies, the appropriate quantities must be used. Let us consider the following example: assume, as usual, $S_{UV}^* = 10^{49}$ s^{-1}, an ambient density of $n = 10^2$cm^{-3}. The radius of Strömgren in the absence of dust is $R_s = 1.06 \cdot 10^{19}$ cm ($= 3.53$ pc) and therefore (Table 11.3):

$$\tau_{H,s}^{(n)} = \alpha_0 n R_s = 7208 \tag{11.60}$$

$$\tau_{s,g} = \frac{\tau_{H,s}^{(n)}}{3100} = 2.16 . \tag{11.61}$$

Interpolating data from Table 11.3 gives $y_g = 0.69$. The radius of the Strömgren sphere in the presence of dust is about 70% of that in the absence of dust. It results $R_{s,g} = 7.31 \cdot 10^{18}$ cm $=2.44$ pc .

11.6 Evolution of the HII Region in a Homogeneous Medium

The radius of Strömgren defines the size of the HII region when the density of the ionized gas is known. As we will see, this density varies over time and therefore the size will also vary. We now attempt to describe this evolution.

The exciting star of an HII region forms in areas of dense molecular gas. To become a source of UV photons it must reach the main sequence and this evolution is so rapid that we can assume that it suddenly " turns on ".

The photons emitted ionize a given region of gas around the star: this region is bounded by a transition layer, in which the gas is only partially ionized and whose thickness is comparable with the average free path of the photons; given its narrowness we can assume that the ionized zone with $x = 1$ is delimited by an ionization front that can be schematized with a mathematical surface. Compared to a reference integral with the star, the ionization front advances with a speed that is determined by the flux of UV photons.

We can describe the situation with the following model: the ionized sphere is separated from the neutral gas by a front that moves outward as new atoms are ionized by the photons. Let R be the radius of this sphere; we consider a cylinder of unitary base area and height dR: the advancement of the front is determined by the condition that the number of UV photons crossing the base of the cylinder in the time dt is equal to the number of atoms contained in the cylinder, i.e.

$$J dt = n dR .$$
(11.62)

The speed of advance with respect to the neutral gas, assumed at rest with respect to the star, is then:

$$\frac{dR}{dt} = \frac{J}{n} .$$
(11.63)

From (11.15) making use of the ionization balance we obtain:

$$\frac{dS_{UV}}{dt} = -4\pi r^2 \beta^{(2)} n^2$$
(11.64)

and integrating up to the generic distance R from the star we obtain:

$$J = \frac{S_{UV}^*}{4\pi R^2} - \frac{1}{3} R n^2 \beta^{(2)}$$
(11.65)

which, with (11.63) gives:

$$\frac{dR}{dt} = \frac{S_{UV}^*}{4\pi R^2 n} - \frac{1}{3} R n \beta^{(2)} .$$
(11.66)

By means of the radius of Strömgren R_s, defined by (11.40) and the characteristic time of recombination t_r, defined by (11.11):

$$\frac{1}{t_r} = \frac{1}{n}\frac{dn}{dt} = n\beta^{(2)}$$
(11.67)

we can introduce two dimensionless variables:

$$\lambda = \frac{R}{R_s} \qquad \tau = \frac{t}{t_r}$$
(11.68)

from which it results:

$$\lambda' = \frac{d\lambda}{dt} = \frac{d}{d\tau}\frac{R}{R_s} = \frac{1}{R_s}\frac{dR}{dt} = \frac{t_r}{R_s}\frac{dR}{d\tau} .$$
(11.69)

With the Eq. 11.66 we obtain:

$$\lambda' = \frac{1 - \lambda^3}{3\lambda^2} :$$ (11.70)

for which the function

$$\lambda = (1 - \exp[-\tau])^{1/3}$$ (11.71)

it is a solution of the equation with $\lambda \to 0$ for $\tau \to 0$. A discussion of this solution and its physical significance is facilitated by a classification of the ionization fronts in relation to their velocity with respect to the neutral gas.

11.7 Classification of Ionization Fronts

The behavior of an ionization front can be studied, in a reference integral with it, in a manner similar to that used for the study of shock waves, making use of the Rankine-Hugoniot conditions.

From the conservation of the flow of matter we have:

$$\rho_1 v_1 = \rho_2 v_2 .$$ (11.72)

From the conservation of momentum, assuming $p_1 = c_1^2 \rho_1$ and $p_2 = c_2^2 \rho_2$, we obtain:

$$\rho_1 (v_1^2 + c_1^2) = \rho_2 (c_2^2 + v_2^2) .$$ (11.73)

We do not need to consider here the third condition, that relating to the conservation of energy, but remember that, given the role of the radiative processes, the temperature behind the ionization front is determined by the condition $G - L = 0$ as in the case of shocks radiative.

By removing v_2 from (11.72) and (11.73) we get an equation in ρ_2/ρ_1 which has for solutions:

$$\frac{\rho_2}{\rho_1} = \frac{(c_1^2 + v_1^2) \pm \sqrt{(c_1^2 + v_1^2)^2 - 4v_1^2 c_2^2}}{2c_2^2}$$ (11.74)

There are real solutions when the discriminant is non-negative, i.e. when

$$c_1^2 + v_1^2 \geq 2v_1 c_2 .$$ (11.75)

This inequality is satisfied when the velocity v_1 of the neutral gas with respect to the ionization front is greater than a higher critical value v_R given by:

$$v_R = c_2 + \sqrt{c_2^2 - c_1^2}$$ (11.76)

Fig. 11.3 Qualitative trend
of ionization front evolution:
radius

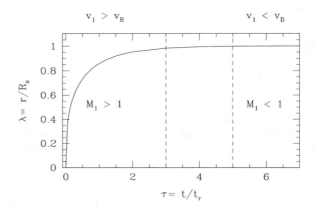

or is less than a lower critical value v_D defined by:

$$v_D = c_2 - \sqrt{c_2^2 - c_1^2} \ . \tag{11.77}$$

Using the condition $c_2 \gg c_1$, which descends from $T_2 \gg T_1$, we can approximate
the two critical values with:

$$v_R \approx 2c_2 \qquad v_D \approx \frac{c_1^2}{2c_2} \tag{11.78}$$

this last expression is obtained from (11.77) by setting $x = c_1^2/c_2^2$ and developing in
series by truncating at the end of the first order.

The classification of the ionization fronts is done by comparing the value of the
velocity v_1 with the critical values, and leads to the following classes.

(1) *Front of type R*: is the one for which $v_1 > v_R$ (the suffix stands for rarefied as
with a sufficiently low density ρ_1 the speed will exceed the critical value). In
this case it is $v_1 > 2c_2 > c_1$ i.e. the ionization front moves supersonically in the
neutral gas.
(2) *Front of type D*: is the one for which we have $v_1 < v_D$ (the suffix stands for dense
because this condition is verified if the density of the neutral gas is high enough).
Since $c_1 < c_2$, the advancement speed of the ionization front in the neutral gas
is subsonic.
(3) When $v_D < v_1 < v_R$ an ionization front of the considered types is not possible.
In this case a shock front develops in front of it and due to the compression
exerted by this the gas, which remains neutral, meets the ionization front with a
velocity $v_1 \leq v_D$. When $v_1 \leq v_D$ the ionization front is called $D - critical$.

A qualitative description of the ionization front progress is sketched in Figs. 11.3 and
11.4.

Fig. 11.4 Qualitative trend of ionization front evolution: velocity

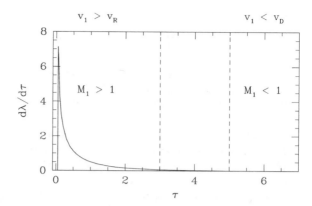

Table 11.4 Evolution of the ionization front

t/t_r	R/R_s	v [km/s]
1.	0.86	604.35
2.	0.95	180.43
3.	0.980	62.33
3.25	0.987	48.17
3.5	0.990	37.29
3.75	0.992	28.91
4.	0.994	22.44

Let us consider a numerical example: an HII region excited by an O8-V star ($S_{UV}^* = 5 \cdot 10^{48}$ s^{-1}) in a molecular cloud with density uniform $n_0 = 10^3$ cm^{-3}. The radius of Strömgren is $R_s = 1.81 \cdot 10^{18}$ cm =0.6 pc and the recombination time is $t_r = 5 \cdot 10^9$ s =167 yr. The evolution of the front is described by (11.71) and the radius of the ionized zone and its expansion rate are given by:

$$R = R_s \left(1 - \exp\left[-\frac{t}{5 \cdot 10^9}\right]\right)^{\frac{1}{3}} \tag{11.79}$$

$$v = \frac{dR}{dt} = \frac{R_s}{1.5 \cdot 10^{10}} \exp\left[-\frac{t}{5 \cdot 10^9}\right] \left(1 - \exp\left[\frac{t}{5 \cdot 10^9}\right]\right)^{-\frac{2}{3}}. \tag{11.80}$$

Taking for the ionized gas $T_2 = 10^4$ K, $\mu_2 = 0.5$, and for the molecular neutral gas $T_1 = 10$ K, $\mu_1 = 2$, it results for the speed of sound $c_2 = 12.86$ km/s and $c_1 = 0.2$ km/s, and for the critical values $v_R = 25.71$ km/s and $v_D = 3 \cdot 10^{-3}$ km/s.

Initially, the advancement speed of the front with respect to the neutral gas is greater than v_R and therefore it is a R type front. At $t \approx 3.88 t_r \approx 1.94 \cdot 10^{10}$ s ≈ 647 yr, v_1 becomes equal to the critical rate v_R.

These results allow us to draw the following conclusions which are generally valid for the early stages of evolution of the HII regions.

- The ionization of the gas increases the pressure of the ionized gas by more than three orders of magnitude, both for the increase in temperature and for the increase in the number of particles (4 particles are obtained from a molecule of H_2); in this way an imbalance is created between the ionized gas and the neutral gas. This imbalance occurs in a time R/v_1 which is less than the time of crossing of the region by the sound signals $t_c = R/c_2$ so that the ionized gas is not able to react to the force that has occurred and retains the initial density. This situation persists until $v_1 \approx c_2$. This first phase is correctly described by the Eq. 11.70 which was deduced by keeping $n_0 = \text{cost}$.
- Initially, basing on shock-front classification, the ionization front is of the R type, but at approximately $t = kt_r$ (with $k = 3 - 5$) the expansion rate becomes equal to v_R and in front of the ionization front a shock front develops which compresses the gas and creates a shell of neutral and dense gas in front of the ionization front. From this moment on, the speed of advancement of the front becomes comparable with the speed of sound of the ionized gas and this begins to expand; the subsequent evolution of the HII region therefore requires a new model.

11.8   HII Region Expansion

We can therefore schematize the evolution of an HII region in three phases.

- *I phase*: the ionization front expands quickly but the ionized gas retains the initial density n_0. We have seen that the first phase is described by the solution of the equation (11.70) which remains valid until the speed of advancement of the ionization front becomes comparable with the speed of sound in the ionized gas.
- *II phase*: a shock front is generated in front of the ionization front, between the two there is a region with density $n_{HI} > n_0$. The ionization front is transformed into a front of the D ($v_1 < v_D$) type. This intermediate phase is difficult to deal with and requires the numerical solution of the dynamics equations, its duration is short and will not be further considered.
- *III phase*: this is the expansion phase itself and will be discussed shortly.
- *IV phase*: conclusion of the expansion. This ends for one of the following two reasons: (a) the ionized gas reaches equilibrium with the neutral gas, (b) the ionizing flux is turned off because the star leaves the main sequence and its temperature becomes too low to produce a significant UV flux.

Let us now consider the expansion phase (phase III). To be able to deal with it in a simple way, we make the following assumptions.

(1) The shocked neutral gas layer is thin due to the compression produced by the shock. With a more detailed analysis it can also be shown that, as long as the

pressure p_2 within the ionized region is much greater than the p_1 of the unperturbed neutral gas, the velocity of the ionization front V_i and that of the shock wave V_{sh} are approximately equal. It can therefore be assumed that the distance of the two edges from the star is the same, therefore, indicating the distance with R, their speed of advance will be given by dR/dt.

(2) Behind the shock, the pressure of the neutral gas and the ionized gas are equal. For this to be true, it is necessary that the crossing time of a sound signal is less than the time required for a significant variation of the structure: this condition is certainly verified for the neutral zone and becomes an increasingly better approximation for the ionized gas during the evolution.

(3) The shock is strong: actually the shock weakens and eventually the strong shock approximation would not be correct.

(4) The neutral gas in front of the shock is stationary, so dR/dt is the rate of advance of the shock with respect to the neutral gas.

It turns out then:

$$p_i = 2n_i kT \tag{11.81}$$

with n_i number of nuclei of H per cm^3 in the ionized region and:

$$c_i^2 = \frac{p}{\rho} \approx \frac{kT}{\mu m_H} = \frac{2kT}{m_H} \tag{11.82}$$

and therefore you can write:

$$p_i = 2n_i kT = n_i m_H c_i^2 . \tag{11.83}$$

The pressure behind a strong shock wave in the adiabatic case is:

$$p_2 = \frac{2\gamma \rho_1 v_1^2}{\gamma + 1} \approx \rho_1 v_1^2 = m_H n_0 v_1^2 . \tag{11.84}$$

This relation is obtained starting from the first two connection conditions in the case $Ma_1 \gg 1$ (see (5.45)): since they also hold in the case of isothermal shock, this relation is of general validity. In the case of the shock now considered, the neutral gas cooling processes are effective: the shock is isothermal and therefore $\gamma = 1$ has been set. From $p_i = p_2$ we get then:

$$\left(\frac{dR}{dt}\right)^2 = \frac{n_i}{n_0} c_i^2 . \tag{11.85}$$

Based on the fact that, neglecting the effect of the dust, the number of recombinations within the ionized region is equal to the flux of UV photons, it can be deduced that:

$$S_{UV}^* = \frac{4\pi}{3} R^3 n_i^2 \beta^{(2)} . \tag{11.86}$$

Table 11.5 Tabulated solution of the equation (11.95)

N	λ	λ'	$\lambda^3\lambda'^2/N$
0.1	1.10	0.93	11.5
0.5	1.43	0.76	3.43
1	1.78	0.65	2.38
5	3.67	0.38	1.41
20	7.75	0.22	1.08
100	19.2	0.11	0.84
250	32.3	0.07	0.74
300	35.9	0.07	0.71

With the definition of R_s and the two previous relations we obtain:

$$\left(\frac{dR}{dt}\right)^2 = c_i^2 \left(\frac{R_s}{R}\right)^{\frac{3}{2}} .$$

(11.87)

By introducing the non-dimensional variables λ and N defined by:

$$\lambda = \frac{R}{R_s} \qquad N = \frac{c_i t}{R_s}$$

(11.88)

the equation becomes:

$$\lambda' = \lambda^{-3/4}$$

(11.89)

with

$$\lambda' = \frac{d\lambda}{dN} = \frac{1}{c_i}\frac{dR}{dt} .$$

(11.90)

With the initial condition $N = 0$, $\lambda = 1$ (we start measuring the time when $R = R_s$) the solution is given by (Table 11.5):

$$\lambda) = \left(1 + \frac{7}{4}N\right)^{\frac{4}{7}}$$

(11.91)

$$\lambda' = \left(1 + \frac{7}{4}N\right)^{-\frac{3}{7}} .$$

(11.92)

Let us briefly mention how the study of the expansion of the HII region should be set up, taking into account the effect of dust. In the absence of dust, the expansion was studied using the Eq. 11.89 obtained from (11.85) and (11.86); now while the (11.85) remains valid, the (11.86) must be replaced with an equivalent equation that takes into account the effect of the dust grains.

The equation for the variation of S_{UV} is:

$$\frac{dS_{UV}}{dr} = -4\pi r^2 n^2 \beta^{(2)} - n_g \sigma_g S_{UV}(r) \tag{11.93}$$

where n_g and σ_g are the numerical density of grains and the cross section. By integrating you get:

$$S_{UV} = S_{UV}^* \exp\left[-\tau_{s,g}\frac{n}{n_0}y\right] + 6\frac{S_{UV}^*}{\tau_{s,g}^3}\left(\frac{n}{n_0}\right)^{-1}\left\{\exp\left[-\tau_{s,g}\frac{n}{n_0}\right] - 1 - \frac{1}{2}\tau_{s,g}^2\left(\frac{n}{n_0}\right)^2 y^2 + \tau_{s,g}\frac{n}{n_0}y\right\}. \tag{11.94}$$

By setting $S_{UV} = 0$ we have an equation in $y = R_{s,g}/R$, which for an assigned value of n gives the edge of the ionized zone. By substituting this quantity, the equation in (11.85) is determined by numerical integration $R = R(t)$.

11.9 End of Expansion

It has been said that the ionized gas expands because its pressure is greater than that of the neutral gas; if the UV radiation source remains active long enough, the final state will be characterized by an equilibrium condition in which the pressure of the unperturbed neutral gas equals the pressure of the ionized gas. The pressure of the neutral gas p_n and the final pressure of the ionized gas p_f are respectively:

$$p_n = n_0 k T_n \qquad p_f = 2n_f k T_f \tag{11.95}$$

where n_f is the final number of nuclei per cm^3 in the ionized region. The temperatures T_n and T_f are respectively 10 and 8000 K. The pressure equilibrium gives:

$$n_0 k T_n = 2n_f k T_f . \tag{11.96}$$

The ionized sphere continues to absorb UV photons emitted by the star:

$$S_{UV}^* = \frac{4\pi}{3} R_f^3 n_f^2 \beta^{(2)} \tag{11.97}$$

where R_f is the final radius.
From (11.40), (11.96) and (11.97) we have:

$$\frac{n_f}{n_0} = \frac{T_n}{2T_f} \tag{11.98}$$

$$\frac{R_f}{R_s} = \left(\frac{2T_f}{T_n}\right)^{\frac{2}{3}} . \tag{11.99}$$

The ratio between the mass M_f of gas contained in the final ionized sphere with respect to the initial M_s is:

$$\frac{M_f}{M_s} = \frac{R_f^3 n_f}{R_s^3 n_0} = \frac{2T_f}{T_n} . \tag{11.100}$$

We have:

$$\frac{n_f}{n_0} = 6 \cdot 10^{-4} \qquad \frac{R_f}{R_s} = 138 \qquad \frac{M_f}{M_s} = 1600 . \tag{11.101}$$

The final density is much lower than the original density. We have seen that at the equilibrium $R_f/R_s = 138$, based on (11.91) a $\lambda = 138$ corresponds to $N_f = 3175$ and the time at which the equilibrium is reached is given by:

$$t_f = \frac{N_f R_s}{c_i} = 1.8 \cdot 10^{10} n_0^{-2/3} [\text{yr}] . \tag{11.102}$$

However, equilibrium is not necessarily reached within the ionized region: in order for this to take place, the time t_f required to achieve equilibrium must be less than or at most equal to the life time t_{MS} of the star in main sequence.

In fact, for a density of 10^2 cm^{-3} the equilibrium time is about $8 \cdot 10^8$ years, much greater than the main sequence life time of the star with $S_{UV}^* = 10^{49}$ s^{-1} (for which we have done the calculations) which is approximately 10^7 years.

Under these conditions, the HII region is deactivated, before reaching equilibrium, when the star leaves the main sequence and progressively all hydrogen recombines. What are the characteristics of the HII region at this moment? For $t = 10^7$ years we have $N = 34.5$ and with this value we get $\lambda = 11.5$ and in correspondence it is $R = 38.3$ pc.

The density of the environment required to allow the HII region to reach equilibrium is obtained by imposing that $t_f = 10^7$ and from (11.102) we obtain $n_0 = 7.5 \cdot 10^4$ cm^{-3}, which is quite high even for a molecular cloud.

Let's now make a few considerations on the energy injected into the ISM. The ISM heating per photon is given by:

$$Q(\nu) = h\nu \tag{11.103}$$

Following the same procedure used in (11.3) we can define a mean energy per photon deposited in the ISM as:

$$\langle h\nu \rangle = \frac{\int_{\nu_0}^{\infty} h\nu S_{UV}^*(\nu)d\nu}{\int_{\nu_0}^{\infty} S_{IV}^*(\nu)d\nu} . \tag{11.104}$$

and can be computed as:

$$\langle h\nu \rangle \approx KT^* \tag{11.105}$$

where T^* is the star effective temperature. Actually, as we have seen, KT^* is a measure of the mean value of $h(\nu - \nu_o)$, but the previous approximation holds is we work with order of magnitudes. If we now consider phase III expansion solution, we found that $\frac{dR}{dt} < c_i$; besides, $(\frac{dr}{dt})^2$ is proportional to the kinetic energy E_k, while c_i^2 is proportional to the thermal energy E_{Th}. As a consequence we have:

$$E_T = E_K + E_{Th} \approx E_{Th} = \frac{4\pi}{3} R^2 2n_i \frac{3}{2} KT_i = 2\pi R^3 n_i H c_i^2 \qquad (11.106)$$

where (11.83) was used. The opposite happens for the neutral gas, because it cools down (isothermal shock):

$$E_n = E_{c,n} + E_{th,n} \approx E_{c,n} = \frac{1}{2} \left(\frac{4\pi}{3} R^3 n_0 m_H - \frac{4\pi}{3} R^3 2n_i m_H \right) \left(\frac{dR}{dt} \right)^2 . \qquad (11.107)$$

In the final stages it is $n_0 \gg n_i$ hence:

$$E_n = \frac{2\pi}{3} R^3 n_0 m_H \left(\frac{dR}{dt} \right)^2 . \qquad (11.108)$$

The comparison of (11.107) and (11.108) yields:

$$\frac{E_n}{E_i} = \frac{2}{3} \frac{n_0}{n_i c_i^2} \left(\frac{dR}{dt} \right)^2 \approx \frac{1}{3} \qquad (11.109)$$

being $(dR/dt)^2 = (n_i/n_0)c_i^2$. Therefore, the energy purchased from the gas is of the order of E_n. The conversion efficiency f is the ratio between E_n and E^*, i.e.

$$f = \frac{E_{c,n}}{E^*} = \frac{2\pi R^3 n_0 m_H}{3\langle h\nu \rangle S_{UV}^* t} \left(\frac{dR}{dt} \right)^2 . \qquad (11.110)$$

Putting $\langle h\nu \rangle = 5 \cdot 10^{-12}$ erg and using the expressions of c_i, $\beta^{(2)}$, and R_s we have:

$$f \approx 5 \cdot 10^{-3} n_0^{-1/3} \frac{\lambda^3 \lambda'^2}{N} . \qquad (11.111)$$

From Table 11.4 a reasonable value of $\lambda^3 \lambda'^2/N$ results to be 1. Adopting $n_0 = 10^2 \text{cm}^{-3}$ we eventually have $f = 10^{-3}$.

11.10 Expansion of an HII Region in an In-Homogeneous Medium

The discussion that we have followed so far assumes that the expansion of the HII region occurs in a homogeneous medium. It may happen that the star forms in an internal region of the molecular cloud so that during its evolution the HII region always remains confined within the cloud. If the cloud is homogeneous, the evolution in the essential lines is that described in the previous paragraphs. Obviously the quantitative aspects of evolution will depend on the density of the environment: size and speed of expansion will be a function of this.

If the star is formed near the edge of the molecular cloud, for part of its evolution the HII region remains confined to the cloud, but starting from a certain instant the shock front and the ionization front partially transshield in the ambient medium which surrounds the cloud and which has a lower density. In this more rarefied environment, the ionization front moves faster than in the cloud, giving rise to a structure like the one represented in Fig. 11.5. This model is called a "blister" or champagne model.

It may also happen that a small cloud or a region of however higher density is engulfed by the HII region. When the shock front meets the cloud, it is progressively destroyed as the gas on its surface is ionized and flows away (photoevaporation). The effect of the presence of a set of these clouds in the middle where the HII region expands has been studied: in contrast to the homogeneous case the radius of the Strömgren sphere contracts, rather than expanding, because photoevaporation increases the density of the medium.

11.11 Radio Frequency Emission

The mechanisms responsible for continuous and line output are different. *Continuous emission* Bremsstrahlung (free-free) radiation is produced when electrons are braked in a Coulomb field. This radiation is emitted over a wide range of frequencies but is more easily observed at radio frequencies.

For this radiation the HII region is partially opaque and therefore it is necessary to use the transport equation. Let I_ν be the intensity of the radiation: it is the energy emitted in the unit frequency interval centered around ν that crosses the unit of area in the unit of time with direction included within the angle solid unit. Let j_ν be the emission coefficient, that is, the energy emitted by the gas in the unit of time from the unit of volume per unit of solid angle in the unit frequency interval. Furthermore, let k_ν be the absorption coefficient. For a plane-parallel geometry the transport equation is:

$$\frac{\mathrm{d}I_\nu}{\mathrm{d}s} = -k_\nu I_\nu + j_\nu . \tag{11.112}$$

In conditions of thermodynamic equilibrium, the Kirchhoff principle applies:

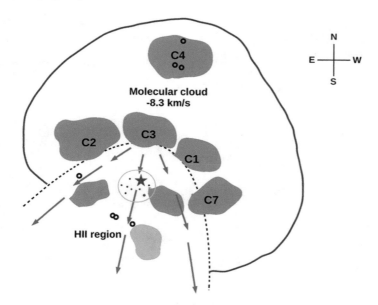

Fig. 11.5 Interaction of an HII region with the surrounding patchy medium

$$\frac{j_\nu}{k_\nu} = B_\nu(T) = \frac{2h\nu^3}{c^2}\left(\exp\left[\frac{h\nu}{kT}\right] - 1\right)^{-1} \tag{11.113}$$

where $B_\nu(T)$ is the Plank function corresponding to the system temperature.

The electronic gas has a Maxwellian distribution with temperature T so the emission and absorption coefficients of the bremsstrahlung are linked through the (11.113). With the optical depth at the frequency ν:

$$d\tau_\nu = k_\nu ds \tag{11.114}$$

the Eq. 11.112 becomes:

$$\frac{dI_\nu}{d\tau_\nu} = B_\nu(T) - I_\nu . \tag{11.115}$$

Assuming $T = $ cost. and $I_\nu = 0$ for $\tau_\nu = 0$ (there are no other sources) we have:

$$I_\nu = B_\nu(T)(1 - \exp[-\tau_\nu]) . \tag{11.116}$$

Let's consider the borderline cases:

$$I_\nu \propto \tau_\nu B_\nu(T) \quad if\ \tau_\nu \ll 1 \tag{11.117}$$

$$I_\nu \propto B_\nu(T)T \quad if\ \tau_\nu \gg 1 . \tag{11.118}$$

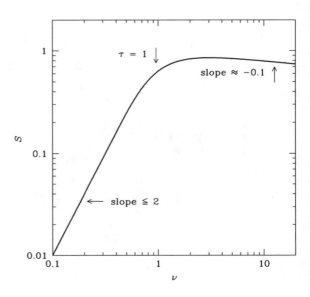

Fig. 11.6 Sketch of continuum spectrum at radio frequency. Courtesy of Valentina Canton

Optical depth is expressed by the relation:

$$\tau_\nu = A\nu^{-2.1}T^{-1.35}n_e^2 L \tag{11.119}$$

with $A = $ cost., and where L is the path length through the HII region. The quantity $E_m = n_e^2 L$ is called the emission measure.

At radio frequencies ($h\nu/kT \ll 1$) the Plank function can be approximated with the Rayleigh function:

$$B_\nu(T) \propto \frac{2kT\nu^2}{c^2} . \tag{11.120}$$

If $\tau_\nu \ll 1$ (based on the expression of τ_ν this implies high frequencies):

$$I_\nu \propto \tau_\nu B_\nu(T) \propto \nu^{-0.1} \tag{11.121}$$

while if $\tau_\nu \gg 1$ (lower frequencies):

$$I_\nu \propto B_\nu(T) \propto \nu^2 . \tag{11.122}$$

The shape of the spectrum is represented in Fig. 11.6; the intersection of the two tangents corresponds to the condition $\tau \approx 1$ and determines the frequency ν_0. Placing $\tau = 1$ and $\nu = \nu_0$ in (11.119), E_m is determined from the knowledge of T and the electron density can be estimated from the estimate of L.

Radio spectrum of lines. The recombination brings the electrons to all the excited levels and therefore cascade transitions bring the atom back to the fundamental level giving rise to a spectrum of lines, in particular in the visible the transitions at the level $n = 2$ originate the lines of the Balmer series.

What transitions give rise to radio domain lines? Let us consider a transition from level n to level n', the corresponding frequency is

$$\nu_{n,n'} = \mathcal{R} \left(\frac{1}{n'^2} - \frac{1}{n^2} \right) \tag{11.123}$$

with $\mathcal{R} = 3.3 \cdot 10^{15}$ Hz (Rydberg constant). The previous relationship can be put in an approximate form as follows:

$$\nu_{n,n'} = \mathcal{R} \frac{(n + n')(n - n')}{n'^2 n^2} = \frac{2\mathcal{R}\Delta n}{n'^3} . \tag{11.124}$$

Setting $\Delta n = 1$, $n' = 100$ we have $\nu_{n,n'} = 6.6 \cdot 10^9$ Hz and $\lambda = c/\nu_{n,n'} = 4.5$ cm. The radii of the orbits corresponding to $n = 100$ are relatively large but the orbits are still well defined, being their radius much smaller than the average distance between the particles.

11.12 Infrared Emission

The bremsstrahlung emission of the ionized gas also concerns the infrared region but the observational data show that in reality the emission of an HII region in this portion of the spectrum is very different, as shown in Fig. 11.7. While the radio emission is in agreement with the theoretical expectations, in the infrared there is an emission maximum at $\lambda = 100\mu$ and in any case the radiation is in excess of the bremsstrahlung emission in the interval $(1\ \mu, 1\ mm)$. The figure also shows a curve very similar to a black body emission with a temperature of 70 K.

The interpretation of this behavior is due to the presence of dust and their emission. Hot stars emit a significant fraction of its radiation in the form of ionizing photons ($\lambda < 912$ Å). Part of this radiation is absorbed by the dust and part by the gas: the absorption by the gas takes place mainly at the edge of the HII region where $x \neq 1$ and where the optical depth due to hydrogen becomes important. The non-ionizing radiation ($\lambda > 912$ Å) is largely absorbed by the dust.

The temperature of the grains is determined by the balance between the radiation absorption process and their thermal emission (see Sect. 10.7). To evaluate it, it is necessary to use a model that describes their properties and the result will depend on these assumptions. We adopt a model consisting of two types of grains:

(1) the first type has relatively large dimensions (radius $a_1 = 10 - 5$ cm) and a high emissivity in the IR (by Kirchhoff's principle this implies a high absorption factor), i.e. $Q_{a,1} = 2\pi a_1/\lambda$;

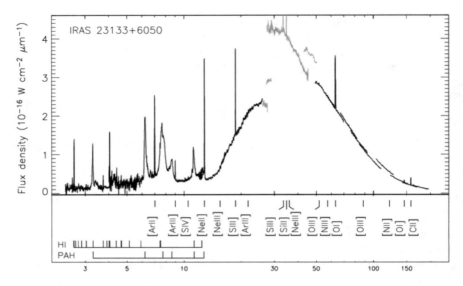

Fig. 11.7 IR spectrum of an HII region

(2) the second type has small dimensions and low emissivity ($a_2 = 2 \cdot 10^{-6}$ cm and $Q_{a,2} = 7 \cdot 10^{-4} a_2/\lambda$) .

The results are reported in Table 11.6 which also indicates the wavelength of the emission maximum of a black body of the same temperature according to Wien's displacement law. It is obvious that, with these temperatures, the emission of the grains takes place mainly in the infrared.

11.13   Calculation of the Lines of the Balmer Series

It is possible to construct a model by which to calculate the flux of the first lines of the Balmer series as a function of the mass of the exciting star. From the atmosphere models the number N_{UV} of UV photons emitted per cm^2 and per second is determined in correspondence to the effective temperature and luminosity; N_{UV} is a function of only the temperature and not of the surface gravity. We the get:

$$S^*_{UV} = 4\pi R^2 N_{UV} . \tag{11.125}$$

By means of the evolutionary traces it is possible to determine the brightness and the temperature at a certain instant t^* and therefore the stellar radius and the total number of UV $S_{UV}(m)$ photons emitted per second.

It has been seen that in a HII region for the recombination of hydrogen atoms the lines of the Balmer series are emitted; the flow of the H β line can be expressed as a function of S_{UV} by means of:

$$\mathcal{F}(H\beta) = 4.09 \cdot 10^{12} \frac{\beta_2}{\beta} S_{UV} \text{ erg s}^{-1} \tag{11.126}$$

where β_2 is the emission recombination coefficient of the H β row and β is the global recombination coefficient. The coefficients β_2 and β depend on the temperature and two cases are considered: the case in which the medium is opaque and the case in which the medium is transparent. The values of β_2/β for the two cases are shown in Table 11.7.

Once $\mathcal{F}(H\beta)$ has been calculated, in the absence of extinction $\mathcal{F}(H\alpha)$ and $\mathcal{F}(H\gamma)$ are given by:

$$\mathcal{F}(H\alpha) = 2.87\mathcal{F}(H\beta) \quad \mathcal{F}(H\gamma) = 0.466\mathcal{F}(H\beta) . \tag{11.127}$$

Note that these relations are independent of the mass of the star and depend, albeit weakly, on the gas temperature. The ionized gas emits some metallic lines during recombination: the most important is the forbidden line [OII] ($\lambda = 3727$ Å). The relationship between the flow of this row $\mathcal{F}(OII)$ and the flow of the row H β is a function of the stellar mass:

$$\frac{\mathcal{F}(OII)}{\mathcal{F}(H\beta)} = h(m) . \tag{11.128}$$

11.14 HII Regions Excited by a Cluster

It often happens that an HII region is excited by the stars of a young cluster. In this case, the UV emission of all the hot stars in the cluster must be taken into account. The geometric configuration can present some complications:

(1) all stars are concentrated near the center of a single large HII region,
(2) the stars are scattered in a region of space large enough so that each of them excites an HII region that does not overlap the others.

The treatment of the first case is relatively simple, because, except for a brief initial phase, the HII region is equivalent to one excited by a single source whose ultraviolet luminosity is the sum of the individual stellar luminosities:

$$S_{UV}(t^*) = \int_{m_i}^{m_f} Am^{-\alpha} S_{UV}(m) dm \tag{11.129}$$

where t^* is the age of the cluster and the stellar mass distribution is given by the Salpeter function.

Of course, it is sufficient to extend the integration to massive stars ($m > 10m_\odot$). Clusters older than a few tens of millions of years do not produce sufficient UV flux.

Table 11.6 In the first column: distance from the star relative to the radius of the HII region. In the second and fourth column: temperatures of cold grains and hot grains. In the third and fifth columns: wavelengths corresponding to the maximum emission

r/R_H	$T_g^{(1)}$	$\lambda_{max}^{(1)}$	$T_g^{(2)}$	$\lambda_{max}^{(2)}$
0.1	180	16.7	370	8.1
0.8	70	42.9	170	17.6
2.0	40	75.0	100	30.0

The flux of the HB line will be given by:

$$\mathcal{F}(H\beta) = \int_{m_i}^{m_f} Am^{-\alpha} f_{H\beta}(m)dm \qquad (11.130)$$

where $f_{H\beta}(m)$, in accordance with what was said previously, is given by a relation like:

$$f_{H\beta}(m) = k S_{UV}^*(m) \qquad (11.131)$$

with $k = \text{cost}$.; introducing this expression in the integral we have:

$$\mathcal{F}(H\beta) = k \int_{m_i}^{m_f} Am^{-\alpha} S_{UV}^*(m)dm = k S_{UV}(t^*) . \qquad (11.132)$$

From the relations (11.127) we get $\mathcal{F}(H\alpha)$ and $\mathcal{F}(H\gamma)$. The flux of the [OII] line is given by:

$$\mathcal{F}(OII) = \int_{m_i}^{m_f} Am^{-\alpha} f_{OII}(m)dm \qquad (11.133)$$

and for (11.128) we have:

$$\mathcal{F}(OII) = \int_{m_i}^{m_f} Am^{-\alpha} f_{H\beta}(m)h(m)dm . \qquad (11.134)$$

A further simplification as in the case of the H β line is not possible here because h depends on the integration variable.

An extension to the case of an entire galaxy can be done taking into account all the HII regions present: a model of this type allows to calibrate a relationship between the intensity of the Balmer lines and the OII line and the star formation rate. This will be discussed in Chap. 16.

Table 11.7 Balmer lines fluxes

T	β_2	β	β_2/β
5000	$3.78 \cdot 10^{-14}$	$6.82 \cdot 10^{-13}$	0.055
10000	$2.04 \cdot 10^{-14}$	$4.18 \cdot 10^{-13}$	0.049
20000	$1.03 \cdot 10^{-14}$	$2.51 \cdot 10^{-13}$	0.041
T	β_2	β	β_2/β
5000	$5.37 \cdot 10^{-14}$	$4.54 \cdot 10^{-13}$	0.118
10000	$3.03 \cdot 10^{-14}$	$2.60 \cdot 10^{-13}$	0.117
20000	$1.62 \cdot 10^{-14}$	$1.43 \cdot 10^{-13}$	0.113

11.15 Synthetic Summary

HII regions are volumes around massive stars where H is fully ionized by the intense stars UV flux. The volume is limited by the so called Stroemgren sphere, whose size depends critically on the amount of dust inside the sphere. Typically, HII regions have sizes of the order of 1 pc. The equilibrium among photo-ionization and collisional recombinations implies that the temperature inside the Stromgren sphere is around 10^4 K. Important cooling processes derive from the recombination of heavy species like O, S, and N. The transition of these elements are important diagnostics of the gas temperature and electron number density. HII regions emit in UV, visual, infrared and also in radio, where free-free transitions are very common, and from their spectrum one can infer the region electron number density. The expansion of an HII region generates a shock waves, and the shock front, which separates the region from the neutral and cold ISM, is significantly thinner than the Stromgren sphere. This shock front is doomed to survive shortly, because the material coming out of the star via stellar wind is going to quickly swipe it. The ionized bubble switches off as soon as the star leaves the main sequence.

Reference

1. Bresolin, F., Gierenn, W., Kudritzki, R., et al.: ApJ **700**, 309 (2009)

Chapter 12
Stellar Winds

12.1 Introduction

The analysis of the spectra of stars of the first spectral types (O and B) reveals that these stars loose mass with a continuous flux: this phenomenon is called stellar wind. The existence of these winds was revealed mostly through the UV spectra obtained with the IUE satellite; the resonance lines of these spectra have P-Cygni profiles (see Fig. 12.1). These profiles consist of an absorption component shifted to shorter wavelengths, and an emission component displaced to longer ones. The most natural explanation is that the star is surrounded by an expanding envelope: the portion of this envelope that moves towards the observer produces the absorbing component, while the portion that surrounds the stellar disk and projects against the envelope of the star generates the emission. The detailed shape of the profile depends on the geometry of the envelope, the velocity trend, the density and the nature of the absorption and emission processes.

Many of the objects studied have high ionization and X-ray emission lines with energy between 0.15 and 4.5 Kev. These phenomena suggest the existence of a non-thermal mechanism responsible for the wind. It is not fully understood, but is probably related to a momentum transfer from the high energy photons to the gas by an absorption process; the moment is redistributed by impacts and the result is an acceleration of the outermost layers of the star in a radial direction.

The parameters that characterise the wind are the ejection speed and the rate of mass loss: both can be evaluated from the spectrum. The terminal wind speed, that is the one at which the wind tends asymptotically if it is not braked by some obstacle, can be determined by the Doppler formula by measuring the wavelength of the blue edge of the profile (this evaluation is easy when the material ejected from the star is abundant and the profile is saturated). This speed is higher than the photospheric escape speed; this depends on both the gravitational potential and the radiation pressure, and is given by:

© The Author(s), under exclusive license to Springer Nature Switzerland AG 2021
G. Carraro, *Astrophysics of the Interstellar Medium*,
UNITEXT for Physics, https://doi.org/10.1007/978-3-030-75293-4_12

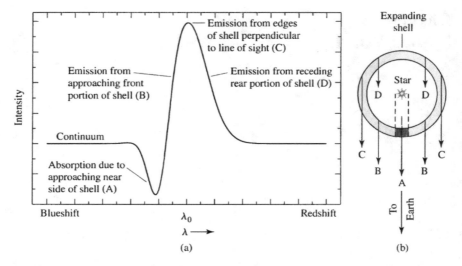

Fig. 12.1 Example of P Cygni profile. From Carrol & Ostlie (2006)

$$v_{\rm esc} = \left[\frac{2\mathcal{G}M(1-\Gamma)}{r} \right]^{\frac{1}{2}}, \tag{12.1}$$

with $\Gamma = L/L_E$ where L is the brightness of the star and L_E is the Eddington brightness, which is the maximum brightness compatible with a radiative flux:

$$L_E = \frac{4\pi\mathcal{G}Mc}{\sigma_e}, \tag{12.2}$$

σ_e being the scattering cross section.

Once the terminal velocity was determined by means of the spectrum analysis, an empirical relationship was obtained between this and $v_{\rm esc}$

$$V_w = a v_{\rm esc} \tag{12.3}$$

where the constant of proportionality has values of about 1 for the cooler OB stars, and about 3.5 for the hottest ones. This relationship can in turn be used to determine the wind speed.

There are different techniques, related to the different spectral regions used, to obtain an estimate of the rate of mass loss: they include radio observations of the free-free emission originating in the wind far from the acceleration region, infrared observations of the free- free which is generated in the acceleration region, optical observations of the Hα line and UV observations of lines with the P-Cygni profile that originate in the accelerating flow. Assuming that:

(a) the emission is isotropic and homogeneous,

(b) the material is at a distance equal to the radius of the star,

we have:

$$\dot{M}_w = 4\pi R^2 \rho V_w \ . \tag{12.4}$$

The column density $R\rho$ is proportional to the equivalent width W_λ of each absorption line:

$$W_\lambda = \tau_\lambda \Delta\lambda = n\sigma R\Delta\lambda \propto \sigma\rho R\Delta\lambda \tag{12.5}$$

and hence:

$$\dot{M}_w \propto R W_\lambda V_w \tag{12.6}$$

where are all quantities can be inferred from the spectrum. Estimates deduced with this method lead to an uncertainty of 50%.

All hot stars with mass greater than $10M_\odot$, show evidence of mass loss in their spectra: this means that this phenomenon is shared by the stars of the main sequence of types O and B, by the supergiants of the types A and B and by the WR stars. From the empirical estimates of the rate of mass loss a relationship valid for OB stars was derived:

$$\dot{M}_w = 1.3 \cdot 10^{-16} \left(\frac{L}{L_\odot}\right)^{1.77} [M_\odot/\text{yr}] \ . \tag{12.7}$$

The uncertainty of the $\dot{M}_w$ estimate deduced from this relationship is a factor of 2.

The speeds V_w are in the range of $1500 - 3000$ km/s. Typical values that we will refer to in the numerical examples can be: $\dot{M}_w = 10^{-6}[M_\odot/\text{yr}]$ e $V_w = 2000$ km/s.

By means of the rate of mass loss and the wind speed it is possible to evaluate the brightness of the wind, i.e. the quantity of mechanical energy transported by the wind in the unit of time:

$$L_w = \frac{1}{2}\dot{M}_w V_w^2 \ . \tag{12.8}$$

With the values previously used, a brightness of 10^{36} erg/s is deduced, while for main sequence stars the brightness of the wind varies between 10^{34} and 10^{37}. If the wind phenomenon is continuous and lasts for a time of τ_w, the total amount of energy associated with the wind is $E_w = L_w\tau_w$. For example, with $\tau_w = 10^6$ years and $L_w = 10^{36}$ erg/s we get $E_w = 3 \cdot 10^{49}$ erg. The thermal energy of the wind is negligible. Abbott [1] derived the energy of the winds as a function of the spectral type and the initial mass on the ZAMS and these data are reported in Table 12.1.

The table shows that the most massive stars, during the main sequence phase, can eject through the wind an amount of mechanical energy equal to that developed in the explosion of a supernova.

The same main sequence stars that exhibit the stellar wind phenomenon, with their UV flux generate the HII regions: a typical value of the number of UV photons emitted is $S_{UV}^* = 10^{49}$ s^{-1}, while the average photon energy is $\langle h\nu\rangle = kT_*$ where T_* is the temperature of the star. A typical value of $\langle h\nu\rangle$ is $5 \cdot 10^{-12}$ erg; therefore the energy produced per second is $E_{UV} = 5 \cdot 10^{37}$ erg/s, greater than L_w. However,

Table 12.1 Energy associated with the wind from different spectral type stars

$M/M_\odot$	Spectr.	Life time [Myr]	E_w [erg]
100	O4-V	3.4	$2.3 \cdot 10^{51}$
60	O6-V	4.2	$7.0 \cdot 10^{50}$
20	O9-V	10.3	$3.6 \cdot 10^{49}$
15	B0-V	11.2	$8.5 \cdot 10^{48}$

it has been seen that the efficiency of converting electromagnetic energy into kinetic energy is very low; to evaluate which of the two processes provides the most energy to the interstellar medium we will have to calculate the efficiency of the wind energy conversion.

The stellar winds from hot stars of the early spectral types are not the only example of injection of energy into the interstellar medium in a continuous and stationary fashion. For example, red giant stars lose mass with a brightness in the range of 10^{32} − 10^{34} erg/s and the ejection velocity is of the order of $100 - 200$ km/s. Stellar winds are also associated with protostars when they are still immersed in the molecular matrix cloud. The mechanical brightness of these winds varies between 10^{31} and 10^{36} erg/s. On a galactic scale, much more energetic phenomena are observed: stellar winds and supernova explosions are linked to the evolution of OB associations and inject matter and energy in an approximately continuous way into the interstellar medium. The mechanical brightness reaches values of 10^{45} erg/s; even more energetic phenomena occur in starburst galaxies ($L_w \geq 10^{45}$ erg/s).

Therefore in the various different astrophysical situations the mechanical luminosity of the winds cover a very wide interval between 10^{31} and 10^{45} erg/s, and also the velocities vary considerably. Furthermore, these winds interact with environments with very different physical characteristics, both in terms of density and temperature. In the following, after a general introduction that will lead us to define the essential aspects of the interaction between winds and interstellar medium and after introducing the distinction between slow winds and fast winds by defining a critical speed, we will deal only with winds produced by the stars of the first spectral types, and thus belonging to the category of fast winds.

12.2   Classification of Winds

Let us consider the interaction of a stationary, spherically symmetric wind with the surrounding environment which is supposed to be homogeneous: this interaction gives rise to an expanding bubble. The wind acts on the environment like a piston compressing it at a supersonic speed and generating a shock (blast) wave S_1 (see Fig. 12.2). Initially, the amount of swept material is small and does not constitute a significant obstacle to the expansion of the wind which will therefore be in an essentially free expansion with the radius of the bubble growing linearly with time. This phase is short-lived because subsequently the material swept by the impact front begins to slow down the wind. The new material ejected from the star then travels at a speed higher than that of the previously ejected wind is partially slowed down, and soon this difference in speed becomes supersonic. As seen in the reference system of the wind exiting the star, it is the previously ejected material that falls supersonically in the direction of the star: therefore, in addition to the divergent shock S_1 that sweeps the interstellar medium, there is a convergent shock S_2 that moves towards the wind and compresses it. From the same Fig. 12.2 one can easily realise that S_2 is convergent. The figure represents the trend of the wind speed with respect to the speed in the vicinity of the star at two close instants: we see that the remote point around S_2 where this speed becomes supersonic tends, instant by instant, to approach the star. However, this does not mean that the front moves towards the star because it also participates in the expansion of the wind and this effect is prevalent. The density of the wind is given, based on the conservation of matter, by:

$$\rho(r) = \frac{\dot{M}_w}{4\pi r^2 V_w} \, . \tag{12.9}$$

Since the wind density is large and the shock velocity S_2 is small, this is initially radiative. However, due to the expansion of the bubble, the density in front of the

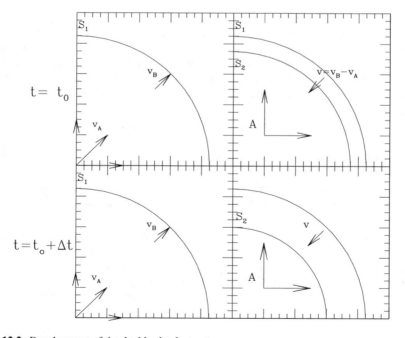

Fig. 12.2 Development of the double shock structure

Fig. 12.3 Comparison of
cooling and dynamical time
during the wind evolution:
the S_2 case

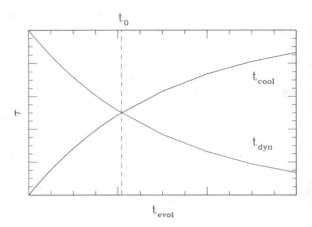

shock decreases while its speed increases and the shock becomes adiabatic; Fig. 12.3 shows, qualitatively, the evolution of cooling time and dynamic time: for $t < t_0$ the shock is radiative, while for $t > t_0$ it becomes adiabatic. The moment in which this transition takes place depends on the speed of the wind: if the wind is fast it takes place while the wind is expanding freely and in this case the radiative phase has no importance from the dynamic point of view. If the wind is slow, the S_2 shock remains radiative even when the free expansion phase has ended and the bubble expansion is slowed down by the action of the swept ambient mass: in this case the effects of the radiative cooling of the shocked wind on the bubble dynamics are important.

These circumstances lead to classify winds according to their speed by introducing a critical speed which is defined based on the comparison between the cooling time of the shocked wind and the time required to sweep an adequate amount of ambient gas. This critical speed is introduced in Appendix A and is expressed by:

$$V_{cr} = 300.5(L_{38}n_{H,0})^{1/11} \text{ [km/s] .} \qquad (12.10)$$

Let us consider as an example the typical wind properties of stars of early spectral types, e.g. $V = 2000$ km / s and $L_w = 10^{36}$ erg/s. In an environment characterized by $n_{H,0} = 1$ cm^{-3}, the value of V_{cr} is 197.7 km/s, and therefore the wind is strong. For the wind of a red giant we assume a mechanical brightness of 10^{34} erg/s and a speed of 100 km/s; the critical speed is 130 km/s and in this case the wind is slow. In general the winds of the stars of the first spectral types are strong, as are those that are generated around the OB associations and in starburst galaxies. The winds of the red giants and protostars (bipolar flows) are instead of the slow type.

Due to the lower density of the environment, the S_1 shock is initially adiabatic because the cooling time is long: during this phase, for a strong wind, we will have to deal with a completely adiabatic bubble. Subsequently when the radiative processes become effective behind the shock S_1 a dense gas shell is formed and we speak of an adiabatic bubble with a radiative shell. Later on, the effects of radiative cooling extend back to the shocked wind and the whole bubble becomes radiative.

12.3 Structure and Evolution of a Bubble Produced by a Strong Wind

Let us now deal in detail with the case of a strong wind expanding in a homogeneous medium. Since, as mentioned, strong winds are generated by stars of the first spectral types we can assume that the wind expands into the ionized gas of the HII region created by the star that emits the wind. The ambient gas, which we assume is still uniform, has a temperature of about 10^4 K, which corresponds to a speed of sound of 10 km/s.

In the bubble we can distinguish different regions starting from the inside (see Fig. 12.4).

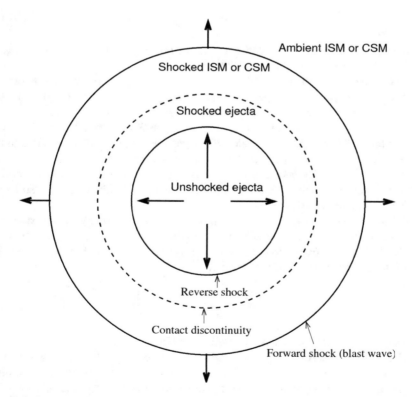

Fig. 12.4 Structure of the shocks created by a fast stellar wind (from Reynolds 2017)

(1) *Region a*: is the innermost and is occupied by the wind not yet reached (un-shocked) by the converging (reverse) front S_2. The wind density is given by the (12.9) and its temperature is 10^4 K.

(2) *Region b*: it contains the stellar wind already hit (shocked) by the converging (forward) shock front. This is a strong shock and, according to what has been said about strong winds, it is adiabatic with the exclusion of a very short initial phase; it is also radiative again at an advanced stage. During the adiabatic phase the temperature behind the shock (see Eq. 5.57)is:

$$T_2 = \frac{3\mu m_H V_w^2}{16k} \approx 4 \cdot 10^7 \text{ [K]} \tag{12.11}$$

where we have put $v_1 = V_w$. In fact, with respect to the star the S_2 front moves very slowly and therefore V_w is approximately equal to the velocity of the front with respect to the gas not hit by the shock. The pressure can be obtained from the second jump condition (related to the conservation of momentum) by neglecting the pressure in front of the shock with respect to the corresponding ram pressure (see Eq. 5.56):

$$p_2 = \frac{3}{4}\rho_1 V_w^2 = \frac{3}{4}\frac{\dot{M}_w V_w}{4\pi R^2} \ . \tag{12.12}$$

The speed of sound associated with the gas behind the shock (i.e. in the *region b*) is therefore very high: $c_b \approx 1000$ km/s.

The expansion of the ambient gas occurs due to the pressure exerted by the hot gas of this region and its speed is reduced precisely due to the work done on the surrounding material. At some point the expansion speed becomes less than the speed of sound c_b, so the time for the propagation of a sound signal becomes less than the expansion time scale and the gas pressure is uniform throughout the region. Due to the dilution resulting from expansion, the density is low and the cooling time, which shortens as the density increases, is large, making cooling negligible for a long time. Since the shock is adiabatic, the compression factor is less than or at most equal to 4.

The total energy of the region is determined by the balance between the expansion work on the surrounding environment and the injection of energy through the S_2 shock. Since the expansion rate is much less than the speed of sound, the energy of this region is practically all thermal:

$$\frac{1}{2}v^2 \ll \frac{1}{2}c^2 = \frac{\gamma}{2}\frac{p}{\rho} = \frac{5}{9}\epsilon \tag{12.13}$$

where ϵ is the internal energy of the unit of mass as usual.

(3) *Region c*: consists of the interstellar gas hit by the external shock S_1. For radiative cooling to be effective it is obviously necessary that the cooling time becomes shorter than the dynamical time, therefore there is a first phase in which the shock S_1 is of the adiabatic type. The gas enters S_1 at a much slower rate than

the stellar wind enters S_2 and therefore the temperature behind S_1 is much lower than that behind S_2 (and cooling is much more effective). Therefore, when the dynamic time is greater than the cooling time, the S_1 shock becomes radiative; the compression behind the shock is very high and the layer is thin. However, very low temperatures are not realised due to the presence of the stellar UV flux: behind the shock the temperature is determined by the balance between cooling and photoionization and is approximately equal to 10^4 K. Since the layer is thin, the crossing time of an acoustic perturbation is short, and also in this region the pressure is uniform.

The stellar wind hit by the shock S_1 (*region b*), and the interstellar gas hit by the shock S_2 (*region c*) are separated by a discontinuity surface C. Through this surface, temperature, density and other quantities of the gas are discontinuous, but not the pressure: in fact there is no flow of matter through C (contact or tangential discontinuity).

(4) *Region d*: is formed by the interstellar ionized gas. It will generally have a speed of the order of 10 km/s, but since the S_2 shock is strong it can be assumed that the gas outside is approximately unaltered and at rest.

The evolution of this structure produced by the wind is analogous to that of a supernova remnant (see Chap. 13). We distinguish four phases:

(A) free expansion phase of the S_2 shock,
(B) phase in which the bubble (*region a*, inside S_2) is completely adiabatic,
(C) phase in which the S_2 shock is adiabatic while the S_1 shock remains radiative,
(D) phase in which both shocks are radiative.

The phase A of free expansion at wind speed lasts some tens of years, during which the wind expands with constant speed as it is practically unhindered by the interstellar gas. When enough material accumulates behind the shock to slow expansion, this phase ceases. Before the conclusion of the free expansion the internal shock S_2 becomes adiabatic.

Phase B during which both the internal S_2 and the external S_1 shocks are adiabatic (the bubble is completely adiabatic) lasts a few thousand years until the cooling processes become important behind the S_1 shock. This phase can be studied by means of self-similarity solutions of the fluid dynamics equations (see Appendix B). In this case, based on the considerations developed in Appendix B, we can construct a single dimensionless quantity ξ:

$$\xi = \frac{r}{R(t)} \tag{12.14}$$

where $R(t)$ can be expressed with t and the parameters ρ_0 (ambient density) and L_w in the form:

$$R(t) \propto \rho_0^a L_w^b t^c . \tag{12.15}$$

The values of a, b and c are determined with dimensional analysis. The result is:

$$R(t) \propto \left(\frac{L_w}{\rho_0} \right)^{1/5} t^{3/5} \,. \tag{12.16}$$

The (12.16) defines the law of variation of the bubble radius, while the variable ξ allows to characterize the position within the *region c* at any time by solving a system of three equations ordinary differentials, as described in the appendix. In particular, these solutions show that for $\xi = 0.86$ ($\xi = 1$ corresponds to the radius of the bubble) there is the surface of tangential discontinuity.

12.4   A Simplified Model for the Radiative Phase

Most of the bubble life produced by wind is covered by phase C, in which the bubble is adiabatic and has a radiative shell, during which the material shock-swept by S_1 gets dense and cold by the radiative cooling processes.

Recalling what was said in the previous paragraph, we can adopt the following approximations:

(1) the *region c* is thin, so its thickness is small and we can assume that c and S_1 are approximately at the same distance R from the star;
(2) in calculating the volume and mass of the wind already hit by the shock (*region b*) it can be assumed that it occupies all the space inside the spherical surface of radius R;
(3) the ambient interstellar gas has a negligible velocity and therefore we can assume that $R' = dR/dt$ measures the velocity of the shock S_1 with respect to the interstellar gas and the expansion velocity of the bubble;
(4) the thin shell of *region c* is compressed and accelerated by the hot wind bubble which has a uniform pressure of P.

The problem can be solved by resorting to conservation laws. Let us consider the conservation of the momentum of the shell, expressed by:

$$\frac{d}{dt} \left[\frac{4\pi}{3} R^3 \rho_0 \frac{dR}{dt} \right] = 4\pi R^2 P \tag{12.17}$$

where ρ_0 is the density of the undisturbed ambient gas. The mass of the shell is equal to the mass originally contained in the sphere of radius R. The previous expression can therefore be put in the form:

$$P = \rho_0 \left(\frac{dR}{dt} \right)^2 + \frac{1}{3} \rho_0 R \frac{d^2 R}{dt^2} \,. \tag{12.18}$$

The conservation of energy in *region b* can be obtained by remembering that the energy of this region is almost completely thermal; the variation of the energy in the unit of time is equal to the rate of energy input by the wind decreased by the work done on the interstellar gas. The thermal energy of the unit of volume is $3P/2$ and we have therefore:

$$\frac{d}{dt}\left[\frac{4\pi}{3}R^2\frac{3}{2}P\right] = L_w - P\frac{d}{dt}\left[\frac{4\pi}{3}R^3\right]. \tag{12.19}$$

By removing P between (12.18) and (12.19) we get:

$$R^4\frac{d^3R}{dt^3} + 12R^3\frac{dR}{dt}\frac{d^2R}{dt^2} + 15R^2\left(\frac{dR}{dt}\right)^3 = \frac{3}{2\pi}\frac{L_w}{\rho_0}. \tag{12.20}$$

We are looking for a solution of (12.20) like:

$$R = At^\alpha \tag{12.21}$$

with A and α positive real constants; this is a physically reasonable solution because for $t = 0$ we have $R = 0$, and for $t \to \infty$ we have R diverges. Substituting (12.21) in (12.20) we obtain:

$$A^5[\alpha(\alpha - 1)(\alpha - 2) + 12\alpha^2(\alpha - 1) + 15\alpha^3]t^{5\alpha-3} = \frac{3}{2\pi}\left(\frac{L_w}{\rho_0}\right). \tag{12.22}$$

Since L_w is constant, the constancy of the second member implies $\alpha = 3/5$. Substituting the found value for α into (12.22) yields the value of A:

$$A = \left(\frac{125}{154\pi}\right)^{\frac{1}{5}}\left(\frac{L_w}{\rho_0}\right)^{\frac{1}{5}}. \tag{12.23}$$

We also have:

$$\frac{dR}{dt} = \frac{3}{5}\left(\frac{125}{154\pi}\right)^{\frac{1}{5}}\left(\frac{L_w}{\rho_0}\right)^{\frac{1}{5}}t^{-\frac{2}{5}} \tag{12.24}$$

and therefore:

$$R = \frac{5}{3}\dot{R}t. \tag{12.25}$$

Using the (12.21) with the appropriate values of the constants it is possible to evaluate the size of the hot bubble created by the wind: after 10^6 years, assuming that the environment has density $n_0 = 10$ cm^{-3}, the radius is several tens of parsecs.

At this point we are able to evaluate the efficiency of converting wind energy into kinetic energy of the interstellar gas in phase C. The kinetic energy of the shell is:

$$E_c = \frac{2\pi}{3}R^3\rho_0\dot{R}^2. \tag{12.26}$$

Taking R from (12.25) and replacing in (12.26) gives

$$E_c = \frac{15}{77} L_w t \ . \tag{12.27}$$

Since the mechanical energy input of the wind up to time t is $E_w = L_w t$, the efficiency is E_c/E_w and is 15/77. Therefore about 20 % of the mechanical energy of the wind is transformed into kinetic energy of the gas: the efficiency factor is much higher than in the process of transforming the UV electromagnetic energy into kinetic energy and amply compensates for the lower emission rate of the wind energy.

Let's analyze the hypothesis we made at the beginning that is that the interstellar gas around the star is ionized. Suppose that an HII region is produced before the wind flows and neglect its expansion; let n_0 be the density of the ionized gas. When the wind " blows ", the interstellar gas is swept into a thin shell of n_s density. This gas cools effectively so that the temperature in the *region c* is approximately equal to that of the *region d* (isothermal shock). The density n_g is related to n_0 according to the following relations, deduced from the shock wave theory. First

$$\frac{\rho_2}{\rho_1} = \frac{u^2}{c_i^2} = \frac{(v_1 - v_2)^2}{c_i^2} \simeq \frac{v_1^2}{c_i^2} \ . \tag{12.28}$$

Therefore

$$\frac{n_s}{n_0} = \frac{\rho_2}{\rho_1} = \frac{\dot{R}^2}{c_i^2} \tag{12.29}$$

from which:

$$n_s = n_0 \frac{\dot{R}^2}{c_i^2} \gg n_0 \ . \tag{12.30}$$

From the conservation of mass in the shell:

$$\frac{4\pi}{3} R^3 n_0 m_H = 4\pi R^2 \Delta R n_s m_H \tag{12.31}$$

we get ΔR:

$$\Delta R = \frac{1}{3} R \frac{n_0}{n_s} \ . \tag{12.32}$$

The recombination rate across the shell is:

$$N_r = 4\pi R^2 \Delta R n_s^2 \beta^{(2)} = \frac{4\pi}{3} R^3 n_0 n_s \beta^{(2)} \ . \tag{12.33}$$

For the phase duration we finally have:

$$t = \frac{77}{30} \frac{c_i^2 m_H S_{UV}}{L_w n_0 \beta^{(2)}} \ . \tag{12.34}$$

With $c_i = 10$ km / s, $S_{UV} = 10^{49}$ s^{-1}, $n_0 = 10^2$ cm^{-3}, $L_w = 10^{36}$ erg/s, we have $t = 2 \cdot 10^{12}$ s ~ 68000 years; also $R = 2$ pc and $\dot{R} = 20$ km/s $\sim 2c_i$.

12.5 Evolution in Phase D

In the course of evolution, the cooling processes behind the internal shock S_2 become important again: we are in the presence of a bubble with both radiative shocks. As a result of the cooling behind the shock S_2, the thermal energy of the gas of the *region b* decreases and the expansion of the outer shell will no longer take place through the work of pressure, as it did in the previous phase, but will happen at the expenses of the wind kinetic energy:

$$\frac{d}{dt}\left[\frac{4\pi}{3}R^3\rho_0\frac{dR}{dt}\right] = \dot{M}_w V_w = \frac{2L_w}{V_w} . \tag{12.35}$$

By developing the operations you get:

$$R^3\frac{d^2R}{dt} + 3R^2\left(\frac{dR}{dt}\right)^2 = \frac{3}{2\pi}\frac{L_w}{\rho_0 V_w} . \tag{12.36}$$

We are looking for a solution such as:

$$R = At^\alpha \tag{12.37}$$

with A and α positive real constants. Substituting this expression of R we have:

$$A^4[\alpha(\alpha-1) + 3\alpha^2]t^{4\alpha-2} = \frac{3}{2\pi}\frac{L_w}{\rho_0 V_w} . \tag{12.38}$$

Since the second member is independent of t, the exponent of t a first member must be null, that is, $\alpha = 1/2$. Substituting the value for α just obtained, the value of A is retrieved as:

$$A = \left(\frac{3}{\pi}\frac{L_w}{\rho_0 V_w}\right)^{\frac{1}{4}} \tag{12.39}$$

while it results:

$$\dot{R} \propto t^{1/2} . \tag{12.40}$$

The transition from the adiabatic to the radiative phase for the S_2 shock occurs at the time t_{rad} determined by the condition:

$$L_{rad} = L_w \tag{12.41}$$

where L_{rad} is the energy lost by radiation in one second from the entire *region b*. For t_{rad} it results:

$$t_{rad} = 3 \cdot 10^6 L_{36}^{0.3} n_0^{-0.7} \text{ [yr]} \tag{12.42}$$

while the size of the bubble at this moment is:

$$R = 50 L_{36}^{0.4} n_0 - 0.6 \text{ [pc]} . \tag{12.43}$$

Let's go back to the solution relating to phase C which can be put in the form:

$$R(t) = 27 \left(\frac{L_{36}}{n_0} \right)^{1/5} t_6^{3/6} \tag{12.44}$$

is

$$V = \dot{R}(t) = 27 \frac{3}{5} \left(\frac{L_{36}}{n_0} \right)^{1/5} t_6^{-2/5} \tag{12.45}$$

from which we obtain:

$$t_6 = 0.6 \frac{R}{V} . \tag{12.46}$$

The solution of phase D can be written:

$$R(t) = 16 \left(\frac{L_{36}}{n_0 V_{1000}} \right)^{1/4} t_6^{-2/5} \tag{12.47}$$

from which we obtain:

$$t_6 = 0.5 \frac{R}{V} . \tag{12.48}$$

12.6   Observations of the Bubbles

The relations (12.46) and (12.48) allow you to estimate the age of the bubble by measuring the size of the bubble and the speed of advancement of the external shock. Since the two expressions are practically the same, one or the other can be used indifferently regardless of whether the bubble is in phase C or phase D. The difference occurs when you want to estimate the mechanical brightness of the wind based on to shell expansion. Recall that L_w can be obtained independently by spectroscopic observations of the star. For this purpose we define two parameters:

$$\epsilon = \frac{\text{kinetic energy of the shell}}{\text{wind energy}} = \frac{M_s V_s^2}{2 L_w t} \tag{12.49}$$

Table 12.2 Parameter values for the phases C and D. See text for details

Parameter	Phase C	Phase D
ϵ	0.2	$\approx 2.2(t/t_{\text{rad}})^{-1/2}$
π	> 1	1

$$\pi = \frac{\text{shell moment}}{\text{wind moment}} = \frac{M_s V_s}{\dot{M}_w V_w t} . \tag{12.50}$$

The solutions relating to phases C and D give different values of these parameters, which are reported in Table 12.2.

In this way it is possible to estimate on the basis of the definitions, the values of the two parameters and, by comparing the values calculated for the two cases, to determine the evolutionary phase of the bubble. In general, in phase D the bubble has a slower expansion rate than in phase C.

The detection of a bubble can be done through the emission of the shell which, being ionized, will emit radiation Hα and the line at $\lambda = 5007\text{Å}$ of the OIII (e.g. M57). The expanding shell can be detected in the spectrum of the central star by absorption lines, e.g. del NII at $\lambda = 1085$ Å.

If there is dense gas in the ambient medium, it may happen that the shell is ionized in the internal part while in the outermost part it is neutral: in this case, as well as with the emission of the ionized gas it can be detected by the emission of the neutral gas and both they measure the same rate of expansion. The hot gas in the interior can be detected by absorption lines in the spectrum of the central star, e.g. del OVI a $\lambda = 1035$ Å (this line is produced in the conduction zone near the surface C of tangential discontinuity, where there is a strong temperature gradient). This emits soft X radiation but, according to the theoretical prediction, its emission is masked by the galactic background.

12.7 Evolution of the Bubble in an In-Homogeneous Medium

The evolution of the bubble is profoundly modified when the wind expands in a non-homogeneous medium due to the presence of clouds. Since the bubble expands, at least for the first part of its evolution, within an HII region, let's first analyze what happens to a cloud that penetrates an HII region. There are two effects on a mass cloud of m by the ultraviolet flux emitted by the star:

- it ionizes and heats the gas on the surface of the cloud and evaporates it by reducing the mass of the cloud (photoevaporation), with a mass reduction rate expressed by:

$$\frac{dm}{dt} = -\frac{m}{t_{ion}} \tag{12.51}$$

where t_{ion} is the characteristic time for evaporation, which is a function of the mass m_0, of the flux S_{UV} and of the distance of the cloud from the star;

- since the material is ejected in the direction of the star, the rocket cloud tends to move away from it.

Suppose there are n_0 clouds per volume unit all with mass m_0. If n_0 is large enough, evaporation increases the density of the material within the ionized region, therefore the number of recombinations per unit of volume increases and the radius R_i of the ionized zone is reduced. The number of clouds within the HII region tends to decrease and the decrease is greater the shorter the distance from the star. If all the clouds present have mass m there is a radius $R_0(m) < R_i$ such that:

(a) for $R < R_0(m)$ the gas is ionized and homogeneous, so all the clouds are eliminated from this region either by photoevaporation or because they are pushed out;

(b) for $R_0(m) < R < R_i$ the gas is ionized but not homogeneous because it still contains clouds;

(c) for $R > R_i$ the gas is neutral and inhomogeneous.

This simplified framework can be applied to an HII region that has formed within a molecular cloud. As will be seen in the chapter on molecular gas, a molecular cloud is not homogeneous and has clumps having densities of about two orders of magnitude greater than the average density and each of them can be considered as a cloud immersed in the middle less dense. These clumps have masses distributed according to a law such as:

$$n(m) \propto m^{-k} \tag{12.52}$$

with $k = 1.5$.

Let us now consider the evolution of the bubble produced by the wind in a medium in which there are inhomogeneities. Photoevaporation of clouds (or clumps) injects gas into the bubble which mixes with the hot gas and increases the effect of the cooling processes, as well as affecting the size of the HII region. With a weak wind the bubble is contained within a homogeneous medium and therefore the theory developed in the previous paragraphs can be applied correctly. With a strong wind the bubble tends to quickly capture a lot of clouds. The cooling processes are accelerated at the expense of the thermal energy of the hot gas and the expansion of the bubble slows down: the dynamics of the bubble produced by strong winds is driven by the photoevaporation of the clouds.

12.8 Synthetic Summary

Stars of any mass loose part of their envelope during their evolution. Pre main sequence stars push away a large amount of the core they are forming in when deuterium is ignited. Red giant and asymptotic branch stars loose mass as well. But the stronger mass loss occurs in OB stars. During their entire main sequence phase they loose mass at the pace of about $10^{-5} M_\odot/yr$. The material escapes the star at velocity of 1000–2000 km/s and shock-heats the ISM. A complicated pair of shocks (forward and reverse) develops, which distorts the HII region, and in the inter-shock region hot gas (10^7 K) can form. The evolution of the bubble is initially adiabatic because of the high gas speed. Then it turns radiative when the cooling time gets short enough. The typical size of the bubble created by the wind is around 4-5 pc, depending on the mass loosing star. The bubble disappears as soon as the star leaves the main sequence.

Appendices

Appendix A: Critical Speed of Winds

In the temperature range between 10^5 and $3 \cdot 10^7$ K, the cooling curve can be interpolated by:

$$L = 1.33 \cdot 10^{-19} n_p n_e T^{-1/2} \ [\text{erg cm}^{-3}\text{s}^{-1}] . \tag{12.53}$$

We assume that the gas has $X = 0.7$ and $Y = 0.30$, but suppose that it is ionized, and completely, only hydrogen, so that $n_p = n_e$.

Therefore the equation of the heat balance for the unit of mass is expressed, on the basis of (4.11), by:

$$T\frac{ds}{dt} = -1.33 \cdot 10^{-19} \frac{n_p^2}{\rho} T^{-1/2} = 6.52 \cdot 10^{-20} \rho m_H^{-2} T^{-1/2} , \tag{12.54}$$

where $n_p = \rho X/m_H$ has been inserted and where s is the entropy per unit of mass:

$$s = \frac{3}{2}\frac{k_B}{\mu m_H} \log \chi \tag{12.55}$$

with μ average molecular weight of the gas particles and with:

$$\chi = \frac{P}{\rho^{5/3}} = \frac{k_B T}{\mu m_H \rho^{2/3}} . \tag{12.56}$$

Equation 12.54 can then be put in the form:

$$\frac{3}{2}\chi^{1/2}\frac{d\chi}{dt} = -\frac{6.52 \cdot 10^{-20}k_B^{1/2}}{(\mu m_H)^{1/2}m_H^2} \, . \tag{12.57}$$

By integrating between the instant τ in which the wind element undergoes the action of the shock and a subsequent instant t we have:

$$\chi^{3/2} = \chi_0^{3/2} - \frac{6.52 \cdot 10^{-20}k_B^{1/2}}{(\mu m_H)^{1/2}m_H^2}(t - \tau) \tag{12.58}$$

which for $t \gg \tau$ can be approximated by:

$$\chi^{3/2} = \chi_0^{3/2} - \frac{6.52 \cdot 10^{-20}k_B^{1/2}}{(\mu m_H)^{1/2}m_H^2}t \, . \tag{12.59}$$

We can define the cooling time t_{cool} so that:

$$\chi(t_{cool}) \approx 0 \, , \tag{12.60}$$

and hence

$$t_{cool} = 3.9 \cdot 10^{-33}\chi_0^{3/2} = 3.9 \cdot 10^{-33}\left(\frac{k_B T}{m\rho^{2/3}}\right)^{3/2} = 5.22 \cdot 10^{-21}\frac{T^{3/2}}{\rho} \, . \tag{12.61}$$

Here T and ρ are the temperature and density behind the internal shock, respectively, and as such they can be derived from the jump conditions of the shocks, i.e.

$$T = \frac{3}{16}\frac{\mu m_H}{k_B}V_w^2 \tag{12.62}$$

is

$$\rho \approx 4\rho_0 \, . \tag{12.63}$$

Therefore we obtain for the cooling time:

$$t_{cool} = 7.91 \cdot 10^{-35}\frac{V_w^3}{\rho_0} = 1.11 \cdot 10^6\frac{V_8^3}{n_{H,0}}[\text{yr}] \, , \tag{12.64}$$

having used:

$$V_8 = \frac{V_w}{10^8[\text{cm/s}]} \, . \tag{12.65}$$

We can assume the transition instant t_s between the free expansion phase and the next phase in which the expansion is decelerating as the one in which the global mass expelled with the wind and the mass of the ambient medium swept by the external shock are equal. This instant is defined by the condition:

$$\dot{M}_w t_s = \frac{4\pi}{3}(V_w t_s)^3 \rho_0 \tag{12.66}$$

From this equation we obtain:

$$t_s = \left(\frac{3\dot{M}_w}{4\pi\rho_0 V_w^3}\right)^{1/2} = \left(\frac{3L_w}{2\pi\rho_0 V_w^5}\right)^{1/2}, \tag{12.67}$$

and, inserting the values of the constants, we finally get:

$$t_s = 1.43 \cdot 10^3 \left(\frac{L_{38}}{n_{H,0} V_8^5}\right)^{1/2} \text{[yr]}, \tag{12.68}$$

with $L_{38} = L/(10^{38}\text{[erg/s]})$. We are therefore able to define the critical wind speed as the one at which the times t_{cool} and t_s are equal:

$$V_{cr} = 300.5(L_{38}n_{H,0})^{1/11}\text{[km/s]} . \tag{12.69}$$

Appendix B: Self-similar Motions

Let us consider, for simplicity, only the case of a one-dimensional motion (which can be a motion with plane, cylindrical or spherical symmetry) and denote the spatial coordinate by r. In general, if the dissipative processes due to viscosity and thermal conduction can be ignored, the equations of fluid dynamics do not contain any characteristic scales of both length and time. The one-dimensional motion of a fluid is described by the functions $v(r, t)$, $\rho(r, t)$ and $p(r, t)$ which depend separately on r and t. These functions contain any parameters that enter the initial and boundary conditions.

A motion of this type is said to be self-similar if v, ρ and p do not depend separately on r and t, but admit expressions like

$$v(r, t) = F(t)\phi(\xi) \qquad \rho(r, t) = K(t)\chi(\xi) \qquad p(r, t) = P(t)\pi(\xi) \tag{12.70}$$

that is, they depend on t and on a dimensionless combination of r and t called similarity variable, i.e. $\xi = r/R(t)$. The distribution of a motion variable evolves over time in such a way as to keep the same shape by changing only the scale, i.e. by keeping $\pi(\xi)$, $\chi(\xi)$ and $\phi(\xi)$, and by contracting or expanding $R(t)$, $P(t)$, $K(t)$ or $F(t)$.

The conditions that must be satisfied for a motion to be self-similar can be identified by dimensional considerations. As has been said, the equations of fluid dynamics do not contain other dimensional parameters besides the dependent variables v, ρ, p and the independent variables r and t. The boundary conditions, on the other hand, can have dimensional parameters. In general we can construct two dimensionless variables r/R_0 and t/t_0, where R_0 and t_0 have dimensions of a length and a time obtained directly from the problem or by suitably combining the parameters of the problem. In this case the problem variables depend separately on r and t. If, on the other hand, with the parameters of the problem it is not possible to construct two quantities with the dimensions of a length and a time, then v, ρ and p cannot depend separately on r and t, but each of them depends only on a dimensionless combination of them and therefore the motion is self-similar.

For a certain number of self-similar motions the form of the function $R(t)$ and therefore the expression of ξ can be determined either by dimensional considerations or by conservation laws: these are the cases that interest us.

As an example we treat the case of a strong explosion in a homogeneous environment, which will be useful in the study of supernova remnants. Let us consider an ideal gas with density ρ_0, in which a short duration explosion takes place which releases a quantity of energy E. As a consequence of the explosion, a shock wave spreads in the environment. We assume that the shock wave is strong, so that the pressure p_0 of the unperturbed gas can be neglected compared to the pressure of the gas behind the shock. This implies that the internal energy of the gas is small compared to the energy E and that the speed of sound in the environment is negligible compared to the speed of the shock.

The parameters that characterise the motion are therefore the density ρ_0 and the energy E, while the speed of sound does not appear as a parameter in our approximation since $c \approx 0$. With them it is not possible to construct two quantities having dimensions of a length and a time, as can be verified by dimensional analysis, therefore the motion is self-similar. To derive the scale expression $R(t)$ we look for a combination of E, ρ and t that has the dimensions of a length, for which we obtain the independent similarity variable:

$$\xi = \frac{r}{R(t)} = r \left(\frac{\rho_0}{Et^2} \right)^{1/5} . \tag{12.71}$$

The shock front will be characterised by a (fixed) value of the similarity variable, ξ_0. The position of the shock as a function of time is given by:

$$R = \xi_0 \left(\frac{E}{\rho_0} \right)^{1/5} t^{2/5} \tag{12.72}$$

while the expansion rate is obtained by deriving the previous relation:

$$V = \frac{dR}{dt} = \frac{2}{5} \xi_0 \left(\frac{E}{\rho_0} \right)^{1/5} t^{-3/5} . \tag{12.73}$$

The physical conditions behind the shock front can be obtained starting from V using the formulas of Sect. 5.4:

$$\rho_2 = \frac{\gamma+1}{\gamma-1}\rho_0 \qquad p_2 = \frac{2}{\gamma+1}\rho_0 V^2 \qquad v_2 = \frac{2}{\gamma+1} V \ . \tag{12.74}$$

The law of expansion can also be deduced from the laws of dynamics: the pressure of the shocked gas is proportional to the density of the thermal energy, which is approximately equal to the total energy E divided by the volume, i.e.

$$p_2 \propto \frac{E}{R^3} \ . \tag{12.75}$$

The wavefront velocity V is related to the pressure behind the shock and the density unperturbed by the (5.87):

$$V \propto \left(\frac{P_2}{\rho_0}\right)^{1/2} \propto \left(\frac{E}{\rho_0}\right)^{1/2} R^{-3/2} \ . \tag{12.76}$$

From the latter through integration we find the (12.72).

The trend of velocity, density and pressure as a function of the distance from the explosion point is obtained by multiplying the time-independent profile by the corresponding scale:

$$v(r,t) = v_2\phi(\xi) \qquad \rho(r,t) = \rho_2\chi(\xi) \qquad p(r,t) = p_2\pi(\xi) \tag{12.77}$$

where the scales v_2, ρ_2, and p_2 are given by (12.74). The profiles $\pi(\xi)$, $\chi(\xi)$, and $\phi(\xi)$ satisfy the system of ordinary differential equations obtained by inserting the (12.77) transformations in the equations of fluid dynamics . This system is analytically solvable.

References

1. Abbott, D.C.: ApJ **263**, 723 (1982)
2. Weaver, R., et al.: ApJ **218**, 377 (1977)
3. Koo, B.C., McKee, C.F.: ApJ **388**, 93 (1992)
4. Zel'dovich, Y.B., Raizer, Y.P.: Physics of Shock Waves and High-temperature Hydrodynamic Phenomena. Dover Publications (2002)
5. McKee, C.F.: Astrophys. Space Sci. **118**, 383 (1986)

Chapter 13
Supernovae Remnants

13.1 Supernovae

In recent times, a very detailed taxonomy of supernovae has been formulated, based mainly on the shape of the light curve and on the characteristics of the spectrum. For our purposes, the following division into two classes is sufficient:

(a) *type II supernovae*, which are the consequence of Population I massive star explosion at the conclusion or in an intermediate phase of the sequence of thermonuclear burnings. They are found in the spiral arms of disk galaxies and are absent in ellipticals;
(b) *type I supernovae*, which are probably a consequence of the accretion of gas in a white dwarf belonging to a binary system, and have no H-lines in their spectra. They are observed in all types of galaxies.

In our galaxy the frequency is one type II supernova every 44 years and one type I supernova every 36 years, but of those accessible to us, one explodes every two hundred years.

The mechanisms of the explosion seem to be understood with sufficient clarity in the case of type II supernovae and depend on the mass of the progenitor star. Let's see what happens to a star that has gone through the entire series of nuclear burnings up to synthesizing iron. As the energy radiated by the star is no longer compensated by the thermonuclear energy, the star begins to contract. The nuclei of iron undergo a disintegration in which helium is produced:

$$Fe^{56} + \gamma \rightarrow 13He^4 + 4n \ . \tag{13.1}$$

This reaction is endothermic and accelerates the contraction which acquires a catastrophic character; the outermost layers of the core, still rich in fuels, fall towards the center and are brought to very high temperatures, at which the reactions proceed with considerable speed, releasing an enormous amount of energy: the outermost part of

G. Carraro, *Astrophysics of the Interstellar Medium*,
UNITEXT for Physics, https://doi.org/10.1007/978-3-030-75293-4_13

the star (the envelope) explodes as a supernova, leaving a residual collapsed central object. Two situations can arise for it:

(a) the collapse of the central zone stops and a neutron star is formed,
(b) the collapse does not stop and a black hole is formed.

This is the fate of stars with mass greater than 12 solar masses.

In stars of lower mass the burning sequence is interrupted before the iron synthesis due to the degeneration of the electron gas. When the degeneration is high, the triggering of the thermonuclear reactions of a certain fuel determines a greater instability the higher the degeneration is. The consequence may be the disintegration of the star which manifests itself as a supernova. This is the fate of stars with mass between 6 and 12 solar masses, and the explosion occurs when carbon or neon or oxygen is ignited. In this case the explosion leaves no residue (these supernovae are called type I 1/2).

In the case of type I (mostly Ia) supernovae it seems that a white dwarf, made up of helium (or carbon and oxygen) receives matter from a companion, typically an RGB star, which tends to expand beyond its Roche lobe, and thereby exceeds the minimum mass for triggering of the reactions of the burning of helium (or those of carbon). The environment in which the combustion takes place has a strong degeneration of the electronic gas and the star explodes without leaving a residue, as in the case of stars with a mass equal to $6 - 12 M_\odot$.

The explosion of a supernova triggers the expulsion of a large amount of material which, traveling at speeds of $10^4 - 10^5$ km/s, produces a blast wave: the energy involved is of the order of $10^{50} - 10^{51}$ erg. The optical phenomenon occurs when the shock reaches the surface of the star and the ejected layers become optically transparent: only 1% of the energy released by the explosion is radiated, however, while 99% is transformed into kinetic energy of the expanding material. The radiated power is equal to that of an entire galaxy, which contains billions of stars. The shock wave, caused by the explosion, sweeps the gas of the surrounding interstellar medium and, displacing it, forms a huge bubble of very high temperature gas (of the order of one million degrees and more) that emits in X-rays and in the ultraviolet: this object is the so-called supernova remnant (supernova remnant, or SNR). The fact that the models are able to reproduce well the fundamental characteristics of the supernova light curve suggests that a good understanding of the physical processes that are important in generating the emission of light have been achieved.

13.2   SNR Evolution

The shock produced by the expulsion of the stellar material propagates in the ambient medium interacting with it. We will first study the evolution of the remnant supposing that a spherical shock propagates in a uniform interstellar medium with density ρ_0 ignoring the effect of the magnetic field on the jump conditions. Although this is an idealised situation, the results obtained reproduce the essential aspects of the phenomenon. We can identify three phases during the expansion:

(a) phase I, called phase of free expansion,
(b) phase II, called adiabatic or Sedov phase,
(c) phase III, called the radiative phase.

Phase I. The density of the expanding material is much greater than ρ_0, so it can be assumed that the expansion occurs in the vacuum and therefore that the expansion rate remains constant. The evolution depends only on the details of the explosive process. In reality, even in this phase the shock can find an obstacle represented by circum-stellar material in the event that before the explosion there was mass loss due to stellar wind; this may be the case with the most massive stars. This phase ends when the mass swept by the shock is equal to the mass M_{SN} ejected by the supernova:

$$\frac{4\pi}{3} R^3 \rho_0 = M_{SN} .\tag{13.2}$$

Assuming $\rho_0 = 1 \cdot 10^{-24}$ g cm^{-3}, $V = 2 \cdot 10^4$km/s, $M_{SN} = 5 M_\odot$, one gets $R = 4.5$ pc for the bubble size, and about 220 years for the duration of this phase ($t = R/V$).

Phase II. The evolution of the remnant is dominated by swept matter, which now exceeds the ejected matter. There are no significant losses of energy by radiation: the shock is adiabatic and the total energy remains constant. We derive in these hypotheses the law of expansion. The study of shocks is done for convenience in a reference frame in which the discontinuity is fixed; with respect to this reference we denote by v_1' and v_2' the gas velocities before and after the collision. With respect to a reference integral with the star, the shock front advances with velocity $V = dR/dt$. Therefore, indicating with v_1 and v_2 the velocities of the gas before and after the shock, we will have that

$$v_1' = v_1 - V \qquad v_2' = v_2 - V .\tag{13.3}$$

Assuming $v_1 = 0$ (the interstellar gas is supposed to be stationary) we have:

$$v_1' = -V \qquad v_2 = v_2' + V .\tag{13.4}$$

In the case of adiabatic shock, the kinetic energy and internal energy of the unit of mass for the gas behind the shock are given by (5.58) and (5.59), and are equal

$$e_k = e_t = \frac{9}{32} \left(\frac{dR}{dt} \right)^2 .\tag{13.5}$$

From energy conservation

$$E_* = \frac{4\pi}{3} R^3 \rho_0 \left[\frac{9}{32} \left(\frac{dR}{dt} \right)^2 + \frac{9}{32} \left(\frac{dR}{dt} \right)^2 \right]\tag{13.6}$$

we have

$$R^3 \left(\frac{dR}{dt} \right)^2 = \frac{4E_*}{3\pi\rho_0} . \tag{13.7}$$

By integrating separating the variables and taking into account that for $t = 0$ we have $R = 0$, we obtain

$$R = \left(\frac{25}{3\pi} \frac{E_*}{\rho_0} \right)^{\frac{1}{5}} t^{\frac{2}{5}} = 1.22 \left(\frac{E_*}{\rho_0} \right)^{\frac{1}{5}} t^{\frac{2}{5}} \tag{13.8}$$

and deriving with respect to time we obtain the expansion speed:

$$V = \frac{dR}{dt} = \frac{2}{5} \left(\frac{25}{3\pi} \frac{E_*}{\rho_0} \right)^{\frac{1}{5}} t^{-\frac{3}{5}} \tag{13.9}$$

which can be rewritten as

$$V = \frac{2}{5} \frac{R}{t} . \tag{13.10}$$

With the self-similarity solution procedure (see Appendix B) Sedov obtained a solution equal to (13.8), with the only difference being the numerical coefficient of value (1.17).

From the solution of (13.7) we get the following properties. First of all, from the conservation of total energy and from the expressions of kinetic and thermal energy, it results that e_k and e_t are conserved separately. From the conservation of thermal energy it follows that the average temperature of the bubble decreases over time, since the mass of gas increases: the reason for this behavior is the adiabatic nature of the expansion.

The temperature of the gas immediately behind the shock is obtained from the equation of state and from (5.87), i.e.

$$T_2' = \frac{\mu m_H}{k_B} \frac{p_2'}{\rho_2'} = \frac{3}{16} \frac{\mu m_H}{k_B} V^2 . \tag{13.11}$$

For a gas with solar composition (reasonable for Pop. I) and completely ionized ($\mu = 0.64$) we have

$$T_2' = 0.45 \cdot 10^{-9} V^2 \text{ K} \tag{13.12}$$

which for $V = 1000$km/s it gives $T \approx 10^7$ K. Combining this relation with (13.9) we have the dependence of temperature on time:

$$T_2' = 3.45 \cdot 10^{-10} \left(\frac{E_*}{\rho_0} \right)^{2/5} t^{-6/5} . \tag{13.13}$$

The temperature behind the shock gradually decreases as the remnant expands. At each instant within the remnant the temperature increases as the distance r from

Table 13.1 Pressure trend behind the shock

r/R	p'/p'_2	r/R	p'/p'_2
1.00	1.00	0.70	0.70
0.94	0.66	0.50	0.31
0.90	0.50	0.40	0.30
0.80	0.40	0.00	0.30

the center decreases because there is no cooling and the temperature T'_2 behind the shock in previous instants was higher.

From (13.8) and (13.13) we get:

$$T'_2 \propto \left(\frac{R^2}{5}\right)^{-\frac{6}{5}} \propto R^{-3} . \tag{13.14}$$

This relation reflects, at each instant, the trend of $T(r)$: from more rigorous calculations it results $T(r) \propto r^{-4.3}$. The pressure behind the shock varies as indicated by Table 13.1, i.e. it decreases moving inwards.

Due to the combined effect of changes in T and p, the density increases even more rapidly as r increases. The consequence of this is the formation of a denser area on the outside; half of the mass is concentrated in a spherical shell with a thickness equal to 6% of the radius, and 3/4 of the mass in a spherical shell equal to 12% of the radius.

As expansion slows down, cooling processes become increasingly important. For a gas of solar composition at temperatures $T > 5 \cdot 10^6$ K, free-free processes prevail but the cooling effect is limited; for $T < 5 \cdot 10^6$ K the bound-bound processes of ions such as C, O, and N become more important, which in the interval $10^5 < T < 5 \cdot 10^6$ produce a cooling as from (12.53):

$$L = 1.33 \cdot 10^{-19} n_H^2 T^{-1/2} \text{ [erg cm}^{-3}\text{s}^{-1}\text{]} . \tag{13.15}$$

Considering the overall rate of energy loss by radiation

$$\frac{de}{dt} = \frac{4\pi}{3} R^3 L = 5.57 \cdot 10^{-19} n_H^2 R^3 T^{-1/2} \text{ [erg/s]} \tag{13.16}$$

which combined with (13.13) and (13.8) gives

$$\frac{de}{dt} = 5.45 \cdot 10^{-14} n_H^2 \left(\frac{E_*}{\rho_0}\right)^{\frac{2}{3}} t^{\frac{9}{5}} \tag{13.17}$$

places $n_H = 4n_0$, $\rho_0 = \mu m_H n_0$ and $\mu = 0.64$, we get

$$\frac{de}{dt} = 3.38 \cdot 10^{-3} n_0^{8/5} E *^{2/5} t^{9/5} . \tag{13.18}$$

By integrating the latter over time we have the overall radiative loss up to time t

$$E_{\text{rad}} = \int_0^t \frac{de}{dt} dt = 1.21 \cdot 10^{-3} n_0^{8/5} E_*^{2/5} t^{5/14} [\text{erg}] . \tag{13.19}$$

We conventionally define the instant at which the adiabatic phase ends and the radiative one begins as that which satisfies the condition that the global radiative losses are equal to half of the total energy E_* released by the explosion. from

$$E_{\text{rad}} = \frac{1}{2} E_* \tag{13.20}$$

it is deduced

$$t_r = 8.6 n_0^{-4/7} E_*^{3/14} \text{ [s] } . \tag{13.21}$$

For $n_0 = 1$ cm^{-3} and $E_* = 10^{50}$ erg, we have $t_r \approx 15,000$ years. At this instant the radius of the remnant is, based on (13.8), $R_r = 3.47 \cdot 10^{19}$ cm, or 11.57 pc, while the forward speed of the expanding shock wave, from (13.10), turns out to be $V_r = 312$ km/s.

III Phase. In this phase, the cooling time is short and the shock can be considered isothermal. In the case of a strong radiative shock (see (5.75)) it results

$$\frac{\rho_2'}{\rho_1'} = \left(\frac{v_1'}{c_1'} \right)^2 = \text{Ma}_1^2 \tag{13.22}$$

and the compression is now high: a thin shell of high-density cold gas is formed, advancing behind the shock front. We speak of a snowplow model because the interstellar gas is swept by the shock in the same way as the snow accumulates in front of the snowplow. The temperature reduction due to the cooling processes makes the effect of the hot gas pressure on the shell less important and roughly it can be assumed that the action of the hot gas pressure becomes negligible and the shell expands to preserving momentum. It is therefore:

$$MV = M_r V_r = \text{cost.} \tag{13.23}$$

when referring to the instant t_r, in which the cold shell is assumed to have already formed.

Almost all of the mass swept by the shock front is concentrated in the cold shell and therefore the previous expression becomes:

$$\frac{4\pi}{3} R^3 \rho_0 \frac{dR}{dt} = M_r V_r . \tag{13.24}$$

Integrating from the beginning of the radiative phase we have:

$$R^4 - R_r^4 = \frac{3 M_r V_r}{\pi \rho_0}(t - t_r) \tag{13.25}$$

or

$$t - t_r = \frac{\pi \rho_0}{3 M_r V_r}(R^4 - R_r^4) . \tag{13.26}$$

In the advanced stages ($t \gg t_r$) we can neglect t_r with respect to t and R_r with respect to R, so we can assume

$$R^4 = \frac{3 M_r V_r}{\pi \rho_0}t . \tag{13.27}$$

Expliciting M_r and expressing V_r as a function of t and R_r with the (13.10) we obtain

$$R = 1.125 R_r \left(\frac{t}{t_r}\right)^{\frac{1}{4}} \tag{13.28}$$

$$t = 0.624 t_r \left(\frac{R}{R_r}\right)^4 . \tag{13.29}$$

The expansion rate is obtained by deriving the (13.27) with respect to time, i.e.

$$V = \frac{dR}{dt} = V_r \left(\frac{R}{R_r}\right)^{-\frac{3}{4}} \tag{13.30}$$

and from this with the (13.28) we obtain the dependence of V on t

$$V = 0.70 V_r \left(\frac{t}{t_r}\right)^{-\frac{3}{4}} . \tag{13.31}$$

In conclusion, these are the evolution pattern in the three phases.

$$\begin{array}{lll} \text{Phase I} & R \propto t & \frac{dR}{dt} = \text{cost.} \\ \text{Phase II} & R \propto t^{0.4} & \frac{dR}{dt} = t^{-0.6} \\ \text{Phase III} & R \propto t^{0.25} & \frac{dR}{dt} = t^{-0.75} \end{array}$$

In the third stage the radius grows slower and the speed decreases more slowly than in the second stage. We have treated the radiative phase assuming that there is conservation of momentum, which implies, as we have said, that the effect of the pressure of the hot gas behind it is negligible. The inclusion of internal pressure changes the expansion law to:

$$R \propto t^\alpha \tag{13.32}$$

Fig. 13.1 A beautiful image of the Crab Nebula, with all the complicated filaments generated by the Rayleigh-Taylor instability. From Wikipedia

for which α must then be intermediate between the coefficient calculated with the conservation of the momentum ($\alpha = 0.25$) and that of the second phase ($\alpha = 0.4$), in which the pressure is not negligible. From numerical calculations performed considering the diminished role of pressure, we obtain in fact $\alpha = 0.31$.

13.2.1   Expansion End

The expansion of the remnant continues at decreasing speed. When this becomes comparable with the speed of the random (thermal or turbulent) motion of the ambient medium ($v \approx 10$ km/s), or in an equivalent way when the pressure of the shell equals the pressure of the medium, the remnant stops expanding and it looses its identity by mixing with the interstellar medium. In the example given, where $E_* = 10^{50}$ erg and $n_0 = 1$ cm^{-3}, we have

- from (13.31) the age at which expansion ends, i.e. $t_f = 2.24 \cdot 10^{13}$ s $= 7.46 \cdot 10^5$ years, for $V = 10$ km/s;
- from (13.28) the final dimension, i.e. $R_f = 8.96 \cdot 10^{19}$ cm, or 30 pc;
- the total swept mass $M_f = (4\pi/3) R_f^3 \rho_0 = 6.84 \cdot 10^{36}$ gr, or $= 3400 M_\odot$.

When the expansion ceases, the remnant structure consists of a cavity of hot, rarefied gas enclosed in a dense, cold shell. Subsequently the shell undergoes a process of gravitational instability which determines its fragmentation with the formation of predominantly filamentary clouds (see Fig. 13.1). The hot gas of the cavity mixes with the external gas tending to form an intra-cloud component (warm component).

Table 13.2 Characteristics of a supernova remnant at the end of each phase

Phase	t [yr]	R [pc]	V [km/s]	M_S [$M_\odot$]
I	220	4.5	20000	5
II	15000	11.6	312	100
III	746000	30	10	3400

The " snowplow " model is very approximate and therefore it correctly describes only the cold shell but is unable to adequately deal with the hot and rarefied gas found in the cavity within the expanding shell. In order to correctly study the behavior of this gas, the fundamental equations of magneto-hydro-dynamics must be taken into consideration. The solution of these equations with numerical methods confirms that the approximation we have considered is adequate for the expansion laws.

13.3 Expansion of a SNR in a Patchy Medium

The considerations developed in the previous paragraph are very idealised and deviate from the real situation for the following reasons:

- the inhomogeneities of the interstellar medium and the lack of spherical symmetry of the remnant,
- the presence of magnetic fields,
- the pressure of cosmic rays within the shell.

Inhomogeneity. Let us now consider how much the evolution of a remnant is modified if the medium in which it expands is not homogeneous. It must be premised that, in the case of supernovae coming from massive stars, the environment in which the remnant expands has been previously modified by the HII region and by the stellar wind. The latter in particular can produce a region of circum-stellar gas capable of making it difficult to identify the remnant. Only in the case of SNRs deriving from type I supernovae the medium in which the remnant expands is, from the beginning, the interstellar medium. As will be seen later (see Chap. 14), this medium is generally inhomogeneous and the more so the greater the quantity of energy that is globally injected into it.

Suppose the shock propagates in an otherwise homogeneous medium except for the presence of a homogeneous, spherical cloud in it. The interaction of the SNR with a non-homogeneous medium has been studied with numerical methods and the outcome is a structure that differs considerably from that described by the simple model presented in the previous paragraph. When the shock front of the expanding SNR encounters a cloud, the rate of advancement of the shock within the cloud will be different from that of the homogeneous medium. We denote by v_a the velocity of the shock front in the medium environment and with v_c that within the cloud, with $\rho_{0,a}$ and with $\rho_{0,c}$ the unperturbed densities of uniform medium and cloud, and both

$$\chi = \frac{\rho_{0,c}}{\rho_{0,a}} \gg 1 \,. \tag{13.33}$$

Let's limit ourselves to consider, for simplicity, the case that the SNR is in the adiabatic phase; nevertheless the shock that propagates in the cloud, due to its high density, will generally be radiative and isothermal. For the pressure behind the shock front in the ambient medium and in the cloud we have respectively:

$$p_{2,a} = \frac{3}{4}\rho_{0,a}v_a^2 \qquad p_{2,c} = \frac{3}{4}\rho_{0,c}v_c^2 \,. \tag{13.34}$$

These relationships are derived from the first two conditions of connection and therefore hold both in the adiabatic and in the radiative case.

The two pressures are equal since the equilibrium between the shocked ambient gas and the cloud gas must be ensured. From the equality of pressures we deduce:

$$v_c = v_a \left(\frac{\rho_{0,a}}{\rho_{0,c}}\right)^{\frac{1}{2}} = v_a \chi^{-\frac{1}{2}} \tag{13.35}$$

from which follows $v_c \ll v_a$. As a consequence of this the shock front curves and eventually the shock will continue in the homogeneous medium while a converging shock will propagate in the cloud from the surface towards the center (see the evolution chain in the panels of Fig. 13.2).

Let's first consider the consequences for the cloud. The interaction between shock wave and cloud is non-linear in nature and its outcome is difficult to determine, however it can be studied with complicated numerical calculations. The analysis, currently incomplete, requires the determination of the role of several parameters and their effect, and for the moment there is no general answer to the problem of the collision between a cloud and a shock. Small clouds are likely to be disrupted by shock compression, while larger clouds are deformed by instability but not disrupted. It is still to be established with precise calculations under which conditions this compression is able to generate a gravitational collapse that gives rise to a star formation process, although there is some observational evidence of this phenomenon (it seems that the association CMa R1 is generated by the compression of the shock wave produced by a supernova remnant). However, it seems that the necessary condition for the cloud to collapse is the duration of the compression

$$\tau_{\text{comp}} = \frac{R_c l}{v_c} = \frac{R_c l}{\chi^{1/2}}v_a \tag{13.36}$$

where R_{cl} is the radius of the cloud, is long enough to prevent the cloud from reexpanding, i.e. the reflection in the center and the reversal of the shock. This condition is expressed by

$$\tau_{\text{ff}} < \tau_{\text{comp}} \tag{13.37}$$

where τ_{ff} is the free-fall time. Under this condition, the possibility of collapse depends on the density and radius of the cloud, the density of the medium and the speed of the shock in the ambient medium.

Let us now consider the effects on the evolution of the SNR. The transfer of matter from the clouds penetrated within the remnant can have a very important role on the dynamics of the SNR and on its final structure. This effect is achieved by thermal conduction. Now the cloud, which has approximately retained its temperature (since the shock within it is radiative), comes into contact with a very hot gas. In the interface between the cloud and the hot gas, heat conduction operates effectively due to the significant thermal gradient. The flow will be

$$\Phi_c = -k \, \text{grad} \, T \tag{13.38}$$

where is it

$$k = 6 \cdot 10^{-7} T^{5/4} [\text{erg cm}^{-1} \text{sec}^{-1} \text{K}^{-1} \tag{13.39}$$

is the conductivity ends (see Sect. 3.2). Consequence of thermal conduction is the evaporation of the cloud into the hot gas. An estimate of the evaporation rate can be made with a hydrodynamic model and the evaporation rate is found to be

$$\frac{dM}{dt} = 1.3 \cdot 10^{-23} R_{cl} T_6^{5/2} [M_\odot/\text{yr}] \tag{13.40}$$

where T_6 is the temperature expressed in millions of Kelvin of the hot medium. The evaporation time can be estimated, i.e. the time required for the complete evaporation of the cloud, such as

$$\tau_{ev} = \frac{M_{cl}}{dM/dt} = 3.23 \cdot 10^{-24} R_{cl}^2 n_{cl} T_6^{-5/2} . \tag{13.41}$$

Putting $R_{cl} = 1$ pc, $n_{cl} = 1$ cm^{-3} and $T_6 = 1$ (data deduced from the UV absorption lines) we obtain $\tau_{ev} = 3 \cdot 10^3$ years.

Evaporation of the cloud leads to an increase in the density of the gas within the bubble. The discussion of thermal conduction requires some caution because the phenomenon of saturation can occur (see Sect. 3.2). In fact, under these conditions the mean free path of the electrons is $\lambda = 2$ pc $> R_{cl}$ (condition for saturation) and the previous formulas overestimate the flow carried. In this case, the flow is instead expressed by (3.20). The phenomenon of evaporation is especially important in the first phase. Radiative cooling works in the opposite direction by condensing gas on the clouds.

Let us now briefly consider other elements which must be taken into account for a realistic description of the evolution of SNRs.

13.3.1   Magnetic Field

With a magnetic field intensity not exceeding $3 \cdot 10^{-6}$ G (equal to the average Galactic magnetic field), the dynamic effects of this field are not relevant during phase II. Behind the shock we have $\rho_2/\rho_1 = 4$, so we also have $B_2 = 4B_1$ if the field is parallel to the shock front; the intensity instead is preserved if the field is normal to the impact front. Assuming an expansion speed of 500 km/s, the hydrostatic pressure behind the shock is given by

$$p_2 = \frac{3}{4}\rho_0 V^2 \tag{13.42}$$

and it amounts to $1.9 \cdot 10^{-9}$ dyn cm $^{-2}$. The magnetic pressure is $B^2/2\mu_0 = 3.6 \cdot 10^{-12}$ dyn cm^{-2}, and therefore is negligible. Therefore the magnetic field is frozen in motion but does not produce relevant dynamic effects. Instead the magnetic field is important in providing the coupling between material ejected from the star and the interstellar medium; in the absence of it the mean free path of the electrons would be very large and the fluid dynamics treatment would probably not be valid, and probably even the shock would not form. In Phase III the role of the magnetic field is important. It has been seen that, in the case of adiabatic MHD shock with **B** field normal to the velocity, the compression, which is however modest, is reduced by the presence of the magnetic field. In the case of radiative shock in which the compression in the absence of a magnetic field can be significant, the magnetic field is significantly reduced: the presence of a transverse magnetic field of $3 \cdot 10^{-6}$ G can reduce the compression to a tenth of what is obtained in the same conditions in the absence of the field. However, if the field is parallel to **v**, it has no effect and high densities can be reached, such as are observed for example in the SNR called Cygnus Loop.

13.3.2   Relativistic Particles

The presence of relativistic particles produced within the remnant, when their energy is comparable to the kinetic energy of the shell, can cause their pressure to tend to reduce the deceleration of the shell.

13.3.3   Spherical Symmetry

Another element that removes realism from the previous model is the assumption of spherical symmetry. Even in conditions of homogeneity of the surrounding medium, the symmetry of the expansion tends to be eliminated when the cold shell is formed, the hot gas inside exerts a pressure on the shell: in the conditions in which a less dense

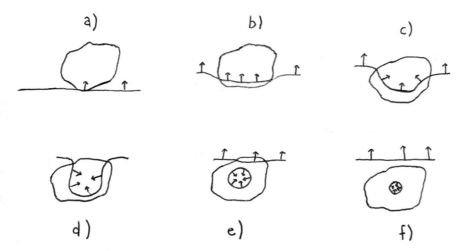

Fig. 13.2   Sketch of the evolution of a SNR in a patchy medium. Courtesy of Agnese Carraro

fluid accelerates into denser one, a (Rayleigh-Taylor) instability develops which breaks the equilibrium and leads to mixing of fluids. The resulting trend is the formation of denser areas behind the isothermal shock front. Both in the front and behind the shell it breaks into fragments which, due to the magnetic field, have the form of filaments.

13.4   Models of Evolution in a Patchy Medium

The study of the evolution of an SNR, taking into account the improvements proposed in the previous paragraph, was carried out through the use of numerical calculations: by interpolating these results, it is possible to give approximate analytical relationships that describe the behaviour of the remnant in the various phases. We present in the following a short summary of these results.

Let us consider a region of the galaxy in which supernovae explode at a rate of r_{SN} per year per unit of volume. The ambient medium is uniform with numerical density n_a and temperature T_a. Spherical clouds with radius a and density n_c are immersed in it, while N_{cl} is the number of clouds per pc ^{3}. Let R_s be the radius of the remnant, T_h the temperature and n_h the density of the coronal hot gas.

Phase I. During free expansion the following relations are obtained:

$$R_s \propto (N_{cl}a)^{-0.1}\tau^{0.6} \tag{13.43}$$

$$T_h \propto (N_{cl}a)^{-0.2}\tau^{-0.8} \tag{13.44}$$

$$n_h \propto (N_{\rm cl}a)^{0.5}\tau^{-1} \tag{13.45}$$

and the phase ends almost instantly:

$$t_1 = 1.9 \cdot 10^4 n_a \Sigma^{-1/2}[\text{yr}] \tag{13.46}$$

where $\Sigma = 5N_{\rm cl}a/8\pi$.

Phase II. During the adiabatic (Sedov) expansion, the relations are:

$$R_s \propto n_a^{-0.2}\tau^{0.4} \tag{13.47}$$

$$T_h \propto n_a^{-0.4}\tau^{-1.2} \tag{13.48}$$

$$n_h \propto n_a[1 + 10^3(N_{\rm cl}an_a - 2\tau^{-2})^{1/3}] \tag{13.49}$$

This set of equations is valid in the interval (t_1, t_2) where t_2 is evaluated by equating the expansion time with the cooling time:

$$t_2 = t_{\rm cool} = 2.09 \cdot 10^4 n_a^{-0.56}[\text{yr}] \tag{13.50}$$

and $t_{\rm cool}$ is inferred from the expression of the cooling rate as in (13.15) .

III Phase. During the radiative phase, finally, the following relations are found:

$$R_s \propto n_a^{-0.26}\tau^{0-29} \tag{13.51}$$

$$T_h \propto n_a^{0.05}\tau^{-0.58} \tag{13.52}$$

$$n_h \propto n_a\tau^{-6/7}[1 + 100(N_{\rm cl}a)^{1/3}n_a^{-0.29}] . \tag{13.53}$$

End of the expansion. The third phase ends when the expansion speed of the remnant is equal to the random speed of the environment (10 km/s), and this happens at the time:

$$t_3 = 1.1 \cdot 10^9 T_a^{-0.7}n_a^{-0-37}[\text{yr}] . \tag{13.54}$$

The shell's temperature and density have value $T_{sh} = 10$ K and $n_{sh} = 10^4 - 10^5$ cm^{-3}, respectively, while the shell is dense and cold, and approximately 1/10 of the SNR radius. The shell becomes gravitationally unstable after a time:

$$t_g = 3.5 \cdot 10^6[\text{yr}] . \tag{13.55}$$

The evolution of one SNR can be influenced by other SNRs if their spatial density is so great that they come into contact during expansion. This circumstance is linked to the supernova explosion rate in a given galactic volume. We denote by t_s the time at which the overlap occurs, which will be a function of the supernova rate $r_{\rm SN}$ which represents the number of supernova explosions per unit of time and per

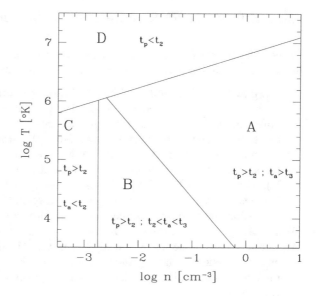

Fig. 13.3 Evolution phases of a SNR in a patchy inter-stellar medium: results from numerical models

unit of volume. The overlap time of remnants is defined as the instant at which the probability $\mathbb{P}$ that any point of the region is internal to a remnant:

$$\mathbb{P} = \int_0^{\tau_0} \frac{4\pi}{3} R_s^3(\tau') r_{SN}(\tau - \tau') d\tau \tag{13.56}$$

is equal to 1. The behavior of the remnants is described in Fig. 13.3 as a function of the density and temperature of the ambient gas; the calculations refer to a constant supernova rate of 10^{-13} SN pc^{-3} yr^{-1}. In addition to the times already defined for understanding the figure, it is necessary to introduce the time t_p which is the instant at which the pressure of the shocked shell, evaluated for the adiabatic phase, is equal to the pressure of the ambient gas:

$$t_p = 10^{10} n_a^{-1/3} T_a^{-5/6} [\text{yr}] . \tag{13.57}$$

In the diagram of Fig. 13.3 four regions are identified: region A represents the positions of those remnants that conclude their expansion during the radiative phase without overlapping with other remnants; the region D on the contrary refers to the systems that stop expanding during the adiabatic phase; region C identifies the environmental conditions in which the remnants overlap before giving rise to the cold shell; finally in region B the superposition takes place during the radiative phase.

13.5   Effects of a SNR on the Interstellar Medium

Beyond the effect of the ISM conditions on the evolution of SNRs, there are also phenomena related to SNRs that affect the interstellar medium.

(1) The gravitational instability of the cold, dense shell that occurs when the remnant stops expanding can produce clouds. There are theoretical considerations and observational indications, however not definitive, that the fragmenting shell gives rise to filamentary structures or to two-dimensional structures (sheets).

(2) The hot gas from the cavities that mixes with the environment after shell fragmentation leads to the formation of coronal gas at very high temperatures ($T \geq 10^6$ K). As we will see later this gas has been highlighted by UV and X observations. If the supernova rate is high, the supernova remnants can overlap during their expansion and give rise to large regions of coronal gas.

(3) The hot gas in the cavities produces X-rays and cosmic rays.

(4) Through the interaction of the SNR with the interstellar medium, part of the energy expelled by the supernova is transferred in the form of kinetic energy to the interstellar medium, similarly to what happens with the stellar wind and the HII region.

(5) Last, but not least, the SNRs inject in the ISM a large amount of chemically processed material.

Let us elaborate a bit more on point 4) and evaluate the efficiency of the conversion of supernova energy into kinetic energy of the shell during the radiative phase. This efficiency will be given by:

$$\varepsilon = \frac{MV^2}{2E_*} \tag{13.58}$$

where M is the mass of the cold shell, V the rate of expansion and E_* the energy released by the explosion. The (13.58) can be written in the form

$$\varepsilon = \frac{4\pi}{3} \frac{R^3 \rho_0}{2E_*} \left(\frac{\mathrm{d}R}{\mathrm{d}t} \right)^2 = \frac{2\pi R^3 \rho_0}{3E_*} \left(\frac{\mathrm{d}R}{\mathrm{d}t} \right)^2 . \tag{13.59}$$

The kinetic energy $MV^2/2$ can be evaluated through the relations of the radiative phase (13.28) and (13.30), so that we have

$$\varepsilon = \frac{2\pi V_r^2 R_r^3}{3(1.125)^3 E_*} \left(\frac{t}{t_r} \right)^{-\frac{3}{4}} . \tag{13.60}$$

With the data from the example we have always considered in this chapter, i.e. $E_* = 10^{50}$ erg and $\rho_0 = 2 \cdot 10^{-24}$ g/cm^3, we have $R_r = 3.25 \cdot 10^{19}$ cm, $V_r = 2.03 \cdot 10^7$ cm/s and $t_r = 6.6 \cdot 10^{11}$ s, and therefore:

$$\varepsilon = 0.42 \left(\frac{t}{t_r}\right)^{-\frac{3}{4}}.$$ (13.61)

At the end of the expansion, when $t = t_f = 2.24 \cdot 10^{13}$ s, we will have $\varepsilon = 0.02$. Therefore $3 \cdot 10^{48}$ erg are transferred, in the case considered, in the form of kinetic energy to the interstellar medium, while the remaining quantity is transformed into thermal energy and electromagnetic energy.

13.6 Superbubbles and Supershells

In recent years, observations have highlighted structures roughly constituted by a large bubble of hot and rarefied gas surrounded by a spherical shell (sometimes only an arc is highlighted) of dense, neutral but sometimes ionized gas. Often, but not in all cases, an OB star association is present inside. Not all of them are able to identify the expansion, but in those in which this is not observed it is probably because the expansion speed is so low that it cannot be detected. They are found in the Galaxy and in other spirals or irregularities. The radii of these structures range from those of normal SNRs (< 100 pc) to more than 1000 pc; in some cases it has been possible to evaluate the kinetic energy of expansion of the set which turns out to be also 10^{54} erg.

The main investigation methods through which these systems are revealed are:

(a) the line at 21 cm, which reveals a neutral gas shell
(b) the optical emission, through which the ionized gas zones are highlighted.

Although other hypotheses have been proposed (UV background radiation pressure, collisions with the galactic disk of high speed clouds coming from the halo) the most natural interpretation for such structures is that they are the combined effect of a large group of massive stars. The idea is suggested by the fact that these arise in clusters and associations and that, during their evolution, they do not move away from the place where they were born by more than $20 - 100$ pc.

Let's consider an association that has a $N_\star$ number of stars destined to end their evolution as supernovae, so they are stars with mass greater than $7M_\odot$ (this limit is uncertain). Once the main sequence is reached, the most massive stars ionize the surrounding medium generating a single large HII region and emit the wind. Virtually all UV radiation comes from O-type stars ($M > 30M_\odot$) and therefore after less than 10^7 years the UV flux decreases considerably and the HII region recombines. After a time of this order also the production of energy by the winds ceases. However, the phase of supernova explosions begins and will continue until $t = 5 \cdot 10^7$ years, equal to the life time of the smallest mass star that explodes as a supernova.

A simple model allows to interpret the main properties of these systems. Let us consider a cluster that contains $N_\star$ stars in the mass range $7 - 100M_\odot$. The life times of the stars are described by the report:

$$\tau = 10^9 M^{-1.53} \text{ [yr]} \tag{13.62}$$

where the mass is expressed in solar masses. Suppose the initial mass function is

$$dN = A M^{-2.35} dM \tag{13.63}$$

where the normalization constant A is determined by imposing that the number of stars with mass in the range of $7 - 100$ is $N_\star$, i.e.

$$N(7 - 100) = A \int_7^{100} M^{-2.35} dM = N_\star \tag{13.64}$$

and whereby $A = 19,203 N_\star$.

Due to the trend of $\tau(M)$ and the IMF, the supernova rate in the first $5 \cdot 10^7$ years (which is the life time of the star of $7 M_\odot$) is constant, in fact

$$r_{SN} = \frac{dN}{d\tau} = \frac{dN}{dM} \frac{dM}{d\tau} = A M^{-2.35} \frac{dM}{d\tau} . \tag{13.65}$$

From (13.62) solving for M and deriving we obtain

$$\frac{dM}{d\tau} \propto \tau^{-1.65} \tag{13.66}$$

and eliminating τ with (13.62) we have

$$\frac{dM}{d\tau} \propto M^{2.52} . \tag{13.67}$$

With (13.67), (13.65) becomes

$$r_{SN} \propto M^{-2.35} M^{2.53} = M^{0.18} \approx \text{cost. .} \tag{13.68}$$

We can schematise the succession of explosions as the input of a continuous flow of energy, the power $\mathcal{P}_{SN}$ being the sum of the energies expelled by all supernovae divided by $5 \cdot 10^7$ years

$$\mathcal{P}_{SN} = 6.7 \cdot 10^{35} (N_\star E_{51}) \tag{13.69}$$

where $E_{51} = E/10^{51}$.

The power produced by stellar winds, referring to the observational data presented in Chap. 12, will be $\mathcal{P}_w = 10^{36} N_\star$ erg/s. We neglect the contribution of the winds and consider that of the supernovae we treat as a continuous flow of energy. Numerical experiments, in which energy is injected in random pulses, indicate that this approximation is good for describing the expansion of super-shells after at least a few supernovae have exploded. The shock wave produced by the explosion of supernovae can be described with the laws of the expansion of stellar wind bubbles, and in dealing with the evolution of the super-bubble that is created we assume that the interstellar medium is homogeneous with density n_0.

The comparison of the cooling time with the dynamic time (equal to the age of the super bubble) allows to evaluate the moment in which the cooling processes become important in conditioning the dynamics of the gas immediately behind the shock. The cooling time is evaluated with the following expression of the cooling rate

$$L = n_e^2 10^{-22} T_6^{-0.7} \tag{13.70}$$

with $T_6 = T/10^6$.

The radiative processes immediately behind the shock front become important starting from the time t_c at which dynamic time and cooling time become equal

$$t_c \approx 2.3 \cdot 10^4 (N_\star E_{51})^{-0.3} n_0^{-0.7} \tag{13.71}$$

therefore the cold shell forms very quickly and the adiabatic phase can be ignored. Using the corresponding results of the evolution of wind bubbles we can derive the expansion law of the super-bubble

$$R = 97.3 \left(\frac{N_\star E_{51}}{n_0} \right)^{1/5} t_7^{3/5} [\text{pc}] . \tag{13.72}$$

The cooling processes, on the other hand, are not relevant inside the bubble, whose evolution is conditioned by the evaporation of the innermost part of the cold shell.

The expansion rate is obtained by deriving the previous relation

$$V = 5.84 \left(\frac{N_\star E_{51}}{n_0} \right)^{1/5} t_7^{-2/5} [\text{km/s}] . \tag{13.73}$$

By setting $E_{51} N_\star = 20$ and $n_0 = 1 \text{ cm}^{-3}$, a $t = 10^7$ years we have that:

(a) the number of supernovae exploded, assuming that the mass of the exploding star at $t = 10^7$ years is deduced from (13.62), i.e. 20 $M_\odot$, is

$$N_{SN} = A \int_{20}^{100} M^{-2.35} dM \approx 4 \tag{13.74}$$

(b) the radius of the super bubble, from (13.72), is $R_s = 177$ pc,
(c) the swept mass in the interstellar medium is

$$M = \frac{4\pi}{3} R_s^3 n_0 \mu m_H \tag{13.75}$$

that with $n(\text{He})/n(\text{H}) = 0.1$ and $\mu = 1.27$ yields $M = 1.39 \cdot 10^{39}$ gr, or $7 \cdot 10^5 M_\odot$,

(d) the speed of advancement of the shock front is obtained by deriving equation (13.72) :

$$V = 6.54 \cdot 10^{11} t^{-2/5} \, [\text{cm/s}] = 10.59 \, [\text{km/s}] \tag{13.76}$$

(e) the kinetic energy of the shell is

$$E_c = \frac{1}{2} M V_s^2 = 8.11 \cdot 10^{50} \, [\text{erg}] \tag{13.77}$$

which corresponds to about 23 % of the energy emitted by the explosions (this efficiency is much greater than that found in Sect. 13.5, but it should be noted that here we are not referring to the end of the expansion)

With respect to the last point, we note that the gas within the shell radiates (radiative shock) and much of its mass is in the form of cold gas ($T < 100$ K), i.e. HI or H_2.

As mentioned, the younger super-shells are also fed by O-type stars suitable for ionizing the material, and therefore the internal surface of the super-shell is characterised by an optical emission typical of the HII regions, in particular it emits the $H\alpha$ line. Older super-shells are only visible through the emission of HI or CO.

The adiabatic character of the hot cavity disappears when the cooling processes also become effective inside it. Starting from this point, evolution slows down and the expansion law can be represented by:

$$R \propto t^\alpha \tag{13.78}$$

with $\alpha \approx 0.25$ if the pressure behind the shell is negligible (see the " snowplow " phase) but more realistically it will have a higher value, since the pressure, although decreased, is not negligible.

The remains of the supernovae that explode when this cavity has now been created find themselves expanding in a hot and refurbished medium and go through an almost free expansion phase with a path of about 30 pcs. They then pass through the dense shell, slow down and eventually merge with the external shock wave. Situations of this type have been observed in the Magellanic Clouds.

An important aspect neglected so far is that the dimensions of these objects are comparable with the size of the disk gas (the scale heights of the HI are 80 pcs in the central areas, 200 pcs in the Sun area and 500 pcs in the outer disk). When the super bubble expands into the rarefied gas, instability occurs and the shell breaks and the hot gas diffuses into the halo. The maximum radius of the bubble is therefore limited by the scale height of the gas in the galactic disc. This is probably the reason why the largest bubbles have been observed in the Magellanic Clouds. Compared to the Galaxy, the Clouds have a lower gravity and the gas has a much larger scale height.

About 20% of these super bubbles have an HII shell; these objects are more compact and expand faster than others. This behaviour is consistent with the model according to which for $t < 10^7$ years there are O stars that are able to ionise a large amount of hydrogen, while after this age and up to $t = 5 \cdot 10^7$ only B stars survive which are unable to ionise a large amount of hydrogen, but with their explosions they fuel the expansion of the shell.

The most relevant Galactic object of this category is the Gum nebula in the direction of the Vela constellation (Galactic coordinate $l \approx 270°$): it has a radius of 400 pc and reveals the existence of an HII region. The observations of the massive stars contained in it confirm the validity of the developed theory.

13.7 Description of some SNRs

Young SNR. Cas A is the youngest galactic SNR. It is exceptionally bright at X and radio wavelengths, but due to the high absorption it is weak in optical and UV. It is associated with a supernova observed only once in 1680; the lack of other observations probably depends on the fact that the supernova was difficult to observe due to the high extinction. Although observations at that time did not allow to reveal the type of supernova, the X and optical emission of the remnant suggest that the supernova was of type II. The optical emission reveals about one hundred small condensations which, according to their motion, are divided into two classes: (a) fast, with $v \sim 6000$ km/s, (b) almost stationary with $v \sim 150$ km/s. From the proper motions of the fastest condensations it can be deduced that the explosion occurred in 1658 but the difference with respect to the date of observation of the presumed supernova could depends on a deceleration. Another interpretation is that the condensations were emitted in the explosion with a velocity lower than the observed one (10 km/s), and therefore are expanding in the already shocked medium and are moving in the low internal pressure zone. The shocks that propagate in these condensations are responsible for the optical emission. In this context the observed velocity is the difference between the ejection velocities and the shock velocity. As for the quasi-stationary condensations, they differ from the previous ones in terms of speed and chemical composition. The interpretation is that they were ejected with low velocity before the explosion.

Cas A has also been studied in radio. The radio surface brightness declines by 1-2% per year. The X emission is 10 erg/s. The X images show a dense shell with radius $102''$ radius and thickness $17''$ (1/5 of the radius); the shell mass is estimated around $20M_\odot$. A second, weaker shell appears within the first. The natural interpretation is that the former is produced by the shocked interstellar medium while the internal one is produced by a shock wave that propagates inwards Reverse shock). X-rays with energy greater than 10 kev have been identified. If the emission is thermal, electrons with an energy of 30 kev are required and therefore temperatures of 10^8 K are implied. Until about ten years ago Cas A seemed unique: there are now others five similar remnants including three in the Magellanic Clouds. The estimated age based on the models is uncertain and varies between 200 and 300 years. Another young remnant is Tycho's supernova remnant SN1592 (SN type I): the X and radio images indicate high spherical symmetry.

SNR of intermediate age. Among the remnants of intermediate age the most studied, due to its beauty, large angular size, high surface brightness and low reddening is the Cygnus Loop. It appears as a shock that propagates at speeds of 400–500 km/s in a medium with a density of 0.2 atoms per cm^3 and in the clouds it induces shocks with a speed of 100 km/s. It has been studied in optics, in X-rays and in radio and in general there is an excellent correlation between the surface brightness at different wavelengths. Observations indicate that this is a fairly old remnant, whose behaviour is independent of the details of the explosion. The age is about 30,000 years.

Old SNRs. They are observed in radio (the radio emission is thermal), with the line at 21 cm, in H α (in emission) and in soft X-rays. Many of these objects, which are quite large in size, are probably powered by more than one supernova (say supershells). An example is the North polar Spur.

13.8   Synthetic Summary

After emitting a strong UV flux and injecting continuously material in the ISM, massive stars explode as SNae. The amount of energy released almost instantaneously matches the energy previously released via UV radiation and stellar winds. All these phenomena are routinely called *stellar feedback*, and are crucial to understand why the ISM is multi-phase. In a single kick the star deposits its entire envelops in the ISM creating a bubble larger than 10 pc on average. Only a small fraction (0.05) of this energy is realised as radiation. It is essentially kinetic energy which creates a strong shock. The material inside the SN bubble gets heated up to 10^7 K. The bubble is surrounded by a cold thin layer of neutral and molecular H. Beside that, SN deposits in the ISM material enriched by the long chain of stellar burning stages, thus becoming critical mechanism to properly understand the chemical evolution of the ISM and of galaxies. The structure left out by the explosion is called SN remnants, and many spectacular examples have been imaged over the years. Taylor instability is the responsible for the fragmentation of the bubble in later stages, when the hot ionised gas mixes with the cold neutral one, giving origin to filaments and sheets whose shape is dictated by the magnetic field.

References

1. Spitzer Jr., L.: Physical Processes in the Interstellar Medium, Chapter, p. 10. Wiley, New York (1978)
2. Tomisaka, K., et al.: Astrophys. Space Sci. **78**, 273 (1981)
3. McCray, R., Kafatos, M.C.: ApJ **317**, 190 (1987)
4. Mac Low, M.M., McCray, R.: ApJ **324**, 776 (1988)
5. McCray, R.: Supernovae and the interstellar medium. In: Physical Processes in Interstellar Clouds, NATO ASI Series. Springer, Berlin (1987)

Chapter 14
The Interstellar Medium and Its Components

14.1 The Structure of the Interstellar Medium

As of 50 years ago our knowledge of the interstellar medium was summarised by Field's theory which included two components: HI clouds and a warm and rarefied medium. This theory is based on the hypothesis of a stationary state in dynamic and thermal equilibrium, the fundamental principles of which have been set out in Sect. 4.5. In the light of the thermal stability analysis, the functional dependence of the cooling rate on the temperature is able to account for the existence of two stable phases (see Fig. 4.3). The thermal equilibrium is achieved with the balance between heating by cosmic rays and cooling consequent to the collisional excitation of ions of C, O and N. Furthermore, there are two components in pressure equilibrium: the clouds of HI ($T \approx 10^2$ K) and a warm ($T \approx 10^4$ K) and rarefied intra-cloud gas. Equilibrium is possible for a narrow range of pressure values which includes those observed in the interstellar medium.

This model of the interstellar medium underwent however a crisis following a series of facts that emerged mainly from observations outside the atmosphere (with balloons or satellites):

(a) the identification of a further very hot ($T = 5 \cdot 10^5 - 1 \cdot 10^6$ K) and rarefied component, responsible for the emission of soft X-rays ($E = 0.25$ kev) and detectable, in addition to this emission, by the interstellar absorption in the UV of the ion OVII; on the basis of the estimates of the pressure of this component, obtained from the observational measures of temperature and density, it appears to be in pressure equilibrium with the two previous components; mass exchanges with these are possible through thermal evaporation and condensation;

(b) the sub-millimeter (mostly from ALMA) and radio observations revealed that the ambient from which stars originate are cold and dense clouds made mainly of molecular hydrogen and detectable through the emission of CO; this component is not in pressure equilibrium with the other three components since they are self-gravitating objects, that is, confined by the force of gravity and not by an

G. Carraro, *Astrophysics of the Interstellar Medium*,
UNITEXT for Physics, https://doi.org/10.1007/978-3-030-75293-4_14

external pressure; molecular clouds seem to contain a relevant fraction of the mass of the interstellar gas;

(c) the observations also revealed that very often the interstellar medium is the site of violent phenomena caused by the propagation of shock waves that are generated by super sonic collisions or by phenomena related to the evolution of massive stars that have been discussed in the last three chapters; often these phenomena appear to be responsible for the generation of new stars.

These facts have led to a drastic change in our view of the interstellar medium and the new picture that emerges envisages a medium that is no longer in dynamic equilibrium, very complex with continuous exchanges between the different components. A fundamental characteristic of the interstellar medium is that very often its components are not in mechanical equilibrium. An essential parameter in the description of this behaviour is the amount of energy injected into the interstellar medium by massive stars. The first models assumed that the most important contribution is that of supernovae and only these take into account. In fact, the evaluation of the relative importance of the different phenomena requires a more detailed analysis which is presented in Appendix A. The inhomogeneous character of the interstellar medium is a direct consequence of the injection of energy by massive stars into the thermally unstable gas.

This behaviour can be studied using a model that considers the stellar component and the different components of the interstellar medium which result from the most recent observations covering the entire spectrum of electromagnetic waves: molecular clouds, HI clouds (cold neutral medium, or CNM), intra-cloud medium up to $T \approx 10^4$ K, which, as will be seen later, can be present in the form of neutral gas (warm neutral medium, or WNM) or in ionised form (warm ionised medium, or WIM), collisional hot ionised medium, or HIM. For each of these components a differential equation can be written which expresses the temporal variation of the mass as a consequence of the interactions between the different components. The rate of energy injected by massive stars is the most important parameter of the model: it is evaluated based on the rate of star formation. In some models the stellar component does not appear and the supernova rate is a variable external parameter.

The results of the model can be summarised as follows.

(A) *One-phase interstellar medium.* If the star formation rate (or the supernova rate) is very small, so that the energy injection rate in the medium is very low, the gas settles on a HI's quiescent disk, with a scale height of only 15 pcs; moreover, this disk is gravitationally stable if its density is low. Obviously with these parameter values the model does not describe a realistic situation.

(B) *Two-phase interstellar medium.* With a higher energy input rate, the gas is present in two phases which can be in dynamic equilibrium. The possible equilibrium pressures are included in an interval (p_{min}, p_{max}): that is, the characteristics of the Field model are found. The gas is confined in a thin cold disk with a thickness of a few tens of parsecs, the gas pressure exceeds p_{max} and therefore only one phase is present. The pressure decreases as it moves away from the galactic plane and becomes less than p_{max} giving rise to a two-phase structure with cold clouds immersed in the warm medium. The scale height of the warm gas is about 15 pcs. At greater heights the gas is all hot and homogeneous. Although this model is more realistic than the previous one, it presents the same difficulties as Field's model: in particular it is quiescent contrary to what is observed in the interstellar medium.

(C) *Multi-phase interstellar medium (violent medium).* If the quantity of energy injected is very high, the hot component appears in large quantities as a consequence of the presence of many massive stars. The scale height of this gas is of the order of kpc. Since most of the cold clouds are confined near the galactic plane, the gas at a great height above the plane appears in two phases: the hot component (HIM) and the warm component (WNM). Onto the galactic plane, the medium is made up of three components: cold clouds, hot gas and warm gas. The agreement of the model with the observational data is reasonable but an important test relating to the volume occupied by the hot component is missing, for which the observations do not provide reliable data at the present time.

We will dedicate the rest of this chapter to the study of atomic components while the analysis of molecular gas, which is of particular importance because it is the environment where the star formation processes take place, is dealt with in the next two chapters.

14.2 Neutral Gas Diagnostic Techniques: The Line at 21 cm

In the hydrogen atom the magnetic moment of the proton interacts with the combined magnetic field produced by the orbiting electron and its intrinsic magnetic moment. This interaction leads to the "splitting" of all energy levels. The line at 21 cm is the result of the "splitting" of the fundamental level which is divided into two sub-levels and of the radiative transition from the upper to the lower level; in the higher level the electron and proton spins are parallel and the statistical weight is 3, in the lower level the spins are antiparallel and the statistical weight is 1. The separation of the levels is $E = 5.9 \cdot 10^{-6}$ eV, and corresponds precisely to the emission of a radiation with a wavelength of 21 cm. The population of the levels is given as:

$$\left(\frac{n_2}{g_2}\right) / \left(\frac{n_1}{g_1}\right) = \exp\left[-\frac{E}{kT}\right] \tag{14.1}$$

and is regulated by the excitation of the hydrogen atoms by collisions with particles, by the cosmic background radiation, and by Lyα photons. When the mechanism at work is the cosmic background radiation, $T = 3$ K, otherwise the temperature is the kinetic temperature of the gas. Since $E \ll kT$, the levels are populated on the basis of statistical weights; three quarters of the atoms are in the upper state. The transition, being a magnetic dipole, is forbidden.

The study of the line at 21 cm must be tackled with the equation of the radiative transport, i.e. (11.112), taking into account also the absorption. Consider an external source, of intensity $I_\nu(0)$, which is on the opposite side to the observer with respect to a HI cloud whose optical depth at the frequency ν is $\tau_{\nu,r}$, being r the spatial coordinate along which τ varies. Using the Kirchhoff principle and the Rayleigh approximation of the Planck function (valid for $h\nu/kT \ll 1$) the solution of the radiative transport equation is given by:

$$I_\nu = I_\nu(0)\exp[-\tau_{\nu,r}] + \frac{2kT\nu^2}{c^2}(1 - \exp[-\tau_{\nu,r}]) . \tag{14.2}$$

The brightness temperature T_B is defined on the basis of:

$$B_\nu(T_B) = I_\nu \tag{14.3}$$

where B_ν is the Planck function. Keeping in mind that this in the radio region of the spectrum can be approximated by:

$$B_\nu = \frac{2kT\nu^2}{c^2} \tag{14.4}$$

we have:

$$T_B = T_B(0)\exp[-\tau_{\nu,r}] + T(1 - \exp[-\tau_{\nu,r}]) . \tag{14.5}$$

The column density of HI, $N(\text{HI})$, along the line of sight is given by the following relationship, which we do not derive:

$$N(\text{HI}) = 1.82 \cdot 10^{13} \int T\tau_{\nu,r}d\nu . \tag{14.6}$$

In the absence of external sources ($T_B(0) = 0$), T can be expressed using the (14.5) and we obtain:

$$N(\text{HI}) = 1.82 \cdot 10^{13} \int T_B(v)\frac{\tau_{\nu,r}}{1 - \exp[-\tau_{\nu,r}]}d\nu . \tag{14.7}$$

If in particular $\tau_{\nu,r} \ll 1$ the previous one reduces to:

$$N(\text{HI}) = 1.82 \cdot 10^{13} \int T_B(v)dv \tag{14.8}$$

where v is the velocity of the atoms emitting the frequency ν and the two quantities are linked by the Doppler effect.

We first consider the case in which there are no external sources, i.e. $T_B(0) = 0$. The solution is then

$$T_B = T(1 - \exp[-\tau_{\nu,r}]) . \tag{14.9}$$

and the following two cases can be distinguished.

(1) *Optically thin case:* $\tau_{\nu,r} \ll 1$. Developing the exponential in series and neglecting the higher order terms, we have:

$$T_B = T\tau_{\nu,r} \tag{14.10}$$

which according to (14.8) is proportional to $N(HI)$.

So in this case the measure of $T_B(v)$ allows us to determine $N(HI)$. This happens because all the photons emitted escape from the cloud without being absorbed. To determine $n(HI)$ it is necessary to have an evaluation of the position of the emitting atoms. Since the gas participates in the galactic rotation, an estimate of the distance from the Sun is obtained if v is attributed to the rotation. Since the gas velocity also has random components, this assumption is valid only approximately and gives correct results only on average. The average hydrogen density found in this way is 0.7 cm^{-3} for $7 < R[kpc] < 11$, while for values of R outside this range $n(HI)$ decreases. Inside the spiral arms around the Sun the average value is between 1 and 2 cm^{-3} (see Fig. 14.1).

(2) *Optically opaque case:* $\tau_{\nu,r} \gg 1$. The approximate solution is:

$$T_B = T \tag{14.11}$$

that is, the brightness temperature is equal to the kinetic temperature of the gas. Almost all the emitted photons are absorbed and only those emitted near the edge from the nearest part, for which $\tau < 1$, are able to leave the emission region. In this case the brightness temperature does not measure the column density and depends only on the gas temperature. We then have from (14.7):

$$N(HI) \propto T_B \tau_{\nu,r} . \tag{14.12}$$

Let us now consider the case of an external source with brightness temperature $T_B(0)$. In reality, what is determined is the difference

$$\Delta T_B(v) = T_B(v) - T_B(0) \tag{14.13}$$

that can be written as:

$$\Delta T_B(v) = (T - T_B(0))(1 - \exp[-\tau_{\nu,r}]) . \tag{14.14}$$

If $T > T_B(0)$ we will have $\Delta T_B(v) > 0$ and the HI is seen in emission, while if $T < T_B(0)$ we will have $\Delta T_B(v) < 0$ and the HI is seen in absorption. In the first case, the emission of the nearest gas exceeds the attenuation of the farthest radiation.

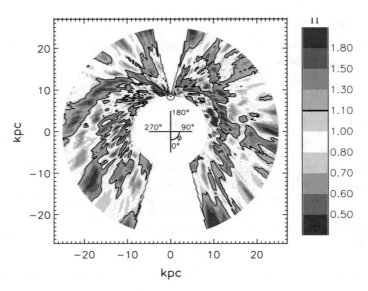

Fig. 14.1 HI distribution in the disk of the MW from Levine et al. [1]

Emission data has been widely used to map the distribution of galactic HI (see Fig. 14.1). The absorption data are needed to separate the cold HI from the hot one and measure the temperature of the hydrogen.

14.3   The Spiral Structure of the Milky Way

Emission data, as mentioned in the previous section, allow us to measure HI column density and, when the distance to the emitter is available, its number density. Emission data have been widely used to map HI distribution in the MW immediately after the discovery of the HI transition at 21 cm, and produced the first MW spiral structure map in 1952 by mean of the very famous so-called Leiden-Sydney survey, whose principal investigator was Jan Oort. Figure 14.1 represents a modern version of the application of this technique. To this purpose, the spectrum in the plane (T_B, v) offers in principle everything we need. From T_B the column density can be derived and therefore the amount of HI in a given direction. From v the distance can be derived assuming that we know, or have derived, a relation between distance and radial velocity. What we instead typically have is a relation between distance and circular velocity around the Galactic center, which is called the Galaxy rotation curve. An example of the application of this technique is sketched in Fig. 14.4. The first assumption is that the clouds orbit the Galactic center along circular orbit. Therefore the distance at which the velocity is maximum (line C in the spectrum) corresponds to the minimum distance to the Galactic center for that specific direction. This point

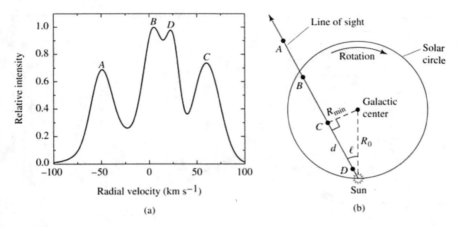

Fig. 14.2 Emission spectrum in a direction toward the Galactic center and distance solution

is called *tangent point* and the velocity is called *terminal velocity*. The second, more subtle, assumption, is that the velocity we measure is entirely due to rotation. This is of course unrealistic, due to the irregular and diffuse shape of the HI clouds, where local velocity irregular pattern can be present due to shocks, random motions, and turbulence.

Basing on this two assumptions, basic trigonometry can be used to infer the cloud distance (Fig. 14.2). This way it is possible to place all clouds which lie along lines of sight for which the Galactic latitude l is smaller than (90°) (first and fourth Galactic quadrants). For other directions (second and third Galactic quadrant, or the Galactic anti-center) this geometrical argument cannot be used, and one has to resort to other distance indicators, such that O,B stars or star clusters in the surroundings. An example of rotation curve is presented in Fig. 14.3. HI was expected to serve as an efficient and ideal tracer of the spiral structure of the Milky Way. However its ubiquitous and diffuse nature together with the insuperable difficulties of deriving meaningful distances, cooled any hope down very soon. Molecular clouds, being discrete and denser, might be better spiral arm tracers (see next Chapter).

14.4 Ionized Gas Diagnostic Techniques

The ionized gas of the interstellar medium can be detected by different types of observations, each sensitive to a combination of electron density and temperature.

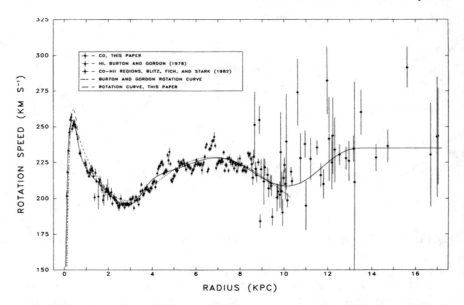

Fig. 14.3 Milky Way rotation curve. Notice the large error bars associated to points in the second and third quadrants. From Clemens [2]

14.4.1 Dispersion of Pulsar Signals

When electromagnetic waves propagate in a plasma of free electrons there is the phenomenon of dispersion. An electromagnetic wave that propagates in a plasma is described by the equation:

$$\frac{\partial^2 E_y}{\partial x^2} - \frac{\omega_p^2}{c^2} E_y = \frac{1}{c^2} \frac{\partial^2 E_y}{\partial t^2} \,. \tag{14.15}$$

The equation represents a monochromatic wave propagating along the x axis, where ω_p is the pulsation frequency of the plasma (see Sect. 7.2). The solution that represents a plane wave is of the type:

$$E_y \propto \exp[i(kx - \omega t)] \,. \tag{14.16}$$

Substituting in (14.15) we obtain the dispersion relation:

$$\omega^2 = k^2 c^2 + \omega_p^2 \,. \tag{14.17}$$

The propagation speed of the disturbance is the group speed obtained from:

$$v_g = \frac{d\omega}{dk} = \frac{d}{dk}[k^2c^2 + \omega_p^2]^{1/2} = \frac{c^2k}{\sqrt{k^2c^2 + \omega_p^2}} \qquad (14.18)$$

where replacing the expression of k derived from the dispersion relation we have:

$$v_g = c\sqrt{1 - \frac{\omega_p^2}{\omega^2}} . \qquad (14.19)$$

If an electromagnetic pulse, such as that emitted by pulsars, moves in a region occupied by a plasma along a path L, the time that the wave with frequency ν takes to travel this distance is:

$$t_\nu = \int_0^L \frac{ds}{c\sqrt{1 - \nu_p^2/\nu^2}} . \qquad (14.20)$$

For characteristic electron densities, e.g. $n_e = 10^{-2}$ cm^{-3}, from (7.31) we have $\nu_p = 1$ kHz, while the radio frequencies at which the pulsars are observed are much higher so $\nu_p/\nu < 1$. Developing in series of $(\nu_p/\nu)^2$ we have:

$$\left(1 - \frac{\nu_p^2}{\nu^2}\right)^{-1/2} \approx 1 + \frac{1}{2}\left(\frac{\nu_p}{\nu}\right)^2 \qquad (14.21)$$

and the time t_ν becomes:

$$t_\nu = \frac{L}{c} + \frac{1}{2c}\int_0^L \left(\frac{\nu_p}{\nu}\right)^2 ds = \frac{L}{c} + \frac{e^2}{8\pi^2\varepsilon_0 m_e c}\frac{D_m}{\nu^2} \qquad (14.22)$$

where is it

$$D_M = \int_0^L n_e ds [\text{cm}^{-3}\text{pc}] \qquad (14.23)$$

is the measure of dispersion. The formula (14.22) in Gaussian units is most often found in the literature:

$$t_\nu = \frac{L}{c} + \frac{e^2}{2\pi m_e c}\frac{D_m}{\nu^2} \qquad (14.24)$$

from which we pass to (14.22) expressing *and* with the transformation relations between the two systems.

The pulsar signals are pulses with periods ranging from milliseconds to seconds (depending on the age of the pulsar) and, as shown by the Fourier analysis, are made up of the superposition of many monochromatic waves of different frequencies. On the basis of the previous relation, for each monochromatic component there is a delay that increases as ν increases and the measure of $dt_\nu/d\nu$ allows to obtain the value of D_m.

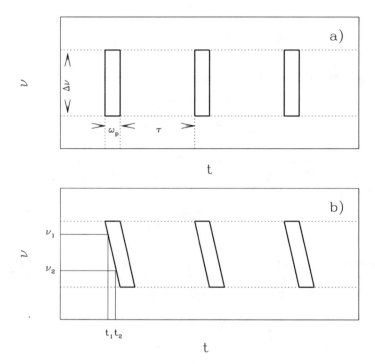

Fig. 14.4 Emission properties of signals from a pulsar

For a certain number of pulsars the distances are known more or less accurately and for these we can evaluate $\langle n_e \rangle$ based on:

$$\langle n_e \rangle = \frac{D_m}{L} . \tag{14.25}$$

Figure 14.4 panel (a) schematises a series of pulse signals emitted by a pulsar: let w_p be the duration of the pulse and $T \gg w_p$ the duration of the inter-pulse. Let $\Delta\nu$ be the frequency range of the monochromatic components of the signal. Figure 14.4 panel (b) instead shows how the signals are received. If we consider an impulse and its components corresponding to the frequencies ν_1 and ν_2 and indicate respectively with t_1 and t_2 the times taken by the two waves to reach the observer, the delay between the two signals will be given by:

$$\Delta t = t_2 - t_1 = \frac{e^2 D_m}{8\pi\varepsilon_0 m_e c} \left(\frac{1}{\nu_2^2} - \frac{1}{\nu_1^2} \right) . \tag{14.26}$$

The amount

$$\delta = \frac{t_2 - t_1}{\nu_2^{-2} - \nu_1^{-2}} \, [\text{Hz}] \tag{14.27}$$

it is called dispersion constant and can be evaluated from the arrival times of the two components of the signal. We then have:

$$D_m = \int_0^L n_e ds = \frac{8\pi\varepsilon_0 m_e c}{e^2}\delta \qquad (14.28)$$

or:

$$D_m = 2.41 \cdot 10^{-16}\delta[\text{cm}^{-3}\text{pc}] . \qquad (14.29)$$

14.4.2 Recombination Optical Emission

There are regions, distinct from the HII regions, where a weak diffuse $H\alpha$ emission is observed due to the recombination of free electrons with protons. The gas is spread over large regions and is almost completely ionized and its temperature is about 10^4 K.

The number of recombinations of H per cm^3 and per second is given by:

$$n_r = n_p n_e \beta \approx n_e^2 \beta \qquad (14.30)$$

If ϵ denotes the average number of $H\alpha$ photons produced by each recombination and ν_α is the emitted frequency, the quantity:

$$h\nu_\alpha \epsilon n_r = h\nu_\alpha \epsilon n_e^2 \beta \qquad (14.31)$$

provides the energy emitted by the unit of volume in the unit of time. Assuming that the emission is isotropic, the emissivity is given by:

$$j = \frac{1}{4\pi}h\nu_\alpha \epsilon n_e^2 \beta . \qquad (14.32)$$

Since the extinction is negligible, the intensity will be obtained by integrating the transport equation along the line of sight by setting the absorption coefficient equal to 0:

$$I = \int_0^L \frac{1}{4\pi}h\nu_\alpha \epsilon n_e^2 \beta ds . \qquad (14.33)$$

Taking into account the temperature dependence of ϵ and β, we obtain:

$$I = A \int_0^L n_e^2 T^{-0.9} ds \qquad (14.34)$$

with $A = \text{cost}$. If we assume that the temperature is constant we have:

$$I = AT^{-0.9} E_m \tag{14.35}$$

where is it

$$E_m = \int_0^L n_e^2 ds \, [\mathrm{cm}^{-6} \mathrm{pc}] \tag{14.36}$$

it is called the emission measure.

From the measure of the intensity of the line Hα it is possible to evaluate the emission measure. If along the line of sight there is only one source at distance L, the mean square electron density $\langle n_e^2 \rangle$ can be determined:

$$\langle n_e^2 \rangle = \frac{E_m}{L} . \tag{14.37}$$

14.4.3   *Absorption at Low Radio Frequencies*

In an ionized gas collisions between electrons and ions can give rise to the emission of bremsstrahlung; the reverse process is called free-free absorption. The optical depth due to this process is expressed by the relation (11.122), i.e.

$$\tau_{ff}(\nu) = C \nu^{-2.1} T^{-1.35} E_m . \tag{14.38}$$

Significant absorption by the interstellar medium occurs at radio frequencies of the order of a few MHz, but usually these frequencies are absorbed by the ionosphere and only exceptionally and in a few locations (for example in Tasmania) the ionosphere becomes transparent to such radiations. Most of the observations are obtained from satellites and balloons. Also in this case the emission measure is determined and, if the distance from the source is known, the electron density as well.

14.5   The Hot Component: Coronal Gas

The existence in the galactic disk of a hot gas with temperatures around $10^5 - 10^6$ K and very rarefied ($n \approx 10^{-3} \mathrm{cm}^{-3}$) had already been postulated in 1965 but was confirmed with observations outside the atmosphere made with the Copernicus satellite (1973).

The hot gas is not in thermodynamic equilibrium; the temperatures referred to are the kinetic ones and in these conditions the ionization equilibrium can only be calculated by equating the rate of creation and destruction of each ion. Processes of interest include collision ionization (as well as radiation) and radiative recombination. Since the production rate of the ions and that of their destruction are both proportional to the square of the electron density, the result is independent of the density and

Table 14.1 Absorption lines used to detect hot gas. We note that the last two species do not emit in the UV. T_{max} is the temperature at which the ion exhibits the greatest abundance

Species	λ [Å]	T_{max}
SiIV	1402.77	$0.6 \cdot 10^5$
CIV	1550.76	$1.0 \cdot 10^5$
SVI	944.52	$2.0 \cdot 10^5$
NV	1242.80	$1.8 \cdot 10^5$
OVI	1037.63	$3.0 \cdot 10^5$
FeX	6374.50	$1.2 \cdot 10^6$
FeXIV	5302.89	$2.0 \cdot 10^6$

depends on the type of atoms and the kinetic temperature. This ionization balance is called collisional (or coronal).

Hot gas can be detected either by absorption measures or by emission measures.

(a) *Absorption measurements.* Absorption lines produced by the gas interposed between the star and the observer can be identified in the spectrum of a star. Absorption measurements can be made in any spectral region as long as there are strong resonance lines in the spectrum of the objects under examination. It is appropriate that the stars considered have a large flux in the region where absorption appears. Particularly suited is the UV region and as a source of the spectrum stars of the first spectral types that have a relevant emission in this region.

(b) *Emission measurements.* The thermal emission of the hot gas can be used to detect it. The radiation produced is mainly localized in the UV and X (soft) regions and is a combination of continuous and lines. The continuum is produced by bremsstrahlung and radiative recombinations.

In the galactic disc, the hot gas was first detected by the absorption lines of the OVI. The correlation between the path length and the column density deduced from the characteristics of the lines confirms that the absorption is interstellar and not circum-stellar. The profiles of the absorption lines with many components indicate that it is a widely distributed gas and that it exhibits a velocity dispersion of about 10 km/s. The temperature obtained from this analysis is $T \leq 4 \cdot 10^5$ K. Other species confirm these conclusions.

The emission X of the hot gas ($E < 1$ kev) has a diffuse character: from the observations we deduce temperatures greater than or equal to 10^6 K, density of 0.004 cm^{-3} and therefore pressures of the order of:

$$\frac{p}{k} = 2n_e T \approx 8000 \, [\text{cm}^{-3}\text{K}] . \tag{14.39}$$

The observations of the two types reveal the presence of this gas also in the galactic halo. We therefore have two independent techniques which reveal the presence of a hot gas. However, there is a discrepancy in the inferred temperature values: the UV absorption observations imply $T \leq 4 \cdot 10^5$ K, while those of the X emission give $T \geq 10^6$ K. This inconsistency can be resolved in the context of the theoretical considerations that had stimulated the search for this component, whose origin was attributed, as we have already seen, to the evolution of SNRs. Their superposition produces a diffuse gas that acts as a substrate for other components of the interstellar medium, at least in certain regions of the disk. According to this theory, the diffuse X emission is determined by the hot substrate while the conductive interfaces between the hot gas and the other components immersed in it are responsible for the observed absorption of the OVI.

It is still uncertain what the filling factor of this component is, i.e. the fraction of the galactic volume occupied by it. The determinations are conflicting, but it seems that a conspicuous part of the interstellar volume (20% or more) is occupied by this gas component.

14.6   The Cold Neutral Component: The Clouds of HI.

Let us now analyze the CNM component made up of low-temperature atomic neutral gas; it has a discrete distribution and its groupings are called clouds. Its temperature can be studied with the absorption line at 21 cm; from these observations it can be deduced that the temperature is not constant, that is, the clouds are not isotherms. The observed "range" is from 60 to 240 K.

Cloud sizes are inferred from the column densities N(HI). There are three methods:

(a) absorption line at 21 cm,
(b) interstellar reddening and relation $E(B - V)/N$(HI),
(c) optical absorption lines.

All methods have drawbacks and favor the identification of the clouds with the largest dimensions. Often these data have been used to deduce the size distribution function under the assumptions that: (1) the clouds are spherical and (2) the density is constant. It will be seen below that the first assumption is not correct and that the smaller clouds have high temperatures if they are in pressure equilibrium they must have a lower density. So such distribution functions don't seem very reliable.

The internal structure of the clouds and their shape can be studied with high resolution HI maps. Even at the smallest scales (with dimensions smaller than or equal to the pc) we observe filaments (structures a in which one dimension is prevalent over the others) or "sheets" (sheet structures in which one dimension is much smaller than the rest) with well marked edges, probably indicators of shocks. Within the filaments and sheets, small condensations (clumps) smaller than the pc are identified. Therefore the shape of the clouds is very irregular and very different from the spherical

one usually used in models. Since evaporation depends on the surface area, shape
is an important parameter in the models and the spherical structure minimises the
evaporation rate. The shape of these clouds confirms what emerges from other observations: diffuse clouds are not gravitationally bound (they are confined by pressure)
and the shape is determined by evaporation, shocks and the magnetic field.

High resolution HI studies lead to draw the following conclusions about the structure and distribution of temperature

- Clumps are numerous in filaments and sheets and are mainly responsible for the
  absorption of HI. They have density $n = 20 - 50 \text{cm}^{-3}$ and temperatures $T = 30 - 80$ K (with an average value of 40 K).
- The filaments and sheets are in turn immersed in warm HI envelopes ($T \geq 500$
  K). These envelopes are responsible for 80% of the HI emission. If we believe
  the theory of thermal instability, the gas at these temperatures cannot exist stably:
  therefore either the envelopes are a transitory phase or the heating agents are
  transient.

Regions containing young objects (OB stars and HII regions) and molecular clouds
are observed on an intermediate scale. Each of these regions is surrounded by a HI
envelope with a high column density. Examples of such regions are in Ophiucus,
Orion and Perseus-Taurus. The average physical properties of these envelopes are:

- $N(\text{HI}) = 10^{21} \text{cm}^{-2}$,
- $\langle n(\text{HI})\rangle = 2.5 \text{ cm}^{-3}$,
- $d = 120$ pc (linear diameter),
- $M(\text{HI}) = 10^5 M_\odot$.

In some cases, the molecular hydrogen of the central zone in which the young stars
and the HII regions are immersed has been studied and it has been found that it has a
mass approximately equal to that of the HI: e.g. in Orion it is $M(\text{H}_2) = 3 \cdot 10^5 M_\odot$.
Using these data to study viral equilibrium it seems that the entire structure is in
equilibrium and both components are essential to determine it: atomic and molecular
hydrogen.

14.7   The Warm Neutral Component

Of the emission lines at 21 cm, the wide ones ($\sigma_v \approx 9$ km/s) are observed in all
directions, while the narrow ones ($\sigma_v < 5$ km/s) appear to come from only one third
of the directions. Only these tight components also exhibit detectable absorption.

Let us consider the situation schematised in Fig. 14.5, bearing in mind the results
of Sect. 14.2. Suppose that R_s is a radio source emitting a continuous spectrum and
that a HI cloud is interposed on the line of sight between it and the observer. If the
observation is made along line (1), a brightness temperature T_B is measured:

$$T_B = T_B(0) \exp[-\tau_{\nu,r}] + T(1 - \exp[-\tau_{\nu,r}]) = a \ . \tag{14.40}$$

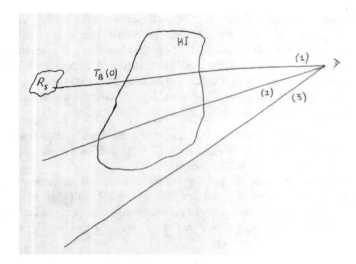

Fig. 14.5 Detection of the WNM component. Courtesy of Agnese Carraro

If you always observe in direction (1) at a wavelength close to but still different from that of the line at 21 cm, you receive the intensity of the source that does not interact with the cloud as the HI does not absorb this radiation. On the other hand, the HI does not emit even at this frequency. We therefore have:

$$T_B = T_B(0) = b \,. \tag{14.41}$$

If we observe in the direction (2) that crosses the cloud but does not intersect the radio source, we have:

$$T_B = T(1 - \exp[-\tau_{\nu,r}]) = c \,. \tag{14.42}$$

Combining the previous relations we have:

$$a = b\exp[-\tau_{\nu,r}] + c \tag{14.43}$$

from which we get $\tau_{\nu,r}$ and from (14.42) we get the temperature T of the gas. This way the thermodynamic parameters of the clouds are obtained.

If we look in the direction (3) that excludes the cloud, we determine the properties of another component, diffuse, warmer and more neutral: the "warm neutral medium" (WNM). Its temperature can be studied using both the 21 cm line and the absorption lines in the UV; currently most of the information comes from the first technique. Since the optical depth is small from the relation (14.10), a high temperature is deduced. In fact, the temperature of most WNM is in the $5000 - 8000$ K range. However, these results cannot be considered definitive because they are based on few observations.

14.8 Synthetic Summary

Feed-back from massive stars, among other perturbations, generates the multi-phase ISM. In this Chapter we have reviewed the various atomic forms in which the ISM appear to be divided: CNM, WNM, WIN, HIM. Besides, we recall the various observational techniques to measure the thermodynamical properties of each component, dividing them according to the ionization degree. HI (the CNM) has played an important role in the past, because of its prominent line at 21 cm, which was discovered in 1951. Much work has been done using this line, especially for Galactic structure, although the dispersed and ubiquitous nature of H1 prevented to reach solid conclusions. Coronal gas (HIM) is found in SN remnants, but also in the solar corona and in the intergalactic medium. Not much is known about the WNM, except that it surrounds HI clouds.

Appendix

Appendix A: The Energy Sources of the Interstellar Medium

In Chaps. 11, 12, and 13 we have analysed the processes by which massive stars in the course of their evolution transfer mechanical or electromagnetic energy to the interstellar medium. Let us now try to evaluate how the conversion efficiency of these energies depends on the mass of the star, and what contribution the HII regions, stellar winds and supernova explosions give respectively, both for individual masses and for a stellar generation.

HII regions. Summarizing what has been said in Chap. 11, the energy emitted by the star and absorbed by the gas is given by (11.107) where for t the life time on the Main Sequence is used, while the average value the photon energy is, based on (11.109), proportional to the effective temperature of the star. The position of the star in the HR diagram is approximated by its position on the zero-age sequence. The efficiency of the conversion is given by (11.111), which can be put in the form:

$$f = 5.75 \cdot 10^{18} n_0^{-1/3} \frac{1}{T_* S_{UV}^{1/3}} \frac{\lambda^3 \lambda'^2}{N} \qquad (14.44)$$

so assuming $n_0 = 10^2 \text{cm}^{-3}$ and adopting 1 as the average value of $\lambda^3 \lambda'^2 / N$ we have:

$$f = 1.24 \cdot 10^{18} (T_* S_{UV}^{1/3})^{-1} . \qquad (14.45)$$

The energy emitted by the star in the form of UV photons during its main sequence life, using the (11.107), is expressed by:

$$\log E(M) = -8.413 + \log S_{UV}(M) + \log T_e(M) + \log t_{MS} . \qquad (14.46)$$

In Table 14.2 we provide the results of the calculation of $E(M)$ and the energy $E_{HII} = fE(M)$ used by the ISM.

The relationship $E_{HII} = E_{HII}(M)$ can be interpolated using the data from previous table, thus obtaining

$$E_{HII} = \begin{array}{ll} 1.1 \cdot 10^{48} M^{0.86} & \text{if } M \geq 17.8 \qquad (14.47) \\ 4.0 \cdot 10^{43} M^{4.5} & \text{if } M \leq 17.8 \qquad (14.48) \end{array}$$

and whose integration yields the energy injected in the ISM by a star cluster with a turn-ff mass M and at birth N_* stars in the mass range $(8, 100) M_\odot$.

Stellar winds. By interpolating observational data or results of theoretical calculations we can derive the following relationships:

(a) brightness of the star on the main sequence, which is approximated with the position at age zero:

$$\log \left(\frac{L}{L_\odot} \right) = -0.148 + 4.62 \log M - 0.747 (\log M)^2 \qquad (14.49)$$

(b) life time in the blue region of the HR diagram:

$$\log t = \begin{array}{ll} 8.29 - 0.97 \log M & \text{se} M \leq 40 \qquad (14.50) \\ 7.41 - 0.43 \log M & \text{se} M \geq 40 \qquad (14.51) \end{array}$$

(c) mass loss rate:

$$\frac{dM}{dt} = 1.36 \cdot 10^{-6} \left(\frac{L}{L_\odot} \right)^{1.77} [M_\odot/\text{yr}] \qquad (14.52)$$

(d) terminal wind speed:

$$\log V_w = 3.28 + 0.178 \log M + 0.138 \log \left(\frac{Z}{Z_\odot} \right) \qquad (14.53)$$

By means of these relations it is possible to calculate the brightness of the wind and the total expelled energy E of the wind, of which we report some values in Table 14.3. The values of E can be interpolated using the relationship:

$$E = 6.46 \cdot 10^{44} M^{3.43} . \qquad (14.54)$$

Taking into account that the efficiency of the transformation of wind energy into kinetic energy of the interstellar medium is equal to 15/77 (see Chap. 12) we have for the energy E_w introduced into the interstellar medium by a mass star M the expression:

Table 14.2 Energy injected in the ISM from an HII region excited by an OB star, as a function of the star mass

$M/M_\odot$	$\log E$	$\log f$	$\log E_{\text{HII}}$
100	53.04	−3.40	49.64
80	52.92	−3.33	49.59
60	52.71	−3.21	49.50
50	52.59	−3.11	49.49
23	51.81	−2.70	49.11
17.5	51.45	−2.50	48.95
8.3	49.34	−1.64	47.70

Table 14.3 Energy injected in the ISM by stellar wind from an OB star, as a function of mass

$M/M_\odot$	$\log L_w$	$\log E$
100	37.71	51.59
80	37.39	51.33
60	36.93	50.96
40	36.21	50.41
30	35.66	49.96
20	34.80	49.27
10	33.15	48.07

$$E_w = 1.26 \cdot 10^{44} M^{3.43} \, [\text{erg}] \,. \tag{14.55}$$

For an age cluster τ (and stars at the mass turnoff point M) initially containing N_* stars in the mass interval $(8, 100)$ $M_\odot$ and with an initial Salpeter mass function, i.e. $\Phi(M) = AM^{-2.35}$, the total energy fed into the medium is:

$$E_{w,\text{tot}} = 23.13 N_* \int_8^{m(\tau)} M^{-2.35} 1.26 \cdot 10^{44} M^{3.43} dM = 1.4 \cdot 10^{45} N_* (M^{2.08} - 271.73) \,. \tag{14.56}$$

Supernovae. Observations seem to suggest an energy of 10^{51} erg in case of a SN of type II explosion, independently from the progenitor mass. Assuming an efficiency of 0.02 fro the conversion of this energy into kinetic energy of the ISM (see Sect. 13.5), we obtain that each SN explosion event contributes with $2 \cdot 10^{49}$ erg, and, hence, for a generation of star we have:

$$E_{\text{SN, tot}} = 2 \cdot 10^{49} N_* \,. \tag{14.57}$$

References

1. Levine, E.S., Blitz, L., Heiles, C.: Science **312**, 5781 (2006)
2. Clemens, D.: ApJ **295**, 422 (1985)

Chapter 15
Molecular Clouds

Molecular clouds are the densest and coldest component of the ISM. They are made of H in its molecular form (H_2), and by many other molecules, although much less abundant than H. Hundreds of species have recently been detected by ALMA in the sub-millimeter regime, where most molecules show rotational emission lines.

15.1 Detection of Molecular Gas

In the last thirty-fourty years, radio observations have revealed that hydrogen in molecular form is the most important component of the interstellar medium: in our galaxy all the regions where star formation is active are associated with molecular clouds. The relationship of the molecular clouds with the other phases of the interstellar medium is quite different from what had been imagined: the molecular clouds are self-gravitating with an internal pressure of approximately an order of magnitude greater than that of the other phases. Therefore these clouds are not in pressure equilibrium with the warmer and more diffused phases.

Since hydrogen is the most abundant element, molecular clouds are mainly made up of H molecules (H_2); nevertheless, the detection of the molecular gas is done not via the emission of H_2 but through the second most abundant molecule, carbon monoxide (CO), which is significantly under-abundant with respect to H_2: n(CO) is in fact only $10^{-4} - 10^{-5}$ n(H_2).

At low temperatures typical of molecular clouds, the thermal energy is such that the rotational states of the molecules are excited by collisions and in fact many of the lines observed are lines of the rotational spectrum.

For linear molecules the permitted levels are those of a rigid rotator:

$$E_J = J(J+1)B \qquad J = 0, 1, 2, \ldots \tag{15.1}$$

© The Author(s), under exclusive license to Springer Nature Switzerland AG 2021
G. Carraro, *Astrophysics of the Interstellar Medium*,
UNITEXT for Physics, https://doi.org/10.1007/978-3-030-75293-4_15

with

$$B = \frac{h^2}{8\pi^2 I} \tag{15.2}$$

where I is the moment of inertia of the molecule. B is called rotational constant, and is measured in Joule. Light molecules have rotational spectra in the far-infrared and microwave regions ($\lambda < 1$ cm).

In the case of the H_2 molecule there is no permanent electric dipole moment (being made of identical nuclei (homo-nuclear) the center of gravity coincides with the charge distribution center) and the transition $J = 1 \to J = 0$ is not allowed . Transitions take place with electrical quadrupole interaction and the selection rule is $\Delta J = \pm 2$. The transition $J = 2 \to J = 0$ ($\lambda = 28\mu$m, visible inn the mid-IR) requires a relatively high temperature ($T = 500$ K) for the upper level to be excited by collision and this temperature is much higher than that normally found in molecular clouds.

Molecular gas is most easily detected through the emission of CO. The CO molecule has a permanent dipole moment and therefore transitions with dipole inter-actions are allowed. In this case, the constant B in (15.2) is $3.83 \cdot 10^{-16}$ erg. The transition $J = 1 \to J = 0$ corresponds to the emission of a line with $\lambda = 2.6$ mm and the probability of spontaneous decay is $A_{1\to 0} = 6 \cdot 10^{-8}$ s^{-1}. The population of the two levels can be evaluated by the balance among the decay rate of the $J = 1$ level and the collisional excitation rate of the $J = 0$ level by molecules of H_2. For this process the cross section is $\sigma = 2 \cdot 10^{-15}$ cm^2. At a temperature of $5 - 10$ K the thermal velocity is of the order of 10^4 cm/s and hence $\langle \sigma v \rangle = 2 \cdot 10^{-11}$ cm^3 s^{-1}.

We then have that:

$$r_{1\to 0} = n_1 A_{1\to 0} , \tag{15.3}$$

and :

$$r_{0\to 1} = n_0 n(H_2) \langle \sigma v \rangle \tag{15.4}$$

where n_1 and n_0 are the densities of the two levels involved. The comparison of these two equation yields:

$$n_1 A_{1\to 0} = n_0 n(H_2) \langle \sigma v \rangle \tag{15.5}$$

from which

$$\frac{n_1}{n_0} = \frac{n(H_2) \langle \sigma v \rangle}{A_{1\to 0}} . \tag{15.6}$$

By imposing $n_1/n_0 > 1$ we have

$$n(H_2) > \frac{A_{1\to 0}}{\langle \sigma v \rangle} = 3000 \text{ cm}^{-3} . \tag{15.7}$$

To obtain, at the temperature of the molecular clouds, a discrete excitation of the rotational level $J = 1$ of the CO, a density greater than 3000 cm^{-3} is required. This estimate is correct only if the excitation of the levels is determined by the balance between spontaneous decay and $CO - H_2$ collisions; in reality the CO emission is optically thick and it is necessary to take into account the induced radiative processes. Taking this into account, we find that the critical density for excitation is reduced by more than a factor of 10: $n(H_2) > 100 \, \text{cm}^{-3}$.

Therefore the CO emission is correlated with the density of the molecular gas (through the excitation of the $J = 1$ level) and therefore it is possible, from the intensity of the CO emission, to evaluate the quantity of molecular gas. In reality what is required is a high number of collisions with CO: the presence of HI could also be fine, but HI at this density should be easily detected by the 21 cm line which is absent; therefore we must think about the molecular form of hydrogen. Through the emission of CO, maps were made in different areas of the sky to highlight the distribution of H2.

The detection of the densest molecular gas regions is done by the emission of different molecules such as NH_3, CS, and HCN, whose excitation requires higher densities.

15.2 Correlation Between CO Emission and Molecular Gas Mass

From the definition of brightness temperature (see Sect. 14.4)

$$I_\nu = \frac{2kT_B\nu^2}{c^2} .$$ (15.8)

The integration of the line at 2.6 mm of the CO yields:

$$I_{CO} = \int_{\Delta\nu} I_\nu d\nu = \int_{\Delta\nu} \frac{2kT_B\nu^2}{c^2} d\nu = \frac{2k\nu_{CO}^3}{c^3} \int T_B(v)dv ,$$ (15.9)

where the integral is made over the velocity profile, taking into account the relationship between frequency and velocity of the source based on the Doppler effect. We can write:

$$I_{CO} = \frac{2k\nu_{CO}^3}{c^3} J_{CO} ,$$ (15.10)

with

$$J_{CO} = \int T_B(v)dv .$$ (15.11)

To derive the brightness one has to integrate over the apparent surface of the source as seen by the observer. The brightness is then:

$$L_{CO} = \int_S I_{CO}d\sigma = \frac{2k\nu_{CO}^3}{c^3} J_{CO}\pi R^2 ,$$ (15.12)

where R is the radius of the cloud, supposed to be spherical. On the other hand we can approximate the integral over the 2.6 mm line as:

$$J_{CO} = \int T_B(v)dv = T_{B,max}\Delta v \tag{15.13}$$

where $T_{B,max}$ is the temperature corresponding to the peak of the emission and Δv can be approximated with the FWHM of the line. We then have:

$$L_{CO} = \frac{2k\nu_{CO}^3}{c^3} T_{B,max}\Delta v \pi R^2 . \tag{15.14}$$

For a cloud in virial equilibrium (we will see later that the molecular clouds are in equilibrium) it is:

$$\frac{\mathcal{G}M^2}{R} = M(\Delta v)^2 , \tag{15.15}$$

from which:

$$\Delta v = \sqrt{\frac{\mathcal{G}M}{R}} . \tag{15.16}$$

Introducing the density to eliminate R we obtain:

$$L_{CO} = \frac{2k\nu_{CO}^3}{c^3} \left(\frac{3\pi\mathcal{G}}{4\rho}\right)^{1/2} M T_{B,max} , \tag{15.17}$$

Eventually, we have:

$$L_{CO} \propto \frac{T_{B,max}}{\rho^{1/2}} M \tag{15.18}$$

where ρ and $T_{B,max}$ are approximately constant, so this equation defines a direct proportionality relationship between L_{CO} and M. Clearly, here M is the virial mass. For clouds with different temperatures and different average densities, the constant of proportionality between L_{CO} and the mass of the molecular gas is scaled as $T_{B,max}/\rho^{1/2}$. Since warmer clouds (e.g. in the Galactic center) tend to be denser, the dependence on $T_{B,max}$ and ρ partially compensate.

In more recent years there have been several attempts to observe the linearity of this relationship observationally and thus to evaluate the constant of proportionality. From the data of Bolatto et al. [1] we obtain the distribution in Fig. 15.1 for clouds with mass between 10% and $2 \cdot 10^6 M_\odot$. α_{CO} is the virial ratio, inferred from the linear relation between the virial mass (M_{vir}, see above) and the CO luminosity, and therefore it reflects the very same kind of dependence. The linear dependence justifies the use of CO as a tracer of the mass of Galactic H_2.

This excellent correlation is equivalent to the relation between the column density of H_2 and J_{CO} and also for this relation, theoretically deduced, there is good agreement with the relation deduced observationally.

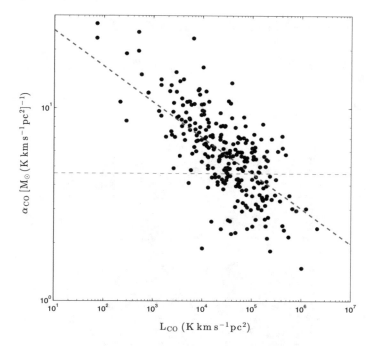

Fig. 15.1 Correlation of L_{CO} and virial ratios (hence masses) for molecular clouds in the Milky Way. From Bolatto et al. [1]

The relationship found is the basis for defining the mass of the molecular gas from the observations of the CO line. By means of it, from the maps of the CO, it is possible to estimate the distribution and the total mass of the H_2 contained in the disk of the Galaxy (see Fig. 15.2).

Besides, the evaluation of the column density of the gas, if we believe it is in molecular form, towards a given line of sight, can also be used to locate dense clouds, making use of the constancy of the relationship $E(B - V)/N$ for the Galaxy.

15.3 The Distribution of H_2 in the Galaxy

The results of various CO surveys (conducted, among the others, with the NANTEN (Chile), Parkes (Australia), and Green Bank (USA) radio-telescopes) have shown that in the Galaxy the emission of CO is confined close to the Galactic center and in a ring of the Galactic plane with an internal radius of 3 kpc and an external radius of 5 kpc. It was also found that most of the molecular material is confined to discrete clouds. These characteristics are completely different from those found with the

Table 15.1 H_2 most prominent concentrations in the Milky Way

	Radius [kpc]	$M_{tot}(H_2)$ [$M_\odot$]	$n(H_2)$ [cm^{-3}]	$M(H_2)/M(HI)$
Galactic Center	$R \leq 0.4$	$2 \cdot 10^8$	$60 - 120$	20
Circum-nuclear ring	$3 < R < 5$	$1.8 \cdot 10^9$	2.7	2
Galactic Disc		$2.3 \cdot 10^9$		1

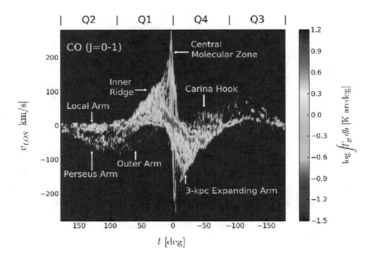

Fig. 15.2 A typical elocity map of CO in the Galaxy. The most relevant emission regions are indicated

surveys of the 21 cm line for the HI, whose distribution is rather uniform and with little tendency to condense into clouds.

The main characteristics of the large-scale distribution of H_2 (Fig. 15.2) are also reported in Table 15.1. The estimate of the mass of the molecular gas in the Galactic center is uncertain because it is possible that the molecular gas has a higher temperature than elsewhere. Overall, the masses of H_2 and HI are almost equal, within the uncertainties of the respective determinations. On the other hand, the radial distributions of the two components are different: approximately 90% of the total mass of H_2 is contained within the solar circle ($R < 8.5$ kpc), while only 35% of HI is found in the same region.

The problem of the association of the emission of CO, and therefore of molecular clouds, to the spiral structure in the Milky Way is important for understanding the origin and evolution of the clouds themselves. If these are relatively short-lived objects that form from HI as a result of the shocks of the spiral arms and are destroyed by leaving the arms, their spatial distribution must follow the design of the arms and there must be a notable contrast between the clouds distributed along the arms compared to those in the areas between the arms. If, on the other hand, the clouds are formed by other mechanisms and survive several passages across spiral arms, they must be more or less uniformly distributed over large regions and should not bear any relation with the arms, as also happens for HI. Recent results indicate that out of a sample of 2,000 molecular clouds, the distribution of hot ($T > 10$ K) and massive clouds resembles the distribution of HII regions, which delineate the spiral structure, while colder and smaller clouds, which represent about 60% of the total CO emission, are more evenly distributed across the disk. The existence of molecular clouds in the regions between the arms was demonstrated by the high-resolution maps of CO in the outer galaxies. In M51 the CO is peaked along the spiral arms, but the intensity of the emission between the arms is equal to 50% of that seen in the arms.

15.4   Distribution of H$_2$ in Other Galaxies

Among spiral galaxies, there is no correlation between the emission of CO and the morphological type. The Magellanic irregulars have a CO emission lower than that of the spirals. There may be two reasons for this difference: (a) a real lower abundance of CO and molecular gas, (b) a lower CO excitation temperature reflecting a less significant heating by cosmic rays. Since this type of galaxies has relatively low metallicity, the lower CO abundance may be due to a longer formation time and a higher photodissociation rate (less obscuring dust). In ellipticals CO was searched for (it may be present due to the mass loss from red giants). It has been found only in NGC 187, although the identification is not entirely certain (NGC 187 is a peculiar elliptical since it shows traces of ongoing star formation): the mass of H$_2$ would be as low as $10^4 \, M_\odot$, equal to 10% of the detected mass of HI. Based on the analysis of the maps of CO in galaxies of different morphological types, the following classification was developed

- *class I*: has a strong central source, and the integrated intensity of CO per unit area decreases monotonously outwards along the entire disk, e.g. M51, NGC6946, IC342;
- *class II*: has a strong central source and one or more rings, e.g. our galaxy;
- *class III*: it has a ring without the central source, and the integrated intensity per unit area in the center is less than 20% of that of the disc, e.g. M31, M81, NGC7331;
- *class IV*: there are only a few isolated regions in the disk, e.g. M33, Large Magellanic Cloud;
- *class V*: CO has not been revealed to date, e.g. NGC6822, Small Magellanic Cloud.

15.5   Clouds Classification and Statistical Properties

It is difficult to set up a classification because in many cases they consist of complex structures. With the exception of single isolated clouds, clouds are generally organised in a hierarchical structure whose parts can be seen as simple clouds or portions of a more complex and larger cloud. There is also a considerable confusion in the terminology used by various authors. Each classification attempt is purely a schematic chart of how the properties of clouds vary according to their size; like any classification, which tends to translate reality into schemes, it is necessarily rough and approximate. Therefore, we avoid considering detailed classifications but in the following we will refer, for practical reasons, to a classification in two categories which is also important for the purposes of studying star formation: giant molecular clouds, supposed to be the formation sites of massive stars, and dark clouds associated with the formation of low-mass stars. The former have masses of the order of $10^5 - 10^6 M_\odot$, temperatures between 10 and 20 K and a velocity dispersion of $5 - 10$ km/s. Examples are the molecular clouds in Orion, M17 and W3. The latter have masses of the order of $10^3 - 10^4 M_\odot$, slightly lower temperatures, about 10 K, velocity dispersion of $1 - 3$ km/s and seem to be the place where only low-mass stars form. The closest and best known clouds of this type is in Taurus.

As we have seen, the excitation of the molecules normally occurs through collisions so that the relative intensity of the emission lines can be used to determine the local density and the temperature. When the excitation is known the total emission in a spectral line gives the column density of the emitting species. The line profile allows you to determine the average speed and dispersion. A map of the emission reveals the internal structure and the cloud edges.

Most of the mass of the molecular clouds is contained in regions of modest extinction, which allows the photons of the interstellar radiation field to maintain sufficient ionization to guarantee the indirect coupling between neutral particles and the magnetic field (freezing of the lines of force at the plasma): in these regions the degree of ionization is $x = 10^{-4}$ (the ionization fraction is measured by molecules of the type of HCO^+, DCO^+, $N_2 H^+$, and $N_2 D^+$) and the ambipolar diffusion time is long with respect to the average life of the molecular cloud ($\tau_{AD} \sim 10^8$ years). Only in the densest portions photons are unable to penetrate and the ionization, due to cosmic rays, is much lower, i.e. $10^{-9} < x < 10^{-7}$. Here, the decoupling between the magnetic field and neutral particles occurs with an ambipolar diffusion time between $2 \cdot 10^4$ and $2 \cdot 10^6$ years.

Molecular clouds, in regions where there is no ongoing star formation, are almost isothermal and are characterised by an average density $n \approx 10^2 - 10^3$ cm^{-3} and are pervaded by a magnetic field. The intensity of it and its direction are important quantities for the study of the physical behaviour of clouds and the star formation process but they are very difficult to measure: the Zeeman effect is used and the results are quite safe for less dense regions, while considerable difficulties are encountered for the denser parts.

The most reliable value is $B \approx 20\mu G$, but values in the range of $10 - 100\mu$ G are considered possible. The spectrum of molecular masses is very broad: at the lower end objects with masses of the order of tens of solar masses are observed: they are considered transient structures that are formed by the action of shock waves. At the other extreme molecular complexes are observed more massive than $10^6 M_\odot$ and with average densities of about 50 cm^{-3}.

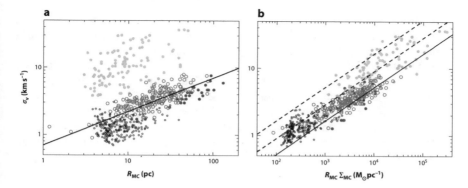

Fig. 15.3 Distribution of velocity dispersions versus size (left panel) and versus column mass density (right panel) of Galactic molecular clouds from different sources, as collected by Heyer et al. [2]

Inside the molecular clouds there are numerous regions with high optical depth, called cores or clumps that appear dynamically isolated from the rest of the cloud and where star formation can be triggered. The width of the NH_3 lines in many of these cores reveals that the turbulence is subsonic and in some of them the thermal agitation is larger than that due to the turbulence.

The analysis of a relatively large number of clouds of various types and sizes made it possible to derive the following statistical relationships.

(1) *Distribution of diameters.* The number of objects having a diameter in the range $(D, D + dD)$ is given by the relation:

$$N(D)dD \propto D^{-2.32\pm0.25}dD \qquad (15.19)$$

where D is expressed in parsecs. This relationship is derived interpolating observational data.

(2) *Diameter-velocity dispersion relationship.* From the data in Fig. 15.3 the following relationship is obtained by interpolation:

$$\sigma_v = (0.31 \pm 0.05)D^{0.55}[\text{km/s}] \qquad (15.20)$$

with D in parsec. A similar relationship also holds for the cores inside molecular clouds: in the Taurus cloud about 40 cores have been identified with an average size of 0.46 pc and an average mass of 23 $M_\odot$; the masses extend over an interval of a factor of 80. These cores can probably be thought of as the lower end of the cloud mass distribution.

(3) *Mass spectrum.* The number of clouds contained in a given volume of the Galaxy having mass in the interval $(M, M + dM)$ is given by:

$$N(M)dM \propto M^{-k}dM \qquad (15.21)$$

with $k = 1.4 - 1.6$. A mass spectrum of this type can also be obtained with a model in which giant clouds form by coalescence in repeated binary collisions between clouds of smaller mass: the upper limit of the mass is determined by the lifetime of the clouds. From the mass spectrum (15.21) it can be deduced that about 50% of H_2 is contained in 1000 clouds with $M > 10^6 M_\odot$, ($D > 50$ pc) and 90% is contained in clouds with $M > 10^5 M_\odot$ ($D > 20$ pc).

(4) *Density-linear dimension relationship.* By interpolating the observational data we find a correlation between the numerical density and the linear dimension (i.e. the diameter) of a cloud; the correlation index is very good (at least 0.9). This reads:

$$n = 4979 D^{-0.9} \tag{15.22}$$

where D is in parsec and n in cm^{-3}. In the literature there are also other similar relationships with slightly different numerical coefficients.

15.6 Mechanical Equilibrium of Molecular Clouds

Cloud masses, inferred from the velocity dispersion and the virial theorem (equilibrium hypothesis) coincides, within the observation errors, with the masses deduced from H_2 column density measurements: this leads us to think that molecular clouds are systems in self-gravitating equilibrium, and not in pressure equilibrium with the ambient medium.

The velocity maps deduced from CO and other molecules line centres indicate that the clouds do not show any indication of global collapse, nor of rapid rotation, although there is a Doppler broadening much greater than the thermal one.

The problem of cloud balance can also be tackled by using the two empirical relations (15.20) and (15.22). The equilibrium condition, applied to a spherical cloud is expressed, according to the virial theorem, by:

$$\frac{2\mathcal{G}M}{D\sigma_v^2} = 1 . \tag{15.23}$$

Since the shape of the cloud deviates from the spherical one and since the density profile is not uniform but is generally a decreasing monotone function of the distance from the center, the value of the second member can be uncertain by an order of magnitude. Using (15.20) and (15.22) we now want to show that the first member of (15.23) has a value not far from 1, in the broad sense just specified. Written in the CGS system, the (15.20) becomes:

$$\sigma_v = 2.1 \cdot 10^{-6} D^{0.55} \tag{15.24}$$

and (15.22), if n is expressed as a function of M and D, becomes:

$$M = 3.8 \cdot 10^{-4} D^{2.1} . \tag{15.25}$$

With (15.24) and (15.25) we have:

$$\frac{2\mathcal{G}M}{D\sigma_v^2} = \frac{2\mathcal{G}3.8 \cdot 10^{-4} D^{2.1}}{D(2.1 \cdot 10^{-6} D^{0.55})^2} \approx 11.7 . \tag{15.26}$$

The numerical result is not quite close to unity; however, for the reasons indicated above, and for the circumstance that if the clouds were not in equilibrium the first member of the previous relation would be much larger, we can confidently concluded that molecular clouds are in virial equilibrium. In addition to these considerations there are other arguments in favour of the state of equilibrium of the clouds: one of these is that the life of the clouds is much longer than the free-fall time. Furthermore, if they were not in equilibrium, the rate of star formation, deduced from the number of clouds present in the disk of the Galaxy, would be considerably larger than that observed. The approximate number of clouds present on the Galactic plane would give a star formation rate of about $200 M_\odot$ per year, while the actual observed rate is lower than $1 M_\odot$ per year.

Once we accept the idea that molecular clouds are in equilibrium, we need to identify which forces are able to balance self-gravitation. Obviously, given their low temperature, the pressure due to thermal motion cannot be responsible for equilibrium; in fact, the widening of the lines observed is much greater than that corresponding to a thermal motion with a temperature of 10 K. There are currently two main conflicting opinions on the type of forces capable of balancing gravity. We will describe both points of view.

(A) Equilibrium is achieved considering the pressure due to turbulence. Several years ago Richard Larson found a relation analogous to (15.20) with exponent 0.38. He noted the remarkable similarity of this relation to the Kolmogorov-Obukhov relation (see Sect. 6.4, Eq. (6.47)) for the turbulence of incompressible fluids. We have already seen what differences exist between the turbulence of incompressible fluids and the behaviour of the interstellar gas for which the interpretation of a relationship of type (15.22) in terms of that theory of turbulence is questionable. However, if we accept that this relationship can be interpreted in terms of the turbulence of a compressible fluid and that the virial equilibrium is due to the action of the turbulence pressure, we still encounter great difficulties. In fact the turbulence decays very quickly and therefore the cloud would be stable only for a time equal to $k\tau_{ff}$ with $k = 2 - 3$, insufficient to explain the actual cloud lifetime. Therefore a turbulence re-feeding mechanism is needed and the most promising is the Galactic differential rotation. The ordered energy of rotation acting on a viscous fluid generates turbulent energy at such a rate and, even if it is subsequently dissipated, it can give rise to a stationary situation.

(B) The self-gravitation forces are balanced by the magnetic forces. More precisely in the interior of the clouds magnetohydrodynamic waves are excited and the velocity field produced is responsible for the observed broadening of the lines;

in reality, as we have already seen, this velocity field is produced by Alfven waves since the other waves decay rapidly. The magnetic pressure produced by these waves guarantees support against the gravitational field. In this case the non-thermal portion of the kinetic energy density $1/2\rho\sigma_v^2$ is related to the energy density of the magnetic field and, within a numerical factor of the order of unity, is

$$\frac{1}{2}\rho\sigma_v^2 = \frac{B^2}{2\mu_0} . \tag{15.27}$$

From this condition we derive:

$$\sigma_v = \left(\frac{B^2}{\mu_0\rho}\right)^{1/2} = v_A \tag{15.28}$$

therefore a cloud, whose energy of non-thermal motions is comparable with its magnetic energy, has non-thermal velocities that are comparable with Alfven's velocity. The non-thermal motions originate from the passage of Alfven waves, that is, the observed turbulent velocity field originates from the propagation of these magnetohydrodynamic waves. Starting from the hypothesis that the cloud is in equilibrium, satisfies the relation (15.27), is spherical with radius R and furthermore has mass M, temperature T, and average molecular weight μ, we show that the observed relationships between dispersion of velocity and size, and between density and size are verified. The virial theorem gives, neglecting thermal energy:

$$\frac{\mathcal{G}M}{R} = \sigma_v^2 \tag{15.29}$$

so by obtaining σ_v from the latter and expressing M by means of ρ and R, we obtain:

$$\sigma_v = \left(\frac{4\pi\mathcal{G}}{3}\right)^{1/2} R\rho^{1/2} . \tag{15.30}$$

Taking ρ from (15.28) and replacing it in (15.30) we finally have:

$$\sigma_v = \left(\frac{4\pi\mathcal{G}}{3\mu_0}\right)^{1/4} R^{1/2}B^{1/2} \tag{15.31}$$

which is of the type $\sigma_v \propto R^{1/2}$. From (15.27), making use of the virial equilibrium condition, we obtain:

$$\rho = \left(\frac{3}{4\pi\mu_0\mathcal{G}}\right)^{1/2} BR^{-1} \tag{15.32}$$

which is of the type $\rho \propto R^{-1}$. The (15.31) and (15.32) should be compared with the (15.20) and (15.22) respectively.

Molecular clouds have been seen to be self gravitating, which means that self gravitation forces are important and determine the confinement of the cloud. For this reason, molecular clouds, although not spherical in shape, have a compact shape. This does not happen for diffuse clouds, in which the self gravitation force is not relevant and the confinement is exerted by external pressure; in this case it is expected that there is no significant density gradient inside and that the shape is irregular.

The distinction between self-gravitating clouds and diffuse clouds can be made using the dimensionless parameter:

$$\beta = \frac{p}{\mathcal{G}\sigma_c^2} \tag{15.33}$$

where σ_c is the mass column density of the cloud.

Let's consider a molecular cloud in equilibrium: we express the equilibrium condition using the virial theorem taking into account the gravitational energy, the kinetic energy of the disordered motion ("turbulence") and the surface pressure:

$$-\frac{\mathcal{G}M^2}{R} + 2\frac{1}{2}M\sigma_v^2 + \int_S p\mathbf{r}\cdot d\boldsymbol{\sigma} = 0 . \tag{15.34}$$

If there is gravitational confinement the surface term can be neglected, so it will be

$$\frac{\mathcal{G}M^2}{R} = M\sigma_v^2 \tag{15.35}$$

and since

$$4\pi p R^3 \ll \frac{\mathcal{G}M^2}{R} \tag{15.36}$$

introducing the density we obtain:

$$\frac{p}{\mathcal{G}\sigma^2} \ll \frac{4\pi}{9} \approx 1 \tag{15.37}$$

ie:

$$\beta \ll 1 . \tag{15.38}$$

In the case of molecular clouds, the value of β is between 0.01 and 0.1.

In the absence of gravitational confinement, it is the external pressure that controls the gaseous structure and results:

$$\frac{\mathcal{G}M^2}{R} \ll 4\pi p R^3 \tag{15.39}$$

from which we obtain:

$$\beta \gg 1 . \tag{15.40}$$

In the case of diffuse clouds, we would have $\beta \sim 50$.

15.7   Why the Clouds Where Stars Form Are Molecular?

In our galaxy, gas from star-forming regions is in molecular form. However, this does not imply that it must be so everywhere. In fact, while star formation takes place when a cloud, or a part of it, collapses under the action of self-gravitation forces, the fact that the cloud is made up of molecules rather than atoms is a consequence of the opacity that shields the ultraviolet radiation that would dissociate the molecules. The condition for collapse can be expressed by the prevalence of gravitational energy density over that of the magnetic field (see Sect. 8.5). Since the gravitational energy density is:

$$\frac{\mathcal{G}M^2/R}{4\pi/3R^3} = \frac{3\mathcal{G}}{4\pi}\frac{M^2}{R^4} \tag{15.41}$$

the condition for collapse becomes:

$$\frac{3\mathcal{G}}{4\pi}\frac{M^2}{R^4} > \frac{B^2}{2\mu_0} \tag{15.42}$$

that is

$$\frac{M}{R^2} > \sqrt{\frac{2\pi}{3\mathcal{G}\mu_0}} B . \tag{15.43}$$

This condition can also be expressed in terms of the mass column density μ. It is:

$$\mu = \int \rho dr = \rho R = \frac{3}{4\pi}\frac{M}{R^2} \tag{15.44}$$

assuming that the cloud is a sphere of radius R. The collapse condition then becomes:

$$\mu > \sqrt{\frac{3}{8\pi\mathcal{G}\mu_0}} B . \tag{15.45}$$

Taking into account the dust-to-gas local value, with reference to the V band,

$$A_V = R_V \frac{E(B-V)}{N(\mathrm{H})} N(\mathrm{H}) \tag{15.46}$$

where R_V is on average equal to 3 in the Milky Way with the exception of particular lines of sight. The study of the extinction leads to the determination of the constancy of the gas/dust ratio which, for our galaxy is $N(\mathrm{H})/E(B-V) = 7.5 \cdot 10^{21}\mathrm{mag}^{-1}\mathrm{cm}^{-2}$, where $N(\mathrm{H})$ is the column density of H in cm^{-2}. It is:

$$A_V = 4 \cdot 10^{-22} N(\mathrm{H}) \tag{15.47}$$

from which, considering that

$$N(\mathrm{H}) = \int n(\mathrm{H}) ds \tag{15.48}$$

we obtain:

$$\mu = 1.4 m_{\mathrm{H}} N(\mathrm{H}) \tag{15.49}$$

Therefore we get the following expression of Av:

$$A_V = 171\mu . \tag{15.50}$$

With (15.50) the condition of gravitational instability (15.45) becomes:

$$A_V > 171 \sqrt{\frac{3}{8\pi G \mu_0}} B = 6.45 \cdot 10^9 B \tag{15.51}$$

where B is measured in tesla. With a value of B of 10^{-6} G $= 10^{-10}$ T, we have:

$$A_V > 0.65 \text{ mag} . \tag{15.52}$$

Why are these clouds (or parts of clouds) in which self gravitation forces prevail also molecular? The molecules, once formed, are kept long enough when the opacity due to the dust is large enough to absorb most of the UV background radiation (stellar and SNR); the transition from atomic gas to molecular gas is usually sudden due to molecular self-shielding. In the case of H_2 and CO, the molecules are present when:

$$A_V > 0.5 \text{ mag} . \tag{15.53}$$

This self-shielding threshold roughly coincides with the condition for gravitational collapse and therefore, in our Galaxy, star-forming regions are molecular. The exclusion of the starlight eliminates the source of heating of the cloud (due to the photoelectric effect and ionisation of weakly bound atoms). The temperature of a cloud is around 80 K in the transparent regions (see Field's model) while it becomes approximately equal to 10 K in the opaque regions. We then explain what we observe, namely that a cloud is articulated in a part essentially consisting of molecular gas and in a warmer envelope where the UV radiation coming from the external stellar light dissociates the molecules and heats the gas (HI envelope). Molecular clouds therefore appear not as isolated clouds but as opaque regions of larger complexes.

The molecular fraction of a cloud, where star formation takes place, depends on the following factors:

- the abundance of heavy elements; it depends on (a) the opacity of the grains that shield UV radiation, (b) the rate of formation of the molecules which is linked to the density of the constituents (eg C and O in CO) and to the surface of the grains (in the case formation of H_2), (c) the rate of formation of molecules through the dependence on temperature, this is linked, in turn, to the metallicity which determines the cooling rate, (d) in part the radiation field can be conditioned by the metallicity of the stars through the opacity of the atmospheres and, possibly, through the shape of the initial function of the stellar masses;
- the local density of stars, which determines the UV flux incident on the cloud;
- the flow of cosmic rays; these have two effects: (a) ionise the cloud and influence the rate of formation of molecules in ion-molecule reactions, (b) as heat sources influence the temperature.

15.8   Mechanisms of Molecular Cloud Formation

Let's consider three fundamental processes that can originate clouds :

(1) accumulation of gas in shells and filaments by shock waves,
(2) formation of massive clouds by collisions between low-mass clouds,
(3) instability of the environment medium or within pre-existing clouds or shells.

These processes can be thought to be operating all together in order to be able to create the complicated spatial organisation observed in clouds. Let's now analyse the processes individually.

15.8.1   Formation of Clouds by the Compression of Ambient Gas Induced by Shock Waves

We have examined processes that can give rise to shells in the interstellar medium due to shocks and it has been said that these phenomena are relatively frequent. The hottest components can be crossed by shocks without having time to cool down before the passage of another shock. Gases that are denser after the shock passage can cool down and give rise to dense layers (called shells) that move behind the shock. It has already been noted that the notion of cloud as a spherical gaseous structure is partially undermined by the most recent observations (in the past with the IRAS or CHANDRA satellites, and now with eROSITA), which highlight the presence of filaments and shells. The filaments are aligned according to the magnetic field and are produced by the passage of a shock with a magnetic field perpendicular to the shock front. Diffuse clouds have sheet shapes indicative of the action of a shock. Along the line of sight the thickness of a cloud can be measured by knowing the column density N and the spatial density. It will be $N = nh$ hence $h = N/n$. Typical values are around 0.1 pc, and this dimension is much smaller than the transversal dimension.

The first model in which shocks are invoked as cloud formation mechanism dates back to the seventies: the HII regions produced by massive stars ionise and heat the gas between the clouds and by expanding compress the gas forming new clouds. From the model one can obtain temperatures, ionisation degree, and internal velocity dispersion of the WNM. Beyond this, also the main characteristics of the clouds, including the mass spectrum, can be predicted. The "hot" component (HIM) was instead not predicted and therefore models could no explain the diffuse X-ray emission or the OVI absorption lines. Models developed later take into account SNRs and stellar winds as well, and could produce better predictions.

15.8.2 Cloud Formation by Agglomeration of Smaller Clouds.

The process of the collision of two clouds is rather complicated and poorly known from a theoretical point of view. The outcome of the collision depends on several parameters: density and size of the colliding objects, collision speed, impact parameter and thermodynamic properties of the gas. There is no global treatment of the phenomenon and therefore there are no general rules to establish the outcome of the collision which could be: the destruction of the two clouds, the fragmentation, their coalescence, the coalescence of one part of the cloud onto the other, coalescence with gravitational instability, etc. With realistic cloud speeds the collision is supersonic and generates two shocks and a compressed middle layer at the interface. The compressed gas tends to expand and extends the shock to non-overlapping regions. If the momentum of the overlapping parts is small, or is reduced if transferred to the non-overlapping parts, coalescence occurs. A criterion for coalescing is that the initial Mach number of overlapping parts versus non-overlapping parts is less than 2. Fast collisions produce fragmentation with two or three fragments. Non-central collisions produce fragmentation or disintegration.

The tendency to fragmentation rather than coalescence when the collision rates are high means that only low velocity collisions are useful to the cloud formation process and this is a significant limitation to this mechanism. A larger scheme predicts that clouds can increase their mass partly by collisions and partly by accretion by passing through the cold SNR shells. A model obtained with these assumptions and assuming that in certain collisions there was fragmentation gave a good reproduction of the observed mass spectrum.

The presence of the magnetic field can favour coalescence: in fact the magnetic lines of one cloud can become entangled with the field lines of the other and this allows a rapid transfer of momentum from one cloud to another. MHD waves can also be produced which slowly dissipate energy and ultimately promote coalescence. The models generally predict that collisions are "random" and that massive clouds are formed by coalescence and smaller clouds by fragmentation. However, when collisions are "random" it is difficult to form massive clouds starting from small ones, because as soon as sufficiently massive clouds can form they get disrupted by the stars that form in them. With the "random" collision models the time to form a giant molecular cloud is of the order of $10^8 - 10^9$ years; the collision time for a small cloud with a giant is between $10^6 - 10^7$ years and the coalescence of 100 diffuse clouds is necessary to obtain a cloud of $100 M_\odot$. A reduction of the collision times can be obtained if the collisions are not "random" as when the clouds are affected by the effect of a gravitational potential or when two cloud currents intersect, as one can think to happen when two gas-rich galaxies collide (mergers).

15.8.3 Cloud Formation Through Medium Instability

We have already seen two important instabilities in the ISM: thermal and gravitational. Both, in principle, can generate clouds. We can anticipate that also the magnetic field may play a significant role.

Thermal instability can give rise to clouds by condensation of diffuse gas, but the flux of cosmic rays, as a heating agent, is too low to trigger any important instability.
  Gravitational instability operates on the Jeans scale:

$$\lambda_J = \frac{c}{\sqrt{\mathcal{G}\rho}} \tag{15.54}$$

and can generate large structures when the density is low and small structures when the density is high. The time of development of the instability is the time of free-fall.
  Another quite interesting instability is the Parker one. This instability is generated by the combined effect of gravitational force and magnetic field. In the Galactic disc, cosmic rays are confined by the magnetic field; the pressure exerted by them is comparable with the magnetic and turbulent ones. On a large scale the interstellar medium is in equilibrium, and this can be expressed by the equation of motion applied to each component. For gas and cosmic rays we have

$$- \operatorname{grad} p_g + \mathbf{j}_g \times \mathbf{B} + \rho_g \mathbf{g} = 0 \tag{15.55}$$

$$- \operatorname{grad} p_{cr} + \mathbf{j}_{cr} \times \mathbf{B} = 0 \tag{15.56}$$

where $\mathbf{g} = \operatorname{grad}\Phi$, and Φ is the gravitational potential. Furthermore, Maxwell's equation holds for the magnetic field:

$$\text{rot}\mathbf{B} = \mu_0(\mathbf{j}_g + \mathbf{j}_{cr}) \tag{15.57}$$

and due to the turbulent pressure of the clouds the:

$$p_g = \frac{1}{3}\rho_g\langle v^2\rangle \tag{15.58}$$

(this relation is derived from the usual relations of the perfect gas). From the previous equations we obtain:

$$-\text{grad}(p_g + p_{rc}) + \frac{1}{\mu_0}\text{rot}\mathbf{B} \times \mathbf{B} + \rho_g\mathbf{g} = 0 . \tag{15.59}$$

$\mathbf{B}$ is contained within the Galactic plane and has a fixed direction: we take the reference so that this direction coincides with the x axis, so it will be $\mathbf{B} = B\hat{\imath}$. We consider the component according to the z axis of the equation (15.59), i.e.

$$-\frac{\partial}{\partial z}\left(p_g + p_{rc} + \frac{B^2}{2\mu_0}\right) - \rho_g g = 0 \tag{15.60}$$

indeed:

$$\text{rot}\mathbf{B} = \frac{\partial B}{\partial z}\hat{\jmath} - \frac{\partial B}{\partial y}\hat{k} \tag{15.61}$$

$$(\text{rot}\mathbf{B}) \times \mathbf{B} = -\frac{1}{2}\frac{\partial B^2}{\partial z}\hat{k} - \frac{1}{2}\frac{\partial B^2}{\partial y}\hat{\jmath} \tag{15.62}$$

$$(\text{rot}\mathbf{B}) \times \mathbf{B} \cdot \hat{k} = -\frac{1}{2}\frac{\partial B^2}{\partial z} \tag{15.63}$$

(note that in these last formulas $\hat{\jmath}$ denotes the versor of the y axis). Assuming g and $\langle v^2\rangle$ constant and setting:

$$\frac{p_{rc}}{p_g} = \beta \tag{15.64}$$

$$\frac{B^2/2\mu_0}{p_g} = \alpha \tag{15.65}$$

we have:

$$\frac{\partial}{\partial z}p_g(1 + \beta + \alpha) + g\rho_g = 0 \tag{15.66}$$

hence with (15.58):

$$\frac{\partial\rho_g}{\partial z} = -\frac{\rho_g}{H} \tag{15.67}$$

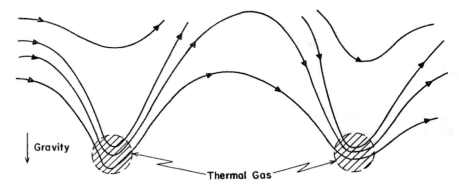

Fig. 15.4 Sketch of the local state of the lines of force of the interstellar magnetic field and interstellar gas-cloud configuration resulting from the intrinsic instability of a large-scale field along the galactic disk or arm when confined by the weight of the gas (Parker, [3])

where is it

$$H = \frac{\langle v^2 \rangle (1 + \alpha + \beta)}{3g} .$$
(15.68)

The solution of this equation is:

$$\rho_g = \rho_g(0) \exp\left[-\frac{z}{H}\right]$$
(15.69)

therefore cosmic rays and the magnetic field increase the height scale of the gas in the Galactic disk. With $\alpha = 0.25$ and $\beta = 0.4$ we have $H = 200$ pc. In such conditions an instability can occur that has analogy with convective instability: it has been shown that a gas immersed in a horizontal magnetic field and in a vertical gravitational field is stable if:

$$-\frac{d\rho}{dz} > -\left(\frac{d\rho}{dz}\right)_{ad} .$$
(15.70)

Note that this condition is exactly the same as the Schwarzschild condition for stability with respect to convection in the interior of stars. The magnetic field does not intervene in the condition of stability but influences the unstable modes. When the (15.70) condition is not verified, instability occurs and the characteristic time for its complete development is of the order of $\langle v^2 \rangle / H = 10^7$ years. Instability tends to alter the lines of force of the magnetic field by rippling them: the gas moves along the lines of force sliding towards the galactic plane and concentrates in clouds and cloud complexes (see Fig. 15.4 for a visualisation of the effect).

15.8.4 Summary

Taking into account all these mechanisms, we can try to provide a picture of the formation of different types of clouds. It has already been said that the interstellar medium appears inhomogeneous. This inhomogeneity has a hierarchical aspect in the sense that the larger complexes contain separate clouds, each of these can contain denser regions (clumps) and even within these, denser and smaller parts can be identified. The statistical relationships considered above indicate that the larger scale structures are virialised and are not in pressure equilibrium with the surrounding gas. Dark clouds tend to be round and at high pressure due to self-gravity and tend to be molecular due to UV self-shielding; transparent clouds are more amorphous and transient because self gravitation is not involved in their confinement. The formation mechanism could be the same in the two types with the same mass; diffuse clouds are probably the low column density part of the mass spectrum.

The smaller scales of the hierarchy are in pressure equilibrium with the surrounding gas and their internal pressure is determined by thermal motions, turbulence or magneto-sonic waves, and magnetic fields. They could be generated by the fragmentation of the SNR shells. On a larger scale there are clusters of clouds held together by magnetic fields and eventually forced to coalescence by some large-scale disturbance such as shocks. Large complexes are formed by coagulation of smaller clouds and by general instability of the medium.

15.9 Synthetic Summary

Molecular clouds are the densest and coldest component of the ISM. They are the sites of star formation. These cloud have different size and masses, although they can be roughly divided in giant and dark molecular clouds. They are much more compact than HI and in principle they can be better tracer of the Galaxy gas structure. These structures are self-gravitating, and not in pressure equilibrium with other ISM components. Molecular clouds are routinely revealed though the CO rotational line in emission at 2.6 mm. The CO luminosity shows a good linear dependence with the H_2 mass, and this way cloud's mass can be inferred. Molecular clouds are in virial equilibrium, and the magnetic forces are the most widely accepted forces to contrast self-gravitation. In the Milky Way, molecular clouds are present everywhere in the Galactic disk, although the major concentrations aree in the bulge and in molecular ring. They are though to form via shocks, collisions, and instabilities, the most relevant of which being the Parker one.

AlCl	NH	AlNC	HCS^+	CH_3	NH_3	NH_4	c-H_2C_3O	c-C_2H_4O	H_3CC_2CN	CH_3C_4H	$(CH_3)_2CO$
AlF	N_2	AlOH	HN_2^+	l-C_3H	CH_4	CH_3O	CH_3CN	CH_3C_2H	$H_2COHCHO$	CH_3OCH_3	$(CH_2OH)_2$
AlO	NO	C_3	HNO	l-C_3H^+	SiC_3	c-C_3H_2	C_2H_4	CH_3NH_2	$HCOOCH_3$	CH_3CH_2CN	CH_3CH_2CHO
ArH^+	NO^+	C_2H	HOC^+	c-C_3H	HMgNC	l-C_3H_2	CH_3NC	CH_2CHCN	CH_3COOH	CH_3CONH_2	CH_3C_5N
C_2	NS	CCN	HSC	C_3N		H_2CCN	CH_3OH	CH_3CHO	H_2C_6	CH_3CH_2OH	HC_9N
CF^+	NaCl	C_2O	KCN	C_3N^-		H_2C_2O	CH_3SH	HC_5N	CH_2CHCHO	C_8H	C_2H_5OCHO
CH	MgH^+	C_2S	MgCN	C_3O		H_2CNH	l-H_2C_4	C_6H	C_7H	CH_3CHNH_2	CH_3COOCH_3
CH^+	NaI	C_2P	MgNC	C_3S		HNCNH	HC_3NH^+	C_6H^-	CH_2CCHCN	CH_3CH_2SH	CH_3C_6H
CN	O_2	CO_2	N_2O	C_2H_2		H_2COH^+	HC_2CHO			NH_2CH_2CN	C_8H_6
CN^-	PN	NH_2	NaCN	H_2CN		C_4H	NH_2CHO			$(NH_2)_2CO$	C_3H_7CN
CO	PO	FeCN	NaOH	H_2CN^+		C_4H^-	C_5H			C_8H_2	C_{60}
CO^+	SH	H_3^+	OCS	H_2CO		HC_3N	C_5N				C_{60}^+
CP	SH^+	H_2C	O_3	H_2CS		HNC_3	C_5S				C_{70}
CS	SO	H_2Cl^+	SO_2	HCCN		HCCNC	C_5N^-				
FeO	SO^+	H_2O	c-SiC_2	$HCNH^+$		HCOOH	HC_4N				
H_2	SiC	H_2O^+	SiCN	$HOCO^+$		NH_2CN					
HCl	SiN	HO_2	SiNC	HOCN		HCOCN					
HCl^+	SiO	H_2S	TiO_2	HCNO		SiC_4					
HF	SiS	HCN		HNCO		SiH_4					
HO	TiO	HNC		HNCS		NH_3D^+					
OH^+		HCO		HOOH		C_5					
KCl		HCO^+		H_3O^+							
		HCP									

Fig. 15.5 An update of organic molecules census in the ISM. Courtesy of Reddavide Matteo

Appendix

Appendix A: Chemistry in the ISM

When ALMA (www.almaobservatory.org) started its operations a few years ago, ISM studies experienced a revolution, and many aspects of molecular cloud structure and evolution were clarified. This went together with a much better understanding of the star formation process (see next chapter). One of the field which better benefits from ALMA observations is ISM chemistry and chemical evolution. ALMA operates in the sub-milli-meter window of the electromagnetic spectrum, where the typical energy is 10^{-3} erg. Rotational transitions of molecules do have this energy, and therefore plenty of new lines have been measured, and new more and more complex molecules discovered. Several of them were detected for the first time outside the solar system. Figure 15.5 provides a recent census of these molecules. In red a few molecules are highlighted. They are called pre-biotic, and are critical for the development of life.

References

1. Bolatto, A.D., Wolfire, M., Leroy, A.K.: ARA&A **51**, 207 (2013)
2. Heyer, M., Dame, T.M.: ARA&A **53**, 583 (2015)
3. Parker, E.N.: ApJ **145**, 811 (1966)

Chapter 16
Star Formation

16.1 Observational Data Related to Star Formation

Bimodal star formation. Observations indicate that the formation of low-mass stars and that of massive stars ($M > 10M_\odot$) involve processes of a different nature. In fact, the places of formation of the two types of stars are spatially separated, more precisely there are regions where only low-mass stars are formed and regions where both low-mass and massive stars are formed. For example, the Taurus region does not seem to contain stars more massive than 2 $M_\odot$, while in Orion both low-mass and large-mass stars are observed. More in detail, considering the "cores", i.e. the densest regions of molecular clouds with $n > 10^2$ cm^{-3}, we can define a subdivision of the "cores" into two categories:

1. *Cores of low mass.* They have been found in regions with high absorption in the visual of nearby dark clouds (e.g. in Taurus) through the molecular emission lines e.g. of NH_3, $C^{18}O$, or HC_5N. The density is $n = 10^4$ cm^{-3}, the mass about 4 $M_\odot$, the temperature about 10 K, while the velocity dispersion deduced from the widening of the lines, that is, the turbulence speed, whatever its nature might be, is subsonic. Several T Tauri stars are associated with these regions. The IR survey carried out by the IRAS and SPITZER satellites revealed that many infrared sources, invisible at optical wavelengths, and of low bolometric brightness, are associated with these "cores": these objects are identified as young stars of low mass.
2. *Cores of large mass.* They are present in giant molecular clouds (e.g. in Orion). The tracers are mainly CS and NH_3 and reveal much larger masses than the "cores" in Taurus: the average mass is about 5000 $M_\odot$. Not much is known about the possible presence of low-mass "cores" in giant clouds. The turbulence, deduced from the widening of the lines, is supersonic. High-mass "cores" are found in the vicinity of stars of early spectral types, ultra-compact HII regions, and infrared light sources with $L > 10^4 L_\odot$.

G. Carraro, *Astrophysics of the Interstellar Medium*,
UNITEXT for Physics, https://doi.org/10.1007/978-3-030-75293-4_16

Table 16.1 Physical parameters of cores in molecular clouds

	Low mass cores	Massive cores
Size [pc]	$0.05 - 0.2$	$0.1 - 3$
Density [cm^{-3}]	$10^4 - 10^5$	$10^4 - 10^6$
Temperature [K]	$9 - 12$	$30 - 100$
Velocity dispersion [km/s]	$0.2 - 0.3$	$1 - 3$
Mass [$M_\odot$]	$0.3 - 10$	$10 - 10^4$
Free-fall time [yr]	$2 \cdot 10^5 - 10^6$	$2 \cdot 10^4 - 2 \cdot 10^5$
Associated Stars	T Tauri	OB

A different classification distinguishes molecular clouds into a cold ($T < 10$ K) and a warm ($T > 20$ K) population. Cold clouds, which do not contain stars of the spectral types earlier than A, are more or less evenly distributed across the Galactic disk. Hot clouds, which tend to be the largest spatially and most massive objects, are associated with HII regions, detected by radio observations, and are predominantly distributed along the spiral arms. The two classifications are almost equivalent.

The idea of a spatial distinction between regions in which only low-mass stars are formed and regions in which large-mass stars are also formed has an increasingly convincing observational confirmation.

Induced star formation. While gravitational instability (Jeans criterion) predicts that the onset of collapse depends on the intrinsic physical conditions of the interstellar gas, there are indications that the formation of OB stars may also or only be induced by processes external to the parent cloud. Mechanisms of this type include shock waves that generate compression in the cloud and cause gravitational collapse to begin. With the exception of two scenarios, the induced star formation process is determined by the massive stars themselves. In fact, the compression can be produced by the following causes:

1. shock waves of SNRs,
2. ionisation fronts of HII regions,
3. stellar winds,
4. collisions between clouds,
5. shock waves produced by the spiral arm perturbations.

The shock waves produced by massive stars can therefore provide a mechanism by which the star-forming process spreads across a region of a galaxy, similar to that by which an infectious disease spreads (self propagating star formation).

While observations seem to confirm the role of these processes in the formation especially of massive stars, from the theoretical point of view their effectiveness is considerably limited by the presence of magnetic fields if these have a relevant role in the mechanical equilibrium of clouds.

A mechanism of this type would be responsible for the formation of discrete groups of stars within a molecular cloud: in fact, stellar groups are observed that, in addition to constituting a spatially ordered sequence, also represent an age sequence. *Efficiency of the star formation process.* The age of Galactic (open) clusters varies in a very wide range from about 10^6 to 10^9 years: less than 30% are older than $2 \cdot 10^8$ years. The presence in the disk of the galaxy of a number of old star clusters (older than 10^9 years) indicates that they are, possibly, gravitationally bound systems. Only 10% of the field stars around the Sun come from the disintegration of bound systems, while the remaining 90% arises in non-bound systems (OB associations). Therefore the star formation process seems to produce unbound systems more easily.

Open clusters are aggregates of stars in a wide mass range with large density, greater than $0.1 M_\odot/\text{pc}^3$ (a critical density) which makes these systems stable against the disintegration by Galactic tidal forces or with respect to collisions with interstellar clouds. On the contrary, associations are stellar aggregates of low density, however higher than that of the field stars: stellar densities are lower than the critical density, and therefore are unstable. Most associations (see Fig. 16.1 for a beautiful example) are younger than 10^7 years. We distinguish OB associations which are groupings of massive stars and T associations which consist of low-mass stars with emission lines (pre-main-sequence stars).

Firstly, an attempt was made to interpret the expansion observed in some associations as due to the fact that the stars of an association are formed out of a gravitationally un-bound cloud, but this hypothesis was soon after denied by the observations of the velocity dispersion. These observations indicate that molecular clouds are gravitationally bound and in virial equilibrium (see Chap. 15). These observations concern clouds of a very wide mass range and also include the clouds where OB stars form. Since the stars acquire the same velocity dispersion of the gas from which they are formed, another phenomenon must obviously intervenes to modify the dynamics of stars.

This phenomenon is the removal of residual gas by the very energetic processes produced by massive stars. The difference between associations and clusters lies in the efficiency of the star formation process: if the conversion efficiency of the gas into stars is low, the removal of the gas greatly changes the mass of the system and therefore its binding energy; the residual star system undergoes a dynamic rearrangement, which relevance increases with the amount of gas removed. In the end, if the quantity of gas removed is large, the system may turn unbound or may have a density lower than the critical one and therefore becomes unstable to whatever perturbation. If, on the other hand, the amount of gas lost is relatively small, the system keeps stable.

In the event of an instability, the star system disperses in less than 10^8 years. This behaviour can be easily studied when gas loss by the system occurs over a time scale lower than the dynamic scale, which approximately corresponds to the time a star takes to travels through the system with the speed that satisfies the theorem of virial (crossing time). Suppose that, before the gas is removed, the system is in equilibrium in a configuration of radius R_i; according to the virial theorem, the velocity dispersion is:

Fig. 16.1 OB associations in the Scorpius/Ara constellations. Credit ESO

$$\sigma_i^2 = \frac{\mathcal{G}M_i}{R_i} . \tag{16.1}$$

The system is not able to immediately adapt after the gas loss and retains both the initial configuration R_i and the dispersion of velocity; its energy, however, has been changed to:

$$E = \frac{1}{2}M\sigma_i^2 - \frac{\mathcal{G}M^2}{R_i} \tag{16.2}$$

where $M = M_i - \Delta M$ is the mass retained, which is also the mass of the formed stars. Subsequently, in a time equal to the dynamic time, the stellar system is rearranged, conserving energy, in a configuration of radius R_f. According to the virial theorem it will be:

$$E = -\frac{\mathcal{G}M^2}{2R_f} . \tag{16.3}$$

This expression is obtained by eliminating in the expression of the viral theorem the kinetic energy using the total energy. From the three previous relations we obtain:

$$\frac{R_f}{R_i} = \frac{1 - \epsilon}{1 - 2\epsilon} \tag{16.4}$$

where $\epsilon = \Delta M / M_i$ is the fraction of mass lost. By introducing instead the efficiency of the star formation $\eta = M_f / M_i = 1 - \epsilon$ we have

$$\frac{R_f}{R_i} = \frac{\eta}{2\eta - 1} . \tag{16.5}$$

From these relations it results that when $\epsilon > 0.5$, i.e. when the quantity of gas expelled exceeds 50% of the initial mass, the system is no longer bound. For smaller amounts of mass loss the radius of the final configuration is in any case much larger than the initial one and therefore the density of the star system can easily fall below the critical value for stability.

Obviously, the case of impulsive loss is an extreme case; another extreme case that is easily treatable is the opposite case, namely when mass loss takes place on a long time scale with respect to the dynamic time: in this case the system dynamically rearranges itself as the gas loss continues and results:

$$\frac{R_f}{R_i} = \frac{1}{\eta} . \tag{16.6}$$

Also in this case the loss of mass by the system leads to a reduction in the density of the resulting star system. The realistic case of a mass loss occurring on a time scale other than those considered is numerically more complicated. Numerical models can explore all intermediate situations. The efficiency is estimated by evaluating the mass of the residual gas and the masses of the stars which formed. From the aforementioned models it is possible to evaluate the efficiency required for a dynamically stable system:

- $\eta = 0.27$ in the case of rapid disintegration ($t_p \ll t_d$),
- $\eta = 0.59$ in the case of slow disintegration ($t_p \gg t_d$).

With numerical models we define an interval for the critical efficiency, i.e. $0.3 \leq \eta \leq 0.55$. For lower values the resulting star system is unstable.

Examples. The λ Ori association is dominated by an O8-III star, λ Ori, which is the source of the UV flux that generates an HII region including most of the members of the association. On the edge of the HII region there is a shell of neutral gas and dust ($M = 10^5 M_\odot$), the last remnant of the parent cloud. The total mass of the stars, including 12 B stars and 83 stars with emission lines, is estimated to be $200 - 300 M_\odot$. The efficiency is $\eta = 0.2 - 0.3$. The history of the association was probably as follows: the star formation process left a large amount of residual gas that was dispersed. Inn turn, the escape velocity of the system dropped to about 0.2 km/s thus triggering the escape of many stars with speed of the order of 2 km/s. The system then re-arranged itself by expanding and regaining virial equilibrium with a stellar density less than $0.1\ M_\odot/\text{pc}^3$.

The dark cloud ρ Ophiuchi is located 160 pc from the Sun on the edge of the Sco-Cen association. The region has a very large extinction, i.e. $A_V = 50$ mag. Infrared sources are identified and, by comparing the characteristics of the stars with emission lines, we conclude that the stellar population is composed of about 40 members having a total mass of $100M_\odot$. The efficiency estimate is $\eta = 0.25$. This fact and the high stellar density of 145 $M_\odot$/pc indicate that ρ Oph is a typical region where a star cluster is forming.

16.2   Star Formation in the Presence of a Magnetic Field

In Sect. 8.5 by means of the virial theorem with the inclusion of magnetic forces we have introduced a critical mass, defined, up to a numerical factor of the order of unity, by:

$$M_{cr}^2 = \frac{8\pi B_0^2 R_0^4}{3\mu_0 \mathcal{G}} \tag{16.7}$$

or, through the magnetic flux Φ_m, by:

$$M_{cr}^2 = \frac{8\Phi_m^2}{3\pi \mu_0 \mathcal{G}} \, . \tag{16.8}$$

This critical mass is independent on the radius R (this is because, in the hypothesis of freezing of the magnetic field lines, both the gravitational and the magnetic energy have the same dependence on R). By eliminating the radius by introducing the density from (16.7) and exploding the values of the constants, we obtain:

$$M_{cr} = 1.9 \cdot 10^4 \frac{B_0^3}{n^2} \, [M_\odot] \tag{16.9}$$

where B_0 is measured in μG.

For a mass $M > M_{cr}$ the virial theorem predicts that gravitational forces prevail over magnetic forces and therefore the cloud undergoes collapse: this situation is called supercritical regime. On the other side, when $M < M_{cr}$, if the lines of force are frozen to the particles, the magnetic forces maintain equilibrium with the gravitational forces and there is no collapse: this is called subcritical regime.

Since the magnetic force is perpendicular to the field, the equilibrium can be exerted by the magnetic forces only in the directions perpendicular to the field, while in the direction of the field itself the magnetic forces have no influence. However, if we interpret the observed "turbulent" velocity field as due to the propagation of variable magnetic fields (magnetohydrodynamic waves), these fields allow equilibrium in the other directions. Given the uncertainty in the evaluation of the intensity of the magnetic field, the estimate of the critical mass is considerably uncertain: from the observational point of view it is hard to establish firmly whether a cloud in its totality does not collapse and therefore it is in a subcritical regime. For instance, with $B_0 = 20\mu$ G and an average density of 50 cm^{-3} we find $M_{cr} = 6.4 \cdot 10^4 M_\odot$. With the same magnetic field and referring to the density of a "core" ($n = 200$

cm^{-3}) the critical mass is 3800 $M_\odot$ which could indicate that there are "cores" in supercritical regime and others in subcritical regime. The former would be unstable and would give rise to a collapse with fragmentation and formation of a group of stars (cluster or association). In the second case, the instability would result from the effect of ambipolar diffusion (or from the decay of turbulence). However, from the observational point of view, there are no firm indications regarding the supercritical or subcritical nature of the "cores" due to the uncertainties of the magnetic field values in the densest areas of the molecular clouds.

In the case that $M < M_{cr}$ and assuming that the magnetic field remains frozen, the cloud remains stable. However, if the flow is not conserved, as occurs in the presence of ambipolar diffusion, there is a contraction that proceeds on a time scale longer than the free-fall time. The slowness of the process probably explains why the efficiency of stellar formation in subcritical clouds is low: the newly formed T Tauri stars have time to disperse the rest of the gas through stellar winds, thus giving rise to poorly bound aggregates.

16.3 The formation of Low-Mass Stars

To improve our knowledge of the star formation process, it is interesting to identify and study the objects immersed in the parent cloud in order to define all their phases. This is relatively simple for the final stages of the process when the stars are not hidden by large amounts of the residual gas from the cloud from which they originated. This is the case with massive hot stars such as OB stars. We have seen that the intense UV radiation field created by them determines the ionization of the surrounding gas. Subsequently the ionized gas expands and in the expansion it drags the gas of the cloud by breaking it up. In this case the observations in the visible and in the ultraviolet allow us to identify the exciting stars as well as the emission of the ionized gas.

Even more important for understanding the star formation process is the study of the phases in which the star is still immersed in the gas from which it originated. In this case, however, we must resort to observational techniques that allow us to somehow penetrate through the clouds. This can be achieved observing in the infrared, at wavelengths of the order of a few up to a few hundreds microns. An abundant amount of data of this type has recently been obtained with the IRAS, AKARI, and SPITZER satellites. These observations give us indirect information about the hidden object through the effect that its radiation produces on the surrounding dust. In fact, its electromagnetic radiation is completely absorbed by the dusts that heat up reaching temperatures ranging between 50 and 500 K depending on the intensity of the radiation field and the distance of the dust from the source. The radiation is, in turn, emitted by dust grains, which can be detected with an infrared telescope. The emission of dust in first approximation is comparable to that of a black body with the same temperature and, according to Wien's displacement law, the radiation is therefore emitted mainly in the far infrared. Only relatively recently this technique has become available, for instance at ESO with instruments like NACO, SINFONI, ISAAC, and CRIRES. Observing in the infrared in fact involves overcoming many difficulties, such as:

(a) due to the weakness of the sources to be detected, very sensitive detectors are required, which can be realized with semiconductors;
(b) carbon dioxide and water vapor from the atmosphere absorb infrared radiation especially at the larger wavelengths (FIR), therefore it is necessary to make observations outside the atmosphere with stratospheric balloons or satellites,
(c) there is always a strong background noise because all objects at ordinary temperature, including the telescope itself, emit radiation of the same type as the signal to be detected, therefore it is necessary to cool the cryogenic structure dow with liquid helium/nitrogen to maintain the temperature constantly around $-270\,^{\circ}$ C.

16.4   Low-Mass Star Formation Scenario

With radio observations of molecular millimeter lines (e.g. of CO) we are able to obtain information on the physical nature of clouds, that is, of objects whose average density is of the order of 100-1000 particles per cm^3. The observations in the infrared reveal the characteristics of the protostars, that is of objects with an average density of $10^{22} - 10^{23}$ particles per cm^3; all the intermediate phases that bring the matter from the density of the clouds to that of the protostars remain, for a large part, inaccessible to our observations and therefore it is necessary to add theoretical considerations to the observations as a guide to our speculations. There are both observational and theoretical indications that low-mass stars, when formed in isolation, originate in a subcritical regime.

Since the picture of the formation of massive stars is still very uncertain, we will only describe the process that gives rise to low-mass stars through the subcritical regime. The cores produced by ambipolar diffusion are initially in equilibrium with gravitational forces balanced by thermal pressure forces and magnetic forces: as the effect of the latter weakens, the equilibrium changes so that the central density increases compared to the average. The model that describes this phase is an isothermal sphere with a density profile $n(r) \sim r^{-2}$. This phase of slow evolution ends with a free fall collapse: this happens because during the slow evolution the ratio between the central density and the average density is progressively increasing and when a critical value is exceeded, equilibrium is no longer possible. The collapse of the core is strongly non-homologous, in the sense that the collapse time of each layer is smaller the larger the density: this causes the outermost layers, less dense, to fall towards the center much more slowly than the core layers that are much denser and this trend increases with time.

Calculations reveal that, in the absence of rotation, a spherical, dense, and opaque central object forms quite soon. The gravitational energy released by the collapse is retained by increasing the temperature of the gas and developing pressure forces that are able to contrast and eventually re-balance the gravitational forces. The balance between pressure forces and gravity forces is, however, continuously modified for two reasons: the increase in mass by the material that continues to fall on the object and the energy loss via radiation through its surface. The overlying material continues to rain on this central object in free fall. If rotation is taken into account, the evolution is the same except for the geometric structure, which changes due to the action of centrifugal forces: in addition to the central object, which has a spherical shape flattened at the poles, a disk develops in the equatorial plane. The surrounding material in free fall feeds both the central object and the disk. This structure is called a protostar.

Before seeing what observational evidence we have supporting the model just described, let's consider what is the subsequent evolution of the protostar. The increase in mass due to the infall of the surrounding material does not cease when this material is exhausted, but due to an intrinsic cause, i.e. the manifestation of a stellar wind, that is, a process of mass loss that blocks and reverses the motion of the infalling material.

The cause that produces the emission of the wind in the protostar is linked to the triggering of the deuterium burning reactions that render the protostar completely convective. Convection in turn, in the presence of rotation, amplifies the pre-existing magnetic fields and the coupling of these with the rotation provides the energy required for the expulsion of the gaseous material.

The stellar wind decelerates the gas falling rate on the protostar and finally determines the removal of the surrounding material which allows, for the first time, to see (in visible light) the protostar, which until now could only be detected in the infrared.

Theoretical calculations show that the growth rate of the material on the central object and on the disk is given by:

$$\frac{dM}{dt} \propto \frac{c^3}{G} \tag{16.10}$$

where c is the sound speed. For a star of $1M_\odot$ this results in $dM/dt = 9.5 \cdot 10^{-6} [M_\odot/\text{yr}]$. The accretion time of a star of the same mass is given by:

$$t_{\text{acc}} = \frac{M}{dM/dt} \approx 10^5 \text{ [yr]} . \tag{16.11}$$

Since the evolutionary time of pre-main-sequence t_{KH} (Kelvin–Hemholtz time) of the same star is approximately 10^7 years, the star of $1M_\odot$ becomes visible optically during the pre-main-sequence phase. Since t_{acc} varies little with M, while t_{KH} decreases rapidly with M, stars with $M > 6M_\odot$ terminate this growth phase when they are already in the main sequence. The locus of the points in the HR diagram where:

$$t_{\text{acc}} = t_{\text{KH}} \tag{16.12}$$

it is called stellar birth-line and is represented in Fig. 16.2.

Fig. 16.2 Different mass
protostars align along an
almost vertical line in the
H-R diagram, the Hayashi,
or birth-line

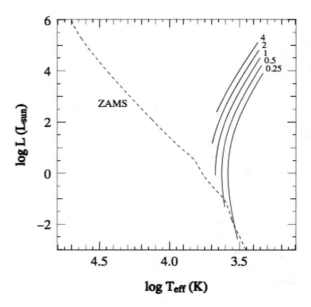

The disc is also destined to disappear: depending on its size and characteristics,
which in turn depend on the physical conditions of the medium from which the
protostar was formed, it can give rise to a planetary system, to another star or to
more than one (multiple system) or it can be swept away. We do not yet know
sufficiently the physics of this process to be able to define under what conditions and
for which masses of the central object a planetary system is formed. The protostar,
now cleaned of the surrounding material, continues to emit stellar wind and has the
characteristics of a T Tauri star. The different phases of the evolution of a protostar
are schematically represented in Fig. 16.4, where we note (a) the formation of the
"cores", (b) the growth of the material on the central object by means of through
an accretion disk, (c) a stellar wind phase, and finally (d) the star released from the
surrounding gas.

The initiation of the hydrogen burning reactions opens the sequence of nuclear
burnings relevant as energy sources for the star. Before the triggering of these reactions, as already mentioned, the star is in equilibrium under the action of gravitational
forces and pressure forces. This balance is constantly changing due to the loss of
energy radiated by the photosphere. This loss tends to cool the star and reduce pressure. This causes the star to contract, increasing its temperature and therefore its
pressure, and in this way the equilibrium is restored. The contraction process will
stop when the increase in the star's core temperature allows enough energy to be
released from nuclear reactions to balance the radiation loss. At this point the star is
on the main sequence.

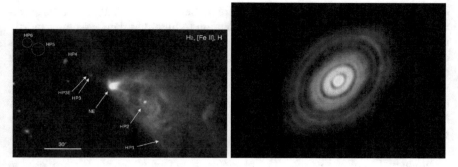

Fig. 16.3 Left panel: bipolar outflow in L1551 from IR emission. Right panel: reconstructed image of L1551 disk from milli-meter interferometry. ALMA and NAOJ credit

16.5   Observational Findings

Collimated stellar winds (bipolar outflows). Optical, radio and infrared observations revealed the rather short phase ($t = 10^4$ years) of mass loss that protostars still exhibit when they are hidden inside the parent cloud.

The phenomenon was revealed through the study of CO emission which highlighted a molecular gas component restricted to a very small area near infrared sources (protostars). One of the most relevant cases concerns the IRc2 source in the molecular cloud of Orion. The spectrum of the CO line reveals, through the broadening of the line by the Doppler effect, that the gas has a velocity range of about 180 km/sec, which is much larger than what results in molecular gas on average due to turbulence. In other objects, the speed range is smaller, but still larger than the average line width.

Considerations on the mass involved in the region where the phenomenon is observed allow one to exclude that the detected motion due to a gravitational collapse or to a rapid rotation, therefore the high speed measured must be attributed to an expansion motion (continuous wind blowing from the protostar).

A relevant aspect that emerged from the study of several of these sources is that the motion is not a uniform emission in all directions but the gas is collimated in two jets in opposite directions (bipolar outflows). The first example was the infrared source L1551, in the molecular cloud of Taurus. Figure 16.3 shows the structure of this source through the emission from the wind: the bipolar character is evident.

The bipolar nature can be revealed by the asymmetric distribution of the CO lines. The degree of collimation varies greatly from object to object and is relatively poor in many cases. Of the seventy objects studied, 75% have a bipolar aspect, 15% appear mono-polar and the remaining 10% have a symmetrical character: it may be that this is a perspective effect (the symmetrical aspect could be presented by bipolar motions in the direction line of sight).

Accretion Disks. The difficulty of imaging disks lies in the fact that most of the observations do not have sufficient spatial resolution to highlight their structure: mostly the entire disk is contained in a single element of spatial resolution (the dimensions of the disks are of the order of thousands of astronomical units) and therefore with a traditional instrumentation we are not able to deduce their existence with certainty, because the same integrated effects could come from a different geometry. However, recently, observations made with HST or with NACO and SPHERE at ESO-VLT have undoubtedly highlighted the presence of a disk around numerous stars within the Orion nebula.

Indirect proof of the existence of the discs comes from observations that have highlighted the collimated mass loss phenomena. In fact, all the models proposed to interpret these observations require the presence of a circum-stellar cloud with axial symmetry (i.e. a structure similar to a disk, more or less flat).

Discs emit mainly at milli-meter wavelengths and therefore interferometric systems (mostly ALMA, but also VLTI in IR) operating at these wavelengths are able to separate the emission of the disc from that of the infalling material. This way the articulated structure of some sources such as L1551 has been highlighted (see Fig. 16.3).

Proto-stars Infrared spectral distributions. From what has been said it can be expected that the protostars are immersed in an absorbing material which is more abundant the younger they are. The emitted radiation is absorbed and reprocessed by the surrounding dust which heats up and, in turn, emits electromagnetic radiation in the infrared part of the spectrum. The analysis of this radiation is fundamental to study the properties of these young objects. The observed spectrum will depend on the brightness of the hidden object and the nature and quantity of the surrounding dust. The first stages, when the protostar is surrounded by large quantities of gas and dust, will give rise to spectral distributions different from those of more evolved objects, when most of the circum-stellar material has been collapsed onto the star or blown away by the wind.

This interpretation is confirmed by the construction of theoretical models of the spectral energy distribution (SED). These models take into account the presence of a central object surrounded by a disk on which matter falls. The radiation emitted by the central object and the disc is progressively absorbed by the surrounding dust. These, in turn, heat up to a temperature which is a decreasing function of the distance from the central object and radiate back. The temperature varies between 800 K for the closest layers and 10 K for the furthest part that is not affected by the action of the central source. The emerging radiation can be evaluated by solving the transport equation

$$\frac{\mathrm{d}I_\nu}{\mathrm{d}s} = -k_\nu I_\nu + j_\nu \ . \tag{16.13}$$

The dust emission coefficient can be expressed using the Kirchhoff principle, i.e. $j_\nu/k_\nu = B_\nu(T)$. The solution of the equation is expressed by:

$$I_\nu = I_\nu(0)\exp[-\tau_{\nu,s}] + \int_0^{\tau_{\nu,s}} B_\nu(T)\exp[-\tau_\nu]d\tau_\nu \qquad (16.14)$$

where $\tau_{\nu,s} = n\sigma_{\nu,s}$ is the total optical depth. The measure of the emerging intensity presupposes that one fixes:

(a) the temperature trend of the grains $T(r)$ as a function of the distance from the central source, which can be determined with an energy balance as described in Sect. 10.7;
(b) the distribution of the density of dust $n(r)$ which must be obtained from a model of collapse of the "core";
(c) the absorbing properties of the dust contained in the cross section σ_ν.

The solution must be compared with the observed spectra and allows to establish constraints on the functions $T(r)$ and $n(r)$.

Using the IRAS, AKARI, and SPITZER satellites, numerous IR sources have been identified and studied which are associated with the dense "cores" of molecular clouds. The IR spectra are usually represented in a $\log(\lambda\mathcal{F}(\lambda))$ versus $\log\lambda$ plot. An analogous representation can be adopted by making use of the frequencies, which we have

$$\lambda\mathcal{F}(\lambda)|d\lambda| = \nu\mathcal{F}(\nu)|d\nu| \qquad (16.15)$$

so that the representation in terms of the wavelength or frequency is the same. We have indeed that

$$\mathcal{F}(\lambda) = \mathcal{F}(\nu)\left|\frac{d\nu}{d\lambda}\right| = \mathcal{F}(\nu)\frac{c}{\lambda^2}. \qquad (16.16)$$

The study of the spectral distributions of a very large number of observed sources made it possible to classify the latter on the basis of the characteristics of the spectrum and this way, originally, three classes were defined.

(a) *Class I*: the source has a wider spectral energy distribution than that of the best fit black body. The difference between the energy of the source and that of the black body best-fitting the observed curve at wavelengths less than 2 μ is called infrared excess. Furthermore, the average slope of the curve at wavelengths larger than 2 μ is positive. This slope defines a spectral index α

$$\alpha = \frac{d\log[\lambda\mathcal{F}(\lambda)]}{d\log\lambda} = -\frac{d\log[\nu\mathcal{F}(\nu)]}{d\log\nu} \qquad (16.17)$$

and a similar index β for the representation in terms of frequency

$$\beta = \frac{d\log[\nu\mathcal{F}(\nu)]}{d\nu} = -\alpha. \qquad (16.18)$$

Sources belonging to this class are not detectable in visible light.
(b) *Class II*: the source has a spectral distribution curve wider than that of the black body but the spectral index α is negative and the infrared excess is now less.

Sources of this class are observable both in infrared and in visual and, when observed in the latter spectral region, they appear as T Tauri stars.

(c) *Class III*: the source has a spectral distribution that resembles that of a black body. Objects of this class are optically visible and have a negligible infrared excess, a sign that the circum-stellar dust has been removed or is possibly present in negligible quantities. These stars have already reached the main sequence or are still in pre-main-sequence but are now very close to the ignition of the hydrogen reactions in the core.

Figure 16.3 shows, in a schematic form, the characteristic spectra of the three classes of sources (Fig. 16.4).

Although, for the sake of simplification, it is convenient to classify the sources into three classes, in reality there is a continuous variation of the spectral index (i.e. of the average slope of the spectrum for wavelengths greater than 2μ) in the range $(-3, 3)$.

Later observations in the sub-millimetric continuum obtained with large radio telescopes provided an interesting complement to the IR observations and revealed the presence of a large number of sources with a maximum emission at longer wavelengths ($\lambda > 300 \mu$ m). Some of these sources also emit in the far IR and have bipolar outflows, an indication that a central object has formed: this objects have been classified as belonging to a class 0.

Other "cores" with sub-millimeter emission do not have any IR sources and are believed to correspond to a stage that precedes the formation of the central object. These objects are called proto-stellar "cores" and are grouped in the −1 class.

In light of the scenario outlined above, the five classes can be interpreted as an evolutionary sequence:

- *class -1* corresponds to already formed "cores" which, however, have not reached the collapse phase and therefore do not have the central object;
- *class 0* represents the next phase during which the central object is being formed but its mass is still small compared to the mass of the envelope in free fall, and the collimated wind phase has already begun;
- sources of *class 1* are still hidden from optical observation by a large amount of dust that the wind has not yet removed;
- sources of the two successive classes represent the progressive liberation of the protostar from the surrounding material.

The agreement between the observational data and the models confirm the validity of the evolutionary scheme outlined and made it possible to define the phase corresponding to each of the three spectral classes identified. The fact that many, if not all, class I objects and many, though not all, T Tauri stars are associated with stellar winds suggests that the wind phase represents the transition between the protostar and the phase in which the star, now freed from the surrounding residual gas, starts to trigger thermo-nuclear reactions.

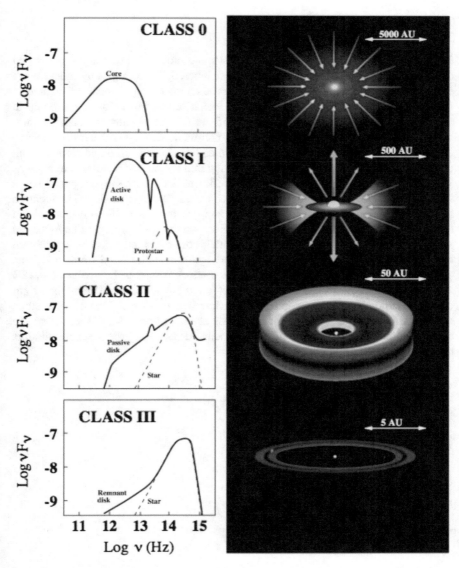

Fig. 16.4 Spectrum and physical state of different class pre-main-sequence stars. Credit to: https://ay201b.wordpress.com/tag/sed-modeling/

16.6   Synthetic Summary

Stars form in the densest regions of the molecular clouds. These regions are called cores. Star formation is seen to be externally triggered, for examples by shocks wave from spiral arms or cloud collisions, or by the energetic phenomena associated to the evolution of massive stars. In this last scenario, star formation keeps going in a self-propagating fashion. The outcome of star formation mostly consists of unbound stellar systems, called OB- or T-association, depending whether they formed in large or small cores. The phenomenon which transform a virialized cloud into an unbound stellar system is mass expulsion triggered by the star formation efficiency. We have devoted this chapter mostly to the formation of small mass stars. These form out of a gravitational instability caused by ampibolar diffusion. The location in the color magnitude diagram where this happens is close to the Hayashi or birth line. Observations in the IR and sub-mm has shown the various stages of a low mass star formation. First, a contraction in a rotating core develops a disk structure with bipolar outflows. When deuterium is ignited, a strong stellar wind starts, and most of the original core mass is wiped out. The stars then stars to be visible in the IR and eventually in the visual when it reaches the main sequence. This is the origin of the classification of pre main sequence stars into classes. As for massive stars, the scenario is not as clear as for low mass stars. In that case, it seems that the accumulation of more cores along gas filaments, designed by the magnetic field, is necessary to put together enough mass .

Printed in the United States
by Baker & Taylor Publisher Services